Guide to Modern Defense and Strategy

GUIDE TO MODERN DEFENSE AND STRATEGY

A complete description of the terms, tactics, organizations and accords of today's defense

DAVID ROBERTSON

Distributed exclusively in the United States and possessions by Gale Research Company

GALE RESEARCH COMPANY
Book Tower, Detroit, MI 48226

First published in 1987 by Europa Publications Limited,
18 Bedford Square, London, WC1B 3JN, as
A Dictionary of Modern Defence and Strategy.
This edition, for exclusive distribution by
Gale Research Company in the United States,
is entitled **Guide to Modern Defense and Strategy.**

ISBN 0-8103-5043-2

Library of Congress Catalog Card Number 87-082651

Made and printed in England by
Staples Printers Rochester Limited,
Love Lane, Rochester, Kent.

for Liz

Preface

Readers of any dictionary or reference book generally require some introductory guidance from the author to gain easy access to the information which they seek. I do not claim that this book is any exception, and offer here a few pointers which should be helpful. This book is a dictionary in so far that it consists of a number of entries arranged alphabetically by their titles. It goes further than a standard dictionary in that the entries are not confined to mere definitions. The main aim of the book is to set out the policy implications and theoretical arguments that lie behind the concepts and the physical specifications. I have consciously tried to avoid being too technical or scientific.

It will quickly become apparent that even the most mundane-sounding topics have profound political and strategic connotations. For example, not to grasp the range of 'correct' definitions encompassed by the single word **Division** could lead to the serious over (or under) estimating of an enemy's military strength. The consideration of such an apparently simple term might be extended by examining the difference between a **Light Division** and a Heavy Division; this, in turn, connects in to what is perhaps the most fundamental political question in American military thinking today—exactly which areas of the globe should be regarded as representing the USA's primary interests abroad? Similarly **Megaton,** in the context of nuclear weapons, cannot be regarded simply as a fixed unit of measure like, for example, a kilometre. When what really counts is the physical damage of a nuclear explosion, to understand fully the fact that a bomb with force of less than one megaton is relatively more effective than a multi-megaton weapon is of vital importance to strategic considerations.

Just from reading so far it will have become clear that, in a field such as this, no entry can be considered as standing alone. Each entry ties in to something else, and most relate to many others. Thus the employment of a thorough cross-referencing system becomes essential. Where a term which itself has an entry devoted to its definition and analysis occurs within the context of another entry, and where a proper understanding of that term is essential to the discussion of the subject in hand, that term is printed in bold type. Examples of this system have already appeared in this introduction. A further characteristic of the subject is the wholesale use of acronyms; in some cases abbreviations have been accepted into the defence language to such an extent that the acronym can even be used by some who do not know the full title. Therefore some definitions will be found headed by an acronym—**ICBM,** for example. Those who still prefer to think in terms of the full title will find a cross reference to the main entry. At which stage the acronym takes over from the full title is obviously open to variations in individual usage. An extensive list of abbreviations follows this Preface and will be of use to those who are looking for an entry under an acronym when I have thought in terms of the full title.

For the purposes of arranging the entries in alphabetical order, where a hyphen joins two words in a title which could alternatively be read as one, that hyphen is ignored. An apparently minor point such as this is mentioned because there are several groups of entries which start with the same word or words. Thus **Air-Launched Cruise Missile** will appear *after* **Air Superiority** and *before* **Airmobile,** but **Nuclear-Free Zone** will appear in its proper place between **Nuclear Fission** and **Nuclear Freeze Movement.** The end result should be that the reader will find an entry positioned where he or she would expect it to be. Where there is an alternative English and American spelling for a word the former has been used throughout this book, except where the word appears in the title of an organization, idea or strategy which is specifically American. The reader will, therefore, generally find **Defence** spelt thus, but will also encounter **Defense Intelligence Agency** and **Strategic Defense Initiative.**

There is, inevitably, an element of overlapping between entries. The **Doctrine** which rules strategic thinking leads to the repetition of the same arguments to support positions in very different subjects. In addition, several concepts will often have core material in common, but have conclusions with significant differences; a group of entries which illustrates this would be **Escalation, Escalation Control** and **Escalation Dominance.** This is also a field where the boundaries between allied definitions can not be regarded as set in stone. Tactical nuclear weapons are usually seen as similar to **Battlefield Nuclear Weapons,** although to be strictly accurate some of the former are certainly verging on the **Theatre Nuclear Forces (TNF)** category. In turn, TNFs *could* be seen as synonymous with the more frequently-used **Intermediate Nuclear Forces (INF).** But does the term INF cover the whole range of weapons which are clearly not 'battlefield' or 'strategic', between about 500 and 5,500 kilometres? Shortly before this book went to press a 'new' distinction between shorter- and longer-range INFs appeared to be emerging. Changes in the boundaries of classes of weapons can occur, as in this case, to 'tune in' with progress, of lack of it, in **Arms Control** negotiations.

This book focuses, unashamedly, on the NATO versus Warsaw Pact confrontation. This is because, in the foreseeable future, any 'world war' would be fought in this context. Similarly, it is with this in mind that the superpowers and their allies make defensive preparations. Regional wars and competitions for local military superiority are in constant session, but appear in this book only so far as they interweave with the theme of global rivalry between the superpowers. Furthermore, the book has had to be written very much from a Western position, as the policies and strategies of the NATO member governments and militaries are far better known than those of the Warsaw Pact. This does not mean that the entries have a pro-Western slant: I have tried to be impartial throughout—which is *not* the same as abstaining from comment entirely! I have done all that can be done to ensure the technical accuracy of the entries. Each of them has been read by a senior member of the US arms control fraternity, and most have also been seen by academics in the relevant fields. I have always listened with interest and concern to any objections.

It is customary to thank one's editors. I wish to make no exception here. Alan Oliver and Paul Kelly have done so much to make this book readable that they are not *really* entitled to the usual waiver of responsibility for error which authors ritualistically offer to their helpers. But I do so waive any shared responsibility.

David Robertson
Oxford
6 August 1987

Abbreviations

AAM Air-to-Air Missile
ABM Anti-Ballistic Missile
A-bomb Atomic bomb
ACE Allied Command Europe
ACM Advanced Cruise Missile
ADM Atomic Demolition Munition
AEC Atomic Energy Commission
AEW Airborne Early Warning
AFCENT Allied Forces Central Europe
AFNORTH Allied Forces Northern Europe
AFSOUTH Allied Forces Southern Europe
AFV Armoured Fighting Vehicle
ALCM Air-Launched Cruise Missile
ANF Atlantic Nuclear Force
ANZUS Australia, New Zealand and United States (Treaty)
ASAT Anti-Satellite Weapon
ASM Air-to-Surface Missile
ASW Anti-Submarine Warfare
AWACS Airborne Warning and Control System

BAOR British Army of the Rhine
BMD Ballistic Missile Defence
BMEWS Ballistic Missile Early Warning System

C^3I Command, Control, Communications and Intelligence
CBM Confidence-Building Measure
CD (Forty Nations) Conference on Disarmament
CDE Conference on Security and Confidence-Building Measures and Disarmament in Europe
CDI Conventional Defence Initiative
CENCOM Central Command
CENTAG Central Army Group, Central Europe
CEP Circular Error Probable
CIA Central Intelligence Agency
CINCHAN Commander-in-Chief Channel
CNA Center for Naval Analyses
CND Campaign for Nuclear Disarmament
CSBM Confidence and Security Building Measure
CSCE Conference on Security and Co-operation in Europe
CTBT Comprehensive Test Ban Treaty
CW Chemical Warfare
CWC Chemical Warfare Convention

DDG Guided Missile Destroyer
DEW Distant Early Warning
DGZ Designated Ground Zero
DIA Defense Intelligence Agency
DOD Department of Defense (US)

DPC Defence Planning Committee (of the North Atlantic Council)

ECM Electronic Counter Measure
EDC European Defence Community
EFA European Fighter Aircraft
ELINT Electronic Intelligence
EMP Electromagnetic Pulse
EMT Equivalent Megatonnage
END European Nuclear Disarmament
ET Emerging Technology
EuroGroup Informal Group of NATO European Defence Ministers
EW Electronic Warfare

FAR Force d'Action Rapide
FEBA Forward Edge of the Battle Area
FOFA Follow on Forces Attack

GCHQ Government Communications Headquarters
GIUK Greenland – Iceland – UK Gap
GLCM Ground-Launched Cruise Missile
GNP Gross National Product
GSFG Group of Soviet Forces, Germany

H-bomb Hydrogen bomb
HLG High Level Group (of NATO)

ICBM Intercontinental Ballistic Missile
IDA Institute for Defence Analyses
IEPG Independent European Programme Group
IISS International Institute for Strategic Studies
INF Intermediate Nuclear Forces
IRBM Intermediate-Range Ballistic Missile

JCS Joint Chiefs of Staff

kg kilogramme(s)
km kilometre(s)
kt kiloton(s)

lbs pounds (weight)
LNO Limited Nuclear Option
LRINF Long-Range Intermediate Nuclear Forces
LRTNF Long-Range Theatre Nuclear Forces
LTDP Long-Term Defence Programme

m metre(s)
MAD Mutual Assured Destruction
MARV Manoeuvrable Re-entry Vehicle
MBFR Mutual and Balanced Force Reductions
MBT Main Battle Tank
MIRV Multiple Independently-Targeted Re-entry Vehicle
MLF Multilateral Force
MLRS Multiple-Launch Rocket System
MOD Ministry of Defence (British)
MRCA Multiple-Role Combat Aircraft
MRV Multiple Re-entry Vehicle
MSBS Mer-Sol Balistique Stratégique
mt megaton(s)
MTE Equivalent Megatonnage
MX Missile Experimental

NAC North Atlantic Council
NATO North Atlantic Treaty Organization
NBC Nuclear, Biological and Chemical Warfare
NCA National Command Authorities
NDB Nuclear Depth Bomb
NFZ Nuclear-Free Zone
NORTHAG Northern Army Group, Central Europe

NPG Nuclear Planning Group
NPT Non-Proliferation Treaty
NSA National Security Agency
NSC National Security Council
NSWP Non-Soviet Warsaw Pact
NTMs National Technical Means

OMG Operational Manoeuvre Group
OMT Other Military Target
OSD Office of the Secretary of Defense

PD-59 Presidential Directive 59
Penaids Penetration Aids
PGMs Precision-Guided Munitions
POMCUS Pre-positioning Of Material Configured to Unit Sets
PPBS Planning, Programming and Budgetary Systems
psi pounds per square inch
PsyOpps Psychological Warfare
PTBT Partial Test Ban Treaty

RAF Royal Air Force
RAND Research and Development (Corporation)
RDF Rapid Deployment Force
RV Re-entry Vehicle

SAC Strategic Air Command
SACEUR Supreme Allied Commander Europe
SACLANT Supreme Allied Commander Atlantic
SALT Strategic Arms Limitation Talks
SAM Surface-to-Air Missile
SDI Strategic Defense Initiative
SHAPE Supreme Headquarters Allied Powers Europe
SICBM Small Intercontinental Ballistic Missile
SIGINT Signals Intelligence
SIOP Single Integrated Operations Plan
SIPRI Stockholm International Peace Research Institute
SLBM Submarine-Launched Ballistic Missile
SLCM Sea-Launched Cruise Missile
SLOCs Sea Lines of Communication
SNLE Sousmarin Nucléaire Lanceur d'Engins
Sonar Sound, Navigation and Radar
SOSUS Sound Surveillance System
SRAM Short-Range Attack Missile
SRINF Short-Range Intermediate Nuclear Forces
SSA Selective Service Administration
SSBN Submarine, Ballistic, Nuclear
SSBS Sol-Sol Balistique Stratégique
SSKP Single Shot Kill Probability
SSM Surface-to-Surface Missile
SSN Submarine, Nuclear
STANAVFORCHAN Standing Naval Force, Channel
STANAVFORLANT Standing Naval Force, Atlantic
START Strategic Arms Reduction Talks

TACAMO Take Charge and Move Out
TNF Theatre Nuclear Forces
TNT Trinitrotoluene
TTBT Threshold Test Ban Treaty
TVD Theatre of Military Operations

UK United Kingdom
UKAEA United Kingdom Atomic Energy Authority
ULF Ultra-Low Frequency
UN United Nations
US(A) United States (of America)
USAF United States Air Force
USSR Union of Soviet Socialist Republics

WEU Western European Union
WP Warsaw Pact

A

Accidental War

Since the early days of the nuclear age there has been a pervasive fear of a total nuclear war triggered by accident. It has, of course, always been possible for wars to begin for trivial reasons, but in the past a mistaken belief by one party that it was under attack from an enemy could be corrected before very much had happened. The reason why accidental war is now seen as so much more likely, and so very much more disastrous, is that automatic systems of alert could theoretically result in the launching of missiles that can not be recalled. Should either superpower launch a major attack on the other retaliation would have to be ordered so rapidly that there would be virtually no time for the two opponents to assess the situation and agree that neither side intended aggression.

The main factor leading to fear of accidental war is that the warning time of a perceived impending missile attack is so short: about 20 minutes for intercontinental missiles, and perhaps half that for submarine-launched missiles. When such an attack appeared on radar screens of the target country, its nuclear forces would immediately be placed at the highest state of alert, its bombers launched and its ground-based missiles held at a state just prior to launch. Because the bombers can be recalled to base there should be no accidental war, because a false alarm would be discovered before the **National Command Authorities** sanctioned the final attack order. In fact the USA has, on several occasions, brought its forces to a very high state of readiness as a result of radar or computer malfunctions, although there has never been any serious danger of a launch order being given. A series of checks is built into the operating system (see **Fail-Safe**), which requires that bombers and missiles return to a passive state unless a verifiable launch order, tested against codes that only the President or his deputies can release, is given.

As long as both superpowers adhere to a fail-safe doctrine based on direct evidence of actual attack, there is no imaginable risk of an accidental nuclear war. However, some problems of ensuring a secure deterrent force have led theorists to consider the doctrine of **Launch on Warning** or **Launch Under Attack,** the adoption of which could drastically increase the chances of an accidental war. Launch on warning and launch under attack suggest that the USA cannot risk postponing retaliation until nuclear warheads actually explode on its territory. This is because it is feared that the US **ICBMs** are now vulnerable to a Soviet **Pre-emptive Strike.** If there was serious evidence of an attack, and the USA did not respond immediately, it might find that up to 90% of its ICBM sites were destroyed before it could retaliate. This problem, sometimes known as the **Window of Vulnerability,** leads to the argument that evidence from radars and satellites, if strong enough, would justify the immediate launching of the ICBM force in order to prevent its virtual destruction. As far as it is known, no such command system has been employed by any nuclear power, neither would there be any justification for it, because the still invulnerable submarine forces (see **SLBM**) would be sufficient to ensure **Assured Destruction** strike level.

The other common worry about an accidentally-started nuclear war is that someone, or some military body, might deliberately launch missiles without authority. Security regulations make it quite impossible for any one person to fire a nuclear weapon without authorization, and a serious attack (such as a squadron of bombers or a group of missiles) would require either collective insanity or criminal conspiracy among a whole group of people. As the military personnel employed in the nuclear services are most carefully chosen and trained for stability and obedience, there is no measurable probability of such collusion, without which no accidental war is credible.

A further theoretical argument that has sometimes lent credence to the fear of accidental war comes from the rarified heights of **Deterrence** theory. In some of the more far-fetched **Scenarios** of this discipline (it has been referred to as nuclear theology), questions have been raised about what a country should do even if it knew that an incoming attack was an accident. Some have suggested that any attack must be treated as intentional, if only because the opponent will not be able to trust you to accept its absence of hostile intention. Fearing you will believe the attack to be deliberate, it will have to follow up as though it *had* intended to make a pre-emptive strike, for fear of your retaliation, and thus you must indeed retaliate, even though you know that the original attack was an accident. No one can really take such logic seriously, but obviously it helps to spread fear. Considerable effort is expended on ensuring communications between the superpowers, for example by the **Hot Line** teleprinter link, and a whole theory of **Crisis Management** has been developed. The latter has recently produced the suggestion by some US Senators for the joint staffing by Soviet and US personnel of an international crisis management centre, probably in Switzerland, to prevent any such scenario becoming a reality.

Action – Reaction

Action – reaction is a phenomenon often used to explain **Arms Races,** international crisis behaviour, and the development of conflicts—whether armed or not. It sometimes implies that each 'actor' in a situation responds to the behaviour of others with pre-planned moves, and that the sequence of actions as a crisis unfolds is not really directed by long-term goals, genuine motivation or, in a sense, even intent. A classic example would be the events of July and August 1914, where the initial response to the assassination of Archduke Ferdinand of Austria set off a chain reaction: mobilization by Russia sparked mobilization by Germany, which in turn led to mobilization by further nations. In each case one participant was *reacting* in a more or less automatic way to another, rather than making a deliberate and thought out move.

With an arms race the action – reaction pattern would occur where the implementation of a new **Weapons System** by one side automatically led to the development of an equivalent or superior system by the opponent. The automaticity is shown by the fact that the first actor most certainly did not deploy the system in order to get its opponent to do likewise, and that the opponent had no intention, originally, of developing such weaponry. Despite appearances this is not an almost tautological explanation for an arms race. It could have been the case that the first nation to act *did* deploy its new system to force the opponent into investing scarce resources in combating it, out of belief that the strain on the opponent's economy would, in the long term, make it weaker. A thesis much like this lies behind the opposition of some US defence analysts to **Arms Control** agreements. Some schools of thought believe that the USA would automatically win an economic competition to afford ever more expensive arms.

Advanced Cruise Missile (see Air-Launched Cruise Missile)

Air Burst

The effects of nuclear explosions vary greatly depending on the altitude at which the warhead is exploded. The principal difference is whether or not the explosion will create a large amount of radioactive **Fall-Out.** If the warhead is detonated at a sufficiently high altitude for the immediate fireball not to touch the ground, fall-out will be relatively slight. Blast damage to buildings and people, however, will actually cover a wider area than it would with a **Ground Burst,** and thus will satisfy most general strategic requirements. Therefore it is expected that, in a nuclear war, most weapons would be set to detonate some thousands of feet above ground level, hence the jargon 'air burst', with ground bursts only planned for use against **Hard Targets.**

Air Superiority

About half-way through the Second World War military organizations became fully familiar with the use of air power, and it was realized that command of the air over a battlefield, and over the area of naval operations, could be crucial for success. Air superiority is the ability to dominate airspace, and consists of two elements. Firstly it means being able to prevent enemy aircraft, especially bombers and reconnaissance planes, from operating over one's own lines; this requires a sizeable interceptor or 'fighter' force. Secondly it implies an ability to fly missions over the enemy's lines, attacking its troop concentrations and supply network, and observing military movements, using what are traditionally thought of as 'fighter-bombers'.

Historically armies, even if stronger on the ground than their enemy, have found it extremely difficult to prevail if they have been inferior in air power. The most recent example, and a powerful one, was the tremendous problem facing the British task force in the Falkland Islands conflict. Although the British troops were superior in training and equipment to the Argentinians, and though the Argentine navy played virtually no active role, the British fleet was provided with very little air support, and no advance warning of air attacks by the Argentine air force, which was able to operate from the mainland. Consequently, serious losses were incurred in ships and men and, had the Argentine air force been better trained, or just a little luckier, it could have prevented the recovery of the islands.

Ever since the naval losses to Axis powers' air forces in the Second World War it has been appreciated that naval forces are extremely vulnerable in the face of enemy air superiority. One consequence of this is that most modern tactical doctrines stress the need for the opposing air forces to fight the battle for air superiority immediately conflict breaks out, and not to engage in close support to their ground forces until it has been achieved. Even this is not new: the importance of the Battle of Britain in 1940 was that the attack by the German *Luftwaffe* was intended to destroy the RAF, in the air and on the ground, as a precursor of an invasion which could not be risked until that goal had been achieved. Only later, having failed in this aim, did the Germans turn their attack towards British cities.

In terms of possible East – West conflict in Europe, air superiority would be sought at almost all costs, because both NATO and the Warsaw Pact are dependent on complex and long-range reinforcement and supply routes.

Interdiction raids on the enemy reinforcements might be decisive; thus the first aim must be to destroy the opposing air forces. A major problem is that air-bases are highly vulnerable not only to attack by aircraft, but also by **Battlefield Nuclear Weapons, Chemical Warfare,** and, possibly, by the new generation of **Emerging Technology** weaponry. At present the air forces of NATO and the Warsaw Pact are probably more evenly balanced than other categories of **Weapons Systems,** especially if technological capacity is considered as well as sheer numbers, and the air battle at the beginning of any conflict is quite likely to determine the final outcome.

Airborne

Airborne troops are, in general, those which can be transported into, or near to, a battle zone by aircraft or helicopter. They should not be confused with **Airportable** troops, which simply can be moved large distances with all their equipment. Various and subtle gradations of airborne troops exist in military manuals. A rough distinction is between those troops which can be dropped by parachute directly over an **Airhead,** and those which can only be delivered by helicopter near to a battle zone. The first category, equivalent to paratroops of the Second World War, is not very important in modern tactical doctrine. Even during the Second World War there were very few completely successful paratroop drops. Apart from the vulnerability of paratroopers while descending, the organizational difficulties and the lightweight equipment necessary for their air mobility caused major problems as soon as they came into contact with ground-based heavy troops. Under the conditions of modern ground-based air defence, the attrition rate for an airborne assault would be very high.

The second category, airborne or **Airmobile** troops flown by helicopter to a point near to a combat area, with long-range support from ground-based artillery or **Tactical Air Forces,** is much more important. Such success as the USA had in Vietnam was very much dependent on this form of troop transport. **Battlefield Mobility** has a very high priority in modern tactical doctrine, especially with the proponents in the USA of **Manoeuvre Warfare.** This is partly because it is the only convincing way of explaining how NATO's ground forces can be a credible deterrent to an invasion by **Warsaw Pact** troops, given that the latter have a considerable numerical superiority.

Nevertheless it is important to realize that helicopter mobility is both expensive and risky. It takes 24 helicopters to deliver 300 troops simultaneously to one dropping zone. Each helicopter can be destroyed relatively easily, not just by anti-aircraft missiles, but even by small arms fire. Only the few, very expensive heavy-lift helicopters in any army corps can carry even a light howitzer or armoured reconnaissance vehicle. Although airborne troops such as the British Parachute Regiment or the

US 82nd Airborne Division remain publicly prestigious élite troops, the **General Staffs** will have some difficulty deciding how they should be used in any future conflict.

Air-Breathing Missiles

A basic distinction between missiles, aircraft, and other vehicles of flight is between those which require an oxygen-laden intake from the atmosphere for operation of their motors and those which do not. Any ordinary aircraft, jet- or propellor-driven, requires an oxygen-rich atmosphere for combustion inside its engines, and is thus described as 'air-breathing'. Missiles designed for flight beyond the atmosphere, whether for use in space or ultimately bound for re-entry into the atmosphere, such as **ICBMs,** cannot be air-breathing. These missiles require oxygen, in one form or another, to be carried as part of their fuel load to enable combustion to take place. The distinction between air-breathing and other airborne objects is vital to radar and other early warning systems, which can detect the difference at some range; air-breathing vehicles can, in principle, be intercepted, while others are not so vulnerable. The most important current example of this distinction is that **Cruise Missiles,** having jet rather than rocket engines, are air-breathing, while ballistic missiles, which otherwise have a similar **Radar Signature,** are not.

Airhead

An airhead, by analogy to a beachhead, is the immediate landing area captured by troops in the first stage of an **Airborne** attack. It would most probably be achieved by means of a paratroop landing, although in principle an airhead could also be gained by helicopter-delivered **Airmobile** troops. The need is to capture, clear and secure an area into which the heaviest equipment and less immediately combat-ready forces can be dropped. Once this activity has ceased the airhead becomes a base from which to advance in pursuit of the attack's objectives. A slightly different example would be where an airborne assault by paratroopers secured an airfield so that the army would have a simple route for delivery of major forces of an **Airportable** nature behind the enemy's lines.

Air-Land Battle Concept

The concept of an 'air-land battle' has been the dominating theme in US army thinking for some years. It is similar to NATO's current official doctrine of **Follow on Forces Attack,** which derived from it. Although it is presented as a doctrine with world-wide relevance, its main drive comes from dealing with conditions on the **Central Front.** The main point of the doctrine is the need to integrate completely the **Tactical Air Forces** with the ground troops in one concerted tactical plan over the entire **Theatre**

of War. In the European context this involves using air-power, and perhaps long-range army weapons, for an **Interdiction** attack on the second echelon Soviet forces while the ground troops hold and try to defeat the first echelon (see **Echeloned Attack**).

One hazard inherent to the air-land battle concept is that it calls for an integration of nuclear, chemical and conventional weapons and might, in fact, lead to early **Escalation** to a nuclear war. Ultimately, the validity of the concept depends on the use of **Emerging Technology** conventional weapons to fight the long-distance part of the war. As these are not yet available, the reliance on nuclear weapons is a matter of concern to many American military thinkers.

Air-Launched Cruise Missile (ALCM)

The air-launched cruise missile is one of three variants of **Cruise Missile** developed by the USA since the mid-1970s. The main purpose of the ALCM is to maintain the utility of the US Air Force bomber fleet, known as the **Strategic Air Command (SAC).** Bombers are now highly vulnerable to the USSR's anti-aircraft defences, and it is improbable that sufficient numbers of the current SAC force of B-52 aircraft could penetrate deep enough into Soviet airspace to deliver a powerful attack. ALCMs, however, could be carried to within range of Soviet targets and launched while the aircraft were still at a safe distance from anti-aircraft defences. The current generation of ALCMs have a range of about 2,500 kilometres, so bombers approaching the USSR from one compass direction or another could target virtually any spot in Soviet territory. American nuclear strategy (like the French, but unlike the British) has always held that missiles launched from piloted aircraft form an indispensable part of the deterrent force, along with **ICBMs** and **SLBMs.** (The three together are known as the deterrent **Triad.**)

To further this policy a major ALCM production programme has been underway since 1976, although the original intention to produce over 4,000 missiles has now been reduced to 1,700. Each bomber can carry at least 12 of these missiles, and plans exist to increase the total capacity of each aircraft to 20. As the new American bombers, the B1-B and the **Stealth Bomber,** come into active service, similar weapon loads will be carried. By the end of 1986 over 1,300 ALCMs were deployed at five USAF bases. As each missile carries a 200-kiloton warhead, the importance of this part of the nuclear triad is evident. It is likely that they would be extremely effective weapons, because they fly at a height of only 30 metres, following closely the terrain from maps loaded into the on-board computer. As a result Soviet radar facilities may not be able to detect them at all. However, once detected they may become vulnerable to ordinary anti-aircraft defences. A major disadvantage is that all piloted aircraft forces are very exposed to attack while on the ground and, particularly as they become more effective, are likely to be made an increasingly prominent target for the Soviet **First Strike** ICBM arsenal. On the other hand the ALCM, because of its relative lack of speed (it travels at about 800 km per hour), may be a **Weapons System** conducive to international stability. As missiles would not reach their targets for several hours they are unlikely to be used as first strike weapons. The bombers themselves can be ordered to take off at any sign of a possible attack, but no irrevocable decision need be taken by the **National Command Authorities.**

The original plan to build 4,000 ALCMs was reduced because of the development of a new generation of cruise missile, the Advanced Cruise Missile (ACM), which has a range of about 4,800 km, and a much more sophisticated navigation capacity; it may even be equipped with **Electronic Counter Measures,** which would make interception even more difficult. Orders for at least 2,800 such missiles have been placed, and they will ultimately equip the next generation of SAC aircraft after the phasing out of the B-52 bombers, which have been the mainstay of the air-launched part of the triad since the mid-1950s. The development of the ACM makes it clear that the USA intends to maintain an extremely powerful air-launched deterrent, despite its vulnerability on the ground, as insurance against any future breakthroughs which might increase the threat to ICBMs in their silos or seriously threaten underwater SLBM forces.

Air-Lift

Air-lift refers to the need to carry troops and equipment over large distances by air if they are to be brought into crisis areas rapidly (see **Force Projection**). Because modern wide-bodied jets have made mass tourist travel so easy, it is often thought that air-lift for military purposes must be relatively easy. In fact the numbers of troops and the weight of equipment needed for any serious military activity are so enormous, compared with the still very restricted ability of aircraft to lift them, that no major power is even close to having sufficient air-lift capacity for its possible strategic needs. A simple calculation will make this clear. With even a moderate amount of luggage, a big civilian aircraft, such as the Boeing 747, can carry only some 400 passengers. This is equivalent to about two companies, only half a battalion, of troops. In fact, with the amount of weapons, equipment, ammunition and supplies needed to keep a soldier active for even a few days, one company is probably the most that could be carried on a single aircraft. In order to deploy an infantry brigade, the size of unit initially landed by ship on the Falkland Islands by the British in 1982, some 20 747s would be required, and any heavy equipment would still have to be left behind.

Air-lift is so expensive, and so inadequate, that forces which depend on it for quick deployment of troops to a distant part of the world could only ever achieve limited results. For example, the US Rapid Deployment Force (see **CENCOM**), intended for intervention in the Middle East, would only be able to establish a small **Airhead** and then hold it for the month which it would probably take to **Sea-Lift** the reinforcements and supplies required for a full campaign. A really major air-lift operation, such as would be needed to get US reinforcements into Europe at the beginning of a NATO – Warsaw Pact conflict on the **Central Front,** would involve the commandeering of most civilian airliners in the USA. One well-known statistic helps put this into context. During the latter days of the Berlin Air-lift the supply of daily necessities for the population of Berlin required the landing of over 1,000 aircraft a day—yet *they* only had to travel the short distance from West German airfields.

Airmobile

An airmobile unit is one that can be delivered directly into a battle zone by helicopters. Inevitably it has to be lightly armed, but will usually be fighting within range of support in the form of artillery and ground attack aircraft. They are to be distinguished from **Airborne** troops in general, which include the traditional paratroops of the Second World War, and **Airportable** troops, which are simply those that can be carried long distances by air. The first extensive use of airmobile fighting units was by the USA during the Vietnam War, where whole regiments of **Infantry** frequently were landed deep within enemy territory on search and destroy missions. The US army even invented a whole new category of regiment, Air Cavalry, especially trained for these missions.

Both the Warsaw Pact and NATO armies have invested in airmobile forces, although the former probably places more faith in them. The major problem is that the helicopters are highly vulnerable, particularly during landing operations. Even against the technologically ill-equipped troops of North Vietnam and the Vietcong the USA is reported to have lost well in excess of 2,000 helicopters on airmobile troop missions. In the context of the **Central Front,** where the armies are very heavily armed with anti-aircraft weapons, it is likely that the attrition rate would be too high to use such troops, except for special purposes. They are also, for obvious reasons, extremely expensive to support. Even modern helicopter designs are unable to lift more than perhaps a dozen armed and equipped soldiers, meaning that about 200 helicopters would be needed to move a three battalion task-force.

Airportable

Airportable forces are those that can be carried considerable distances by aircraft, along with all their equipment. Thus they are typically light forces, either purely **Infantry** or at least accompanied with only light armour and artillery. They can be brought close to a battle area, as long as there are suitable air-bases, but cannot be inserted directly into the battlefield. Thus they are distinguished from **Airborne** and **Airmobile** forces. Airportable troops are particularly valued by military planners whose priority is **Force Projection** into areas where **Forward Based** forces cannot be relied upon. The limitations on their use can be severe because of the restriction on their equipment and resupply, and because of the reliance on available landing sites. Because of the huge expense of maintaining the transportation squadrons, even the superpowers can only afford limited amounts of airportable capacity.

All Volunteer Force

During the Vietnam War a major reason why the American public was hostile to the war effort was the use of conscript troops. Furthermore, the need to minimize this opposition led to a variety of militarily inefficient practices, and often to over-cautious tactics aimed primarily at cutting American casualties rather than defeating the enemy. As a result the US military establishment was almost as eager as the anti-War elements in Congress to abolish **Selective Service,** as the **Conscription** system was known. The drafting of civilians into military service during peacetime was abolished in 1973, under President Nixon, for the first time since the Second World War, and the American military became known as the All Volunteer Force.

Initially the services experienced great difficulty recruiting the required numbers. The overall quality of troops was poor, with very low educational standards, morale problems leading to high rates of absenteeism, and a low re-enlistment rate among trained technicians and senior non-commissioned officers. By the late 1970s Pentagon inspectors were forced to regard many divisions as below combat readiness. However, for a variety of reasons, not the least among which have been unemployment and the general swing to the right of public opinion under President Reagan, this problem has since been abated. The US services are finding recruitment of highly-motivated and well-educated people surprisingly easy. Re-enlistment rates have risen, and the army, for example, now has the highest proportion of high school graduates in its ranks in history.

Allied Command Europe (ACE)

Allied Command Europe is one of the three major commands into which NATO's forces are organized. (The other two are Allied Command Atlantic, and Allied Command Channel.) ACE is, therefore, the main land and air war structure for NATO, and comes under the command of **SACEUR** (Supreme Allied Commander Europe), who is always an American general. It covers an enormous land area, extending from Norway's North Cape south to the Mediterranean, and from the Atlantic to Turkey's eastern border. The total number of troops and equipment in this area under SACEUR's command is somewhat difficult to define, partly because of the ambiguity of France's position (it is not formally a member of NATO's military organization), and also because portions of other country's forces, particularly the UK's, are not NATO-dedicated. The following table, which excludes French forces, shows some approximate totals:

Total ground forces	2,900,000
Main battle tanks	over 20,000
Land attack aircraft	37,300
Divisions deployed (in peacetime)	33
Total divisions (full mobilization)	132

Source: *The Military Balance 1986/87,* IISS, London.

The headquarters of this command, known as **SHAPE** (Supreme Headquarters Allied Powers Europe) used to be at Versailles in France but, after French withdrawal from the integrated command structure of NATO, moved to Casteau, near Mons, in Belgium. The principal activities of NATO military planning are focused here, with the Permanent Military Representative from each of the 14 military powers in NATO (that is, excluding France, and also Iceland, which has no defence forces of its own) liaising between SACEUR and their respective **Chiefs-of-Staff.** Apart from full war, the most important function of ACE is the provision of a stand-by force, consisting of units in various countries, for instant intervention in trouble spots. Known as the ACE Mobile Force, and under the direct control of SACEUR, it could be of particular importance in the vulnerable Norwegian and Turkish border areas (see **Northern Flank** and **Southern Flank**).

Anti-Ballistic Missile

An anti-ballistic missile is designed to intercept a ballistic missile, or one of its **Warheads** (or, to be more technical, one of the **Re-entry Vehicles**), at some stage during the later part of its trajectory. The only system of anti-ballistic missiles currently in operation is that set up to defend Moscow from nuclear attack which is, in NATO circles, known as **Galosh.** During the 1960s both the USA and the USSR worked on developing such a defence against **ICBMs,** and both installed systems. For a variety of reasons, neither superpower wished to commit itself to the inevitable **Arms Race** which would follow were these systems to be expanded, and it was relatively easy for them to come to an agreement, the 1972 **Anti-Ballistic Missile Treaty,** restricting ABM deployment and development. The treaty allowed each signatory two ABM systems, to cover the national capital and an ICBM site. In the event, neither the USSR nor the USA deployed ABMs to the agreed limit, and the USA actually decommissioned their one existing system, the Nike-Hercules batteries protecting Washington, DC. The USSR did not build its second system, nor did it bring its existing one up to the allowed maximum of 100 interceptor missiles, leaving it at the 64 or so that it had in place before the treaty was signed. It is believed that 32 of the Galosh missiles have been removed, although this may just be preparation for their replacement with more modern missiles.

Intercepting an incoming ICBM is an enormously demanding task, which was basically well beyond the technological capacity of the 1960s, and is even now extremely challenging, as the research problems of the **Strategic Defense Initiative** show. To a large extent the Soviet and American willingness to negotiate was because of serious doubts that the existing systems could work at anything but a very low level of efficiency, combined with a fear that the cost of an even remotely effective system would be crippling. It is generally thought that the USSR retained its Moscow ABM shield principally as protection against the very small nuclear arsenals of France and the UK, rather than the larger volume of US missiles, which could swamp the limited Galosh system even if it worked with complete accuracy and reliability. (See also **Ballistic Missile Defence.**)

Anti-Ballistic Missile Treaty

The Treaty between the USA and USSR on the Limitation of Anti-Ballistic Missile Systems was signed in Moscow on 26 May 1972. It was part of the Strategic Arms Limitation Talks **(SALT I)** which had started in Vienna in April 1970, and was the most clear-cut success of the process. The treaty had two main points. It restricted the two parties to having relatively small **Anti-Ballistic Missile (ABM)** systems. The two superpowers were both to be allowed to set up two ABM systems, each of no more than 100 interceptors. One could be used to protect the national capital and one to guard **ICBM** sites. The technological limitations of that period meant that neither country believed a cost-effective and efficient ABM system could be constructed, and were perfectly happy with the restrictions. In fact, the USA subsequently demolished the only ABM site it did have, consisting of Nike-Hercules batteries around Washington, DC, and the USSR neither built its second **Galosh** system nor equipped the one it already had, protecting Moscow, with its full complement of 100 interceptor missiles.

The second aspect of the ABM Treaty later became much more important. Various clauses severely restricted research on more advanced ABM systems by banning full-scale testing, although allowing purely laboratory research. These restrictions covered both the interceptor missiles and the radar systems needed to operate them. When, in 1983, President Reagan announced a major **Strategic Defense Initiative (SDI)** research programme it became a matter of considerable debate how far the USA could proceed without breaching the ABM Treaty. It is generally agreed that research of some form is legitimate, if only because of the impossibility of preventing theoretical and small-scale laboratory research. It is also generally accepted that actual deployment of any device that is a clear development of the old systems, based on interceptor rockets that would be fired into the path of an incoming ballistic missile, would be within the bounds of the treaty. The problems arise from an ambiguous area. Does the treaty preclude the testing of new methods of destroying ballistic missiles? The USSR and many voices in the West, including, for example, Senator Sam Nunn, the respected and powerful Democratic Chairman of the Senate **Armed Services Committee,** think such moves should be prevented. Under this 'narrow' interpretation, serious testing or deployment of *any* product of the SDI programme would put the USA in breach of the ABM Treaty.

The US administration began, in 1986, to insist that the correct interpretation of the treaty (one never thought of for the first 14 years of its life) actually allowed systems based on physical principles or 'new concepts' that had not been in operation in 1972. Although there *is* some textual ground on which to base this claim in part of the treaty, it is clear to most of those who took place in the original negotiations that the intent of the treaty would be contradicted by insisting on this new interpretation. The American counter-attack tends to revolve around claims that the USSR is itself in breach of the treaty. The main argument here is that the Soviet Union has built a very large **Phased Array Radar** system of the type best suited for handling an ABM system. Again there is a great interpretative problem. The USSR has *indeed* built such a radar station, at Krasnoyarsk. It *could* run an ABM system, but it could also have other uses—and the USSR insists that it does. Does the ABM Treaty prohibit the construction of systems with ABM potential, or only systems actually intended for such use? There is, again, no clear answer. The whole controversy is a good example of how hard it is to restrict activity that the superpowers actually want to engage in by treaty. The ABM Treaty was agreed, and abided by, only because it forbade the superpowers to do something they did not want to do anyway. As soon as conditions changed, the treaty came under great strain, as would any other treaty under similar circumstances.

Anti-Satellite Warfare (ASAT)

Anti-satellite weapons, often referred to as ASATs, are not as new a development as is sometimes thought. The US Air Force started planning for an anti-satellite strategy almost before the space age began, and tested a weapon successfully as early as 1963. It is believed that the USSR had carried out tests by 1968. However, it seems that no working system has ever been fully deployed, and the new phase of testing and research does represent a major technological and strategic development.

ASATs are seen by many as extremely dangerous because they threaten a destabilization of the nuclear balance. Satellites are now crucial to the military command and control functions, intelligence gathering, navigation and targeting operations of both superpowers (see **C³I**). A clear and unrivalled capacity for one superpower quickly to destroy the other's orbiting satellites would give it a considerable advantage, and would produce a very real fear on the part of the vulnerable power that it could be subject to a **First Strike** attack. If one of the superpowers suddenly lost a large portion of its satellites it would be unable to detect signs of an impending nuclear attack, and might not even know that a strike had been launched until a few minutes before it suffered the resulting damage. Among other risks, this might prevent the use of a **Launch on Warning** response. In a non-nuclear warfare **Scenario,** loss of satellites would seriously impair both communications with one's own forces, and the tracking of the enemy's, resulting in a major strategic, and probably tactical, advantage to the enemy.

So far it has not become clear what real capacity will be developed, in part because of the considerable technical problems, but more because it is an area in which it may be in the interest of both superpowers to reach an **Arms Control** agreement. Successful anti-satellite weaponry would inevitably start a classic **Arms Race,** with measure and countermeasure following each other at great cost to both the USA and the USSR. The USSR is believed to have developed a cumbersome technique which involves launching a special satellite, more or less an orbiting nuclear bomb, and guiding it close to a target satellite before exploding it. The USA, on the other hand, has succeeded in testing a method in which a fighter aircraft flies to its upper range of about 30 kilometres, and fires a missile with a solid warhead which homes in on the target satellite, destroying it by a direct hit as a **Kinetic Energy Weapon.**

It appears that the USA is ahead in the anti-satellite sector of the arms race, but this is not necessarily a permanent lead. There is considerable feeling against the full development of an ASAT capacity in the US Congress, and the research programme may never be completed. Its future is, to a large extent, connected to the progress of the **Strategic Defense Initiative.** Ironically, as any form of complex **Ballistic Missile Defence** system is certain to be heavily dependent on satellites, it would hardly be in the American interest to spark a technological arms race in ASATs.

Anti-Submarine Warfare (ASW)

The origins of anti-submarine warfare are in the Battle of the Atlantic of the Second World War, and modern technology has only caused methods to change quantitively rather than qualitatively. Submarines are still very serious threats to surface fleets and, with the deployment of **Strategic** missiles in submarines, have become the single most important nuclear weapon carrier to both superpowers. Although the technology of submarine detection has developed, it is probable that improvements in submarine design have made the threat from this quarter even more serious than it was originally. The development of nuclear-powered submarines has removed the need for frequent surfacing to recharge batteries; a submarine can, if necessary, remain on patrol, submerged, for months. Weapons developments have also enabled submarines to attack with missiles from a range of thousands of miles, thus increasing their chance of evading either premature discovery or destruction after an attack.

Of course, there have also been considerable improvements in detection methods. Airborne search for submarines is a very complex method, with aircraft such as the RAF's Nimrod able to patrol for more than 12 hours at long range, searching for submarines by a variety of methods. These include not only dropping **Sonar** buoys which can detect submarines by sound, but also, among others, scanning for infra-red signs, and even for the anomalous patterns in the earth's magnetic field created by underwater craft. Helicopters carried by virtually all naval ships can carry out similar screening in a large area around a fleet, and some submarines themselves, known as **Hunter-Killer Submarines,** can detect and attack enemy underwater vessels. Finally, there are static detection systems, for example the US networks of listening devices fixed to the sea bed, known as **SOSUS,** which are able to detect any submarine passing through vital marine **Choke Points,** such as the **Greenland – Iceland – UK Gap (GIUK).**

More than anything else, each superpower seeks for a method of detecting and monitoring the other's strategic missile-carrying submarines (see **SSBN**), but it is generally accepted that neither side yet has this capacity, or is anywhere near to developing it. Although it is obviously very difficult to be sure, most commentators agree that the USA and its allies, particularly the UK, retain a considerable lead over the USSR in ASW, in part because the USSR has only relatively recently acquired a **Blue Water Navy,** even though their submarine fleet is large and advanced.

ANZUS

ANZUS stands for the Australia, New Zealand and United States Treaty, which was signed in September 1951, and came into effect in April 1952. It is an agreement for defence co-operation between the three countries in the Pacific area. Exactly which threats the three countries face in the Pacific has never been clear, although Australia did send troops to support the USA in Vietnam. The treaty itself, unlike, for example, the **NATO** treaty, does not involve fixed troop deployments or a common military staff. Since the election, in 1984, of the Labour government of Prime Minister David Lange in New Zealand, with its commitment to a **Nuclear-Free Zone** policy, ANZUS has effectively ceased to exist. The USA is not prepared to meet the non-nuclear demands of the New Zealand government, and has withdrawn intelligence-sharing arrangements. Although relations between the USA and Australia have not suffered to the same extent, there is inevitable strain due to the latter's regional loyalty towards New Zealand, and a treaty that was always more symbolic than actual has ceased to have any real meaning.

Appeasement

Appeasement was the term given to the policy taken by the British and French governments towards the Fascist dictatorships of Germany and Italy during the 1930s. It was argued, both at the time and later, that this policy, and perhaps that of Britain in particular, so encouraged aggression by Hitler and Mussolini that it made the Second World War inevitable. Appeasement consisted largely of weakness in the face of territorial expansionism and international aggression; it was a policy of not opposing firmly the behaviour of these countries, and even of giving in to their demands and acquiescing in their invasions of weaker countries. Although the politicians responsible for appeasement policies claimed that the military weakness of the democracies prevented their taking firm action against Hitler, the very failure to rearm in response to obvious expansionist ambitions is itself seen as part of appeasement.

The main argument is that a rearmament programme, combined with a firm condemnation of international lawlessness and a manifest preparedness to intervene militarily, would have stopped both German territorial expansion in central and eastern Europe and Italy's Abyssinian imperialism. It is further pointed out that much of the adventurism by both dictatorships was carried out *before* they were fully militarily prepared, and that appeasement not only encouraged them, but gave them time to complete their own armament programmes. Thus, if France and Britain had taken military action when Hitler moved troops into the officially demilitarized Rhineland in 1934, the German army would have been utterly unable to resist. Even as late as the invasion of Czechoslovakia in 1938 the apparently powerful German army was overstretched and highly vulnerable. However, on these and other occasions, including their aid to the anti-Republican forces in the Spanish Civil War, bold and provocative acts by Hitler and Mussolini were met by weak protests and ineffective diplomacy. It is thought that this policy instilled a belief in the minds of the Fascists that the democracies would never fight, allowing their ambitions to grow unfettered. However, on the invasion of Poland, Britain and France were ultimately forced to declare war, but on much less advantageous terms than would earlier have been possible.

Appeasement has become a powerful criticism of any foreign policy in which aggressive acts by a potential enemy are allowed to pass unpunished. It is feared that the same cycle of tolerating increasing aggression and failing to prepare militarily could lead to over-confidence in the enemy. Furthermore, it is not just aggression aimed at one's own vital interests which can be regarded as a long-term danger. If a state is thought to have an insatiable appetite for expansion of its power, a potential enemy, even if not the target of the current aggression, may only be able to avoid the long-term threat inherent to appeasement by acting in defence of a third country.

Some American critics frequently charge Western Europe of appeasement towards the USSR when its governments fail to join the USA in vigorous protests and diplomatic retaliation against what they see as Soviet imperialism. Western Europe's failure, for example, to accept President Reagan's economic sanctions over Poland has been seen as appeasement. More plausibly, perhaps, both Western European governments and the US government itself are charged with appeasement for not having taken more vigorous action after the Soviet invasion of Afghanistan. Some right-wing theorists go so far as to equate **Détente** with appeasement.

The problem with the concept of appeasement, as much today as in the 1930s, is that stable international relations and peaceful coexistence often require an acceptance of a nation's legitimate security interests within a sphere of influence. In addition, there is very little to be gained by taking a high moral tone over the almost inevitable expansionism of an emerging power. Indeed, classic **Balance of Power** theory actually requires that rising powers be accepted into the international system on an equal basis with existing major powers. Any government which seeks to keep international stability by listening sympathetically to claims of emerging powers, or which doubts the legitimacy of resorting too readily to sanctions, is in danger of being labelled an appeaser simply because of the chauvinistic interests of an ally.

Area Defence

In discussing defence against attack from the air, or from missiles, it is usual to distinguish between area defence and **Point Defence.** A typical and important use of the distinction is seen in the example of the defence of a squadron of naval ships from air attack. In any such squadron one or more ships will have the full-time task of anti-aircraft defence for the whole squadron. They will be armed to shoot down aircraft coming from any direction at long range, whether threatening themselves or other ships. Each ship in the squadron will be armed in a more limited way so that they can destroy, at short range, enemy aircraft which have evaded the area defence screen, and present a direct threat to that ship. The technical problems of the two tasks are very different. It is impossible to give each ship an area defence capacity because this would leave no room for its primary purpose. Yet the short-range point defence interception task is itself so complex that an area defence ship, entirely dedicated to this task, cannot also be expected to defend specific ships from an attacker.

An associated concept is that of **Layered Defence.** The most appropriate current conventional warfare example is again naval. The new plans for the US navy developed under President Reagan call for the use of large **Carrier Battle Groups** north of Norway and close to the Soviet coast to fly offensive missions against USSR military bases and ports. This task is so hazardous that a multiple set of 'layers' of anti-aircraft defence is planned, starting with long-range fighters flying from ground bases in Norway, followed by the aircraft carriers' own fighters, then area defence by anti-aircraft cruisers, and finally point defence by anti-aircraft guns on board the carriers themselves.

Area defence, layered defence and point defence have all become important in the arguments about planning **Ballistic Missile Defence** in the US **Strategic Defense Initiative.**

Armed Services Committees

Both the Senate and the House of Representatives of the US Congress have permanent standing committees on the armed services. The committee system in Congress is where all serious debate, analysis and amendment of legislation, and most of the work of overseeing the executive, takes place. As a result the committees have become very powerful, both in internal congressional politics, and in the whole policy-making process of American government. The armed services committees are among the most influential, and the growth in their power has accelerated since the early 1970s. Their primary function is to vet the military budget and, in particular, procurement plans. More recently they have also used this power to influence military doctrine and strategy. The US army's new doctrine of the **Air-Land Battle Concept** (which has been a great influence on the NATO doctrine of **Follow on Forces Attack**) was, for example, developed partly at the prompting of the Senate Armed Services Committee. Growing anger with the inefficiency of the US Department of Defense's budgetary and planning procedures and, perhaps more serious still, doubts over the military hierarchy's capacity for strategic planning and even adequate command and control, has forced both committees into an investigative mood.

In 1985 both committees, independently, started hearings aimed at producing potentially radical reforms of the whole defence structure. The House of Representatives Armed Services Committee has been considering proposals to change the military command structure to make it much more like a Western European **General Staff,** hoping to create a system which would give the President much better military advice. The Senate proposals, while not ignoring this aspect, concentrate more on the civilian hierarchy and, indeed, the failings of the Congress itself in its supervisory role. This latter aspect is crucial, because the way in which the Armed Services Committees have behaved has certainly contributed to the mismanagement of defence expenditure. The membership of both committees has always been heavily unrepresentative of their respective legislative houses. They tend to be dominated by Representatives and Senators from the southern States who, whether Democrat or Republican, are notably more conservative than other members of Congress. Therefore the committees have had a general ideological inclination to be sympathetic towards **Pentagon** requests for high defence expenditure. Furthermore, the members typically have major military bases or defence industry plants in their constituencies, and the tendency not only to accept any Pentagon request benefiting those areas, but even to *increase* procurement where it will help their constituents, is well established.

The overall consequence has been an absolute inability of the committees to form consistent long-range policy, or to deal with defence budgets comprehensively, rather than on a piecemeal basis. As these committees could play a vital role in ensuring the rational development of strategy and in designing a force structure to match it, the absence of which they themselves lament, and future reform in the Armed Services Committees could have an effect on Western defence quite out of proportion to its apparent importance.

Armoured Fighting Vehicle (AFV)

An armoured fighting vehicle is used for transporting **Infantry** into combat. It is somewhat like a tank, being both armoured to protect the occupants from machine-gun and shell fire, and moving on tracks rather than wheels so that it can traverse rough ground. However, it is usually only lightly armed with a small calibre cannon, or sometimes only with machine-guns, and the plating will not be strong enought to protect against armour-piercing weapons. The advent of such vehicles has transformed the nature of infantry combat, as well as considerably increasing the pace of battle by removing the need for infantry to move at a walking pace.

Originally AFVs were intended only to transport troops to or around the battlefield, when they would disembark for actual combat. Increasingly the tactics of the Warsaw Pact and NATO armies are geared to the idea of fighting from inside the vehicles themselves, making the entire battle one between moving armoured vehicles. The Soviet army is particularly well-equipped with AFVs, as is the US army in Europe. Little practical experience has been gained by any army in the use of these modern AFVs, and some military analysts doubt their value. There has for some years, for example, been a controversy over the US army's most modern version, the Bradley AFV, partly on the grounds of cost, and partly because it is thought to be too vulnerable. However, modern tactical thinking, especially the current US doctrine of the **Air-Land Battle Concept,** and even more so the fashionable **Manoeuvre Warfare** theories, so emphasize rapidity of battlefield movement that extensive use of armoured fighting vehicles seems inevitable.

Arms Control

Arms control is the policy of securing negotiated limits or reductions on the deployment of **Weapons Systems.** As such it has a very high priority in defence policy for all members of **NATO,** and probably of the **Warsaw Pact.** Arms control can usefully be divided into two types: one is the control over existing weapons systems, the other, sometimes called pre-emptive arms control, tries to prevent the original deployment of some new or potential weapon. The most important international successes in arms control have been the two Strategic Arms Limitation Talks treaties between the USA and the USSR, (see **SALT: the General Process**), signed in 1972 and 1979, which put limits on the numbers and types of **Strategic** nuclear weapons either side could deploy. It could also be argued that the **Anti-Ballistic Missile Treaty** which formed part of **SALT I** was a successful application of pre-emptive arms control. In limiting the number of ABMs which the USA or the USSR could build, and setting firm restrictions on the testing and development of new methods of **Ballistic Missile Defence (BMD),** it was essentially preventing the emergence of a weapons system rather than controlling the deployment of one that had already been achieved with any degree of reliability.

The history of arms control negotiations since the late 1970s has been complex and chequered. At the level that most people see as the highest priority, which is strategic nuclear arms control, there has been no progress since SALT II, though both the Reagan and Gorbachev administrations appeared eager to make some sort of progress in this area. Next in importance to strategic nuclear controls, most would place the limitation of **Intermediate Nuclear Forces (INF).** When the Soviet Union began to modernize its INF forces in the late 1970s with the **SS-20** intermediate range rockets, NATO responded with what was called the **Twin-Track** policy. This was to begin preparations to offset the SS-20 threat by deploying both **Ground-Launched Cruise Missiles (GLCMs),** and a new generation of the existing **Pershing** missile, and at the same time to pursue an arms control agreement with the USSR which would either ban or severely limit the numbers of either side's modernized **Theatre Nuclear Forces.** These negotiations, held in Vienna, finally collapsed and the USA went ahead with GLCM and Pershing deployment in Europe, while the Soviet Union increased the numbers of its SS-20 missiles. However, after a lengthy period of complete failure to make any progress towards arms limitation in this area, negotiations took on renewed energy and flexibility following the Reykjavik Summit of October 1986. Further initiatives from Secretary-General Gorbachev during the first half of 1987 raised hopes of an agreement on INF reductions being achieved. (See also **Double Zero Option, Walk in the Woods** and **Zero Option.**)

While the greatest media emphasis is placed on these nuclear arms control efforts, major effort has also gone into attempts at **Conventional** arms control, in a series of negotiating fora over a long period of time. Briefly, there are three fora and areas of attempted arms limitations. The longest running, in their current form, are the talks in Vienna on **Mutual and Balanced Force Reductions (MBFR),** which have been in progress since 1973—with virtually no success. The main aim here is to withdraw proportionate numbers of Soviet and US troops from the European theatre, to be followed by equivalently proportional numbers of troops from other Warsaw Pact and NATO members.

A completely separate approach to increasing international safety by negotiated agreements in the conventional area has been the attempt to agree on what are known as **Confidence-Building Measures (CBMs),** or sometimes Confidence and Security-Building Measures (CSBMs). This means that both the Warsaw Pact and NATO would inform each other of planned troop exercises and war games, possibly limit the number of troops engaged in them or observe restrictions on where in Europe they could be held, and allow observers from the other alliance to monitor the exercises. The idea is to remove, as far as is possible, the fear of surprise attack under cover of major troop movements for training purposes. Some very limited progress has been made in notification of such exercises, and arms control experts in Western Europe and the USA have some degree of hope that further progress can be made in this area. The main work on confidence building has been conducted under the auspices of the **Conference on Confidence and Security-Building Measures and Disarmament in Europe (CDE),** itself part of the **Conference on Security and Co-operation in Europe (CSCE),** in Stockholm since 1984.

The final major area of conventional arms control is the attempt to achieve a globally valid and enforceable complete ban on the production and stockpiling of **Chemical Weapons** in a **Chemical Warfare Convention.** Negotiations for such a convention have been in process, in one form or another, since 1967, and the most recent form of negotiations, which have been running since 1982 in the **Forty Nations Conference on Disarmament** at Geneva, were thought to be capable of some degree of success towards the end of the 1980s.

These several arms control fora, with their bewildering array of initials, and various and overlapping membership, make even the experts confused. However, it is generally thought better to have a wide range of bodies trying to deal with relatively specialized topics, than to risk obscuring possible agreements by combining issues on which one or other side might be intransigent with others where mutual self-interest may be discovered. What all the issues and fora have in common is a central problem, that of

Verification. Mutually acceptable limitations can often be arrived at by negotiation, only for the two sides to become deadlocked because of mutual distrust.

Typically, the Western countries have demanded that they be allowed ways of checking that Warsaw Pact countries really are observing the limitations. The USSR has traditionally responded by saying that no form of verification could be accepted which involved intrusive inspections of sites in Warsaw Pact territory. Instead they have argued that formal commitments should be enough—that nations should trust each other. Furthermore, they argued that adequate means existed to carry out remote checks without entering another nation's territory. There were suggestions in 1987 that the so-called Glasnost (openness) atmosphere in the Soviet Union might lead to increased willingness to accept verification procedures. If so, this would place the Western countries' position on inspection under closer scrutiny.

External techniques for verification are usually referred to as **National Technical Means (NTMs)** and involve, for example, observation by satellites to check on deployment of missiles or radar stations, or long-range seismographical equipment to monitor underground nuclear testing. The problems are obvious: satellites cannot see through the roofs of buildings, and seismographs may not be able to distinguish categorically between an earthquake and a nuclear explosion. It is impossible for those without access to classified material to be sure how often the West's complaints about the inadequacy of NTMs are real, and how often they are just excuses not to ratify an arms control agreement which might be against its interest. Similarly, one cannot know the extent to which Soviet objections to on-site inspections have stemmed from a legitimate fear of the security risks, and a belief that their international status would be impugned, or from a calculated intention to cheat.

The consequence is that, even on the existing agreements like SALT I and **SALT II,** and the ABM Treaty, both sides have periodically accused the other of having broken the agreement, but can only offer evidence that is usually less than fully convincing. There is little doubt that the USA has committed technical breaches of SALT II (which has never been ratified by the US Senate, in any case) since late 1986. However, it is unclear whether these were made for reasons of defence policy or as arms control tactics. Verification remains the biggest single restriction on arms control progress, and often leads to impressive sounding, but relatively hollow, proposals.

Arms Race

Properly used, the term 'arms race' is applied to a process where potential enemies gear their arms procurement to each other's military development, with the intention of gaining a specific level of comparative military strength.

That one or more countries, or a pair of hostile countries, are simultaneously engaging in development or modernization of their forces does not necessarily mean that an arms race is taking place. A country's aim may simply be to gain, or retain, either **Parity** or superiority with its potential enemy. On the other hand, the policy can be more subtle, and the objective may be a certain proportion of another country's military capacity. Thus naval treaties after the First World War actually set legitimate naval strength for some countries as one-half of that enjoyed by the UK. Had Britain sought to increase its naval strength, an automatic multiplier would have led to a world-wide increase in naval forces.

A more complex situation is where a country links its arms policy to the ability to achieve a certain set of tasks, rather than specifically to the force size of its enemies, but has increasing difficulty in matching the equation as other countries increase their armaments. For part of the post-war period the USA, for example, defined its military security as the ability to fight and win two-and-a-half major wars. Obviously, as the countries against whom this target was set armed themselves, the US arms requirement increased. In fact, in this case the USA did not engage in an arms race, but instead reduced its definition of security to only one-and-a-half wars.

The subject of an arms race need not be restricted to technological development, or to material procurement, although these usually define the context. The successive modifications and extensions to the **Conscription** laws in 19th-century France and Prussia had all the hallmarks of an arms race. Both countries tried to increase their stocks of the prime war material of the time—trained infantrymen. Both set their targets entirely in terms of maximizing superiority over the other, and this was a continuing process, with conscription targets changing as estimates of the other's success increased.

The first truly technological arms races were the competitions for naval supremacy from the mid-19th to early 20th centuries, initially between France and Britain, and later between Britain and Germany. In this example it was not just the number of ships which was of importance, as the race between France and Britain began with the exploitation of new technologies, such as steam power and armour-plating, to try to outdo each other in developing ships of greater speed and invulnerability.

As imperial Germany began to desire its own foreign colonies to rival other European nations, and especially Britain, it shifted from its natural position as a continental land power and acquired ambitions of having a powerful fleet. Britain, on the other hand, had always based its own security on more or less unchallenged control of the seas. Thus, while the German naval programme was a direct threat to Britain, overtaking the Royal Navy's paramount position became the German planner's target. The competition was typical of arms races, with both quantity and quality important—more and more, and heavier and heavier battleships. The result was one nearly inevitable in an arms race between countries of approximately equal wealth and technological stature—parity. Both countries built huge and expensive fleets which, by the time of the First World War, were so evenly matched that they achieved virtually nothing. The German fleet stayed at anchor for most of the war, while the British fleet was almost equally tied by the need to stay in harbour in wait in case the Germans put to sea. Ultimately they did so, and the two fleets fought the great Battle of Jutland. This ended in a draw. As has happened subsequently in the nuclear arms races, the forces finally fulfilled no purpose other than to deter each other.

The arms races of the post-Second World War era have tended in the same direction. As one side has achieved an advantage by using a new technological development, the other has caught up. Any breakthrough, as for example in **MIRVing** submarine-carried missiles, has been rapidly matched by the other. However, the logic of nuclear deterrence does, unlike that of naval competition, offer some opportunity of controlling the race. Particularly with ideas such as **Mutual Assured Destruction,** there is a limit to the force any side needs in order to be able to deter the other. Until they became worried by the **Scenario** usually known as the **Window of Vulnerability,** the USA had not really raced against the USSR for missile strength since US Secretary of Defense Robert **McNamara** built the basic **Triad** in the 1960s. Both sides, furthermore, have been prepared to accept the **SALT** treaties which enshrined general parity, there being no point in pursuing **Strategic Superiority.**

What could unsettle, or destabilize, the nuclear balance and rekindle the arms race is some technological breakthrough that might allow for a genuine 'superiority'. Thus, for example, the American **Strategic Defense Initiative,** or a real capacity on either side to destroy the other's satellites, vital both for intelligence gathering and communications (see **Anti-Satellite Warfare**), would inevitably spark an attempt to match the advance, whether or not either side actually sought advantage over the other.

Arms races today are so clearly tied to the economic resources of the competing parties that they may have begun to take on the main functions of actual conflict, instead of being preparatory to them. One quite common American perspective, on the political right, is that the USA should accelerate armaments development, not with any intention of using the weapons, even if superiority was achieved, but because the much weaker Soviet economy would ultimately collapse under the strain of military procurement. The USA would thus win a victory, without the hazards of either conventional or nuclear war. More plausibly, it seems that the futility of competition in any one area shifts attention to those where it remains logically meaningful. Although neither side has really been involved in a strategic nuclear arms race since SALT, both **Conventional** armaments and **Intermediate Nuclear Forces** have seen considerable growth, first by the Warsaw Pact and then by NATO. Even these are gradually being brought to a position of parity. The principal restraint on ever-increasing competition in these areas is that the relevant technologies grow exponentially more expensive as efforts are made to find **Force Multipliers.** In the USA certainly, and possibly in the USSR as well, the conviction is growing that military procurement budgets simply cannot keep pace with inflation in arms costs. (See also **Arms Control** and **Mutual and Balanced Force Reductions.**)

Assured Destruction

Assured destruction is a major goal of all forces of nuclear **Deterrence,** and is usually encountered as part of the doctrine of **Mutual Assured Destruction.** The idea is very simple: a nation can deter an attack on itself by having the retaliatory ability utterly to destroy organized society in its attacker. For this to be possible a country must possess forces which cannot be eliminated by an opponent, hence the 'assured' part of the concept. The only current type of nuclear force that can give this guarantee is an **SLBM** system, such as the US **Poseidon** or **Trident,** or the British **Polaris** forces, because, unless **Anti-Submarine Warfare** techniques improve dramatically, these vessels are virtually undetectable.

A definition of the total destruction of a society is obviously highly arbitrary. The criteria used by most analysts were given by the US Secretary for Defense, Robert **McNamara,** in the early 1960s when he advocated the doctrine of mutual assured destruction. He set the guaranteed damage required for full deterrent effect against a potential aggressor as the deaths of one-third of its population and the destruction of about two-thirds of its industrial capacity. In fact these figures are not derived a priori, or even sociologically. They occur naturally on a graph of the effects of a nuclear strike against the USSR when **Megatons** used are plotted against cities destroyed, where a peak is reached at a level of around 400 megatons. The population structure of the USSR is such that dramatically reduced return for yield is encountered above this level. While 400 megatons destroy about 60 – 65% of industry, 800 megatons would only add a further 5% or so. The USA estimates that it would suffer equivalent, or greater, levels of destruction in a similar attack.

These calculations are highly conservative, anyway, because they omit the likely effects of social dislocation, and even of long-term risk to life from disease, nuclear **Fall-Out** and other secondary hazards. It is very probable that the delivery of much less than even 200 megatons would have an utterly destructive impact. This is a significant point because, while only the USA or the USSR can currently deliver a 400-megaton strike, both the British and French forces will, by the mid-1990s, be able to launch an attack with over 100 megatons. This means that, within the foreseeable future, the superpowers will be vulnerable to assured destruction from second rank powers.

The real question about assured destruction is whether it has the other ingredient of a true deterrent—**Credibility.** It is one thing to have the capacity to do unacceptable damage to an enemy, but if it is simply not believed that the threat will be carried out there is no deterrent value. Clearly there are actions by the USSR that can be almost guaranteed to provoke the full destructive strike; a **Pre-emptive Strike** on US cities would surely be one. But what else would lead an American President to cause the deaths of as many as 100 million people, quite apart from the fact that it would surely be inviting the same fate for the US population? The conclusion is that it is possible that the threat of assured destruction deters nothing except its equivalent. For the USA this questions the validity of the whole idea of **Extended Deterrence,** and for Western Europe it raises NATO's nightmare of **De-Coupling** of the US deterrent from Western European security, which is still in theory based on the NATO doctrine of **Flexible Response.**

Atlantic Nuclear Force (ANF)

The Atlantic Nuclear Force was one of the several suggestions developed in the early 1960s for a way in which NATO's nuclear guarantee, upon which its conventional strategy in Europe has always ultimately rested, could be put in the hands of all member states. This would have reduced the reliance on US **Extended Deterrence.** As such it was a rival to the more serious American plan for a **Multilateral Force (MLF).** All such plans came about initially because of the unhappiness that many Western European powers, and particularly West Germany, felt about their security ultimately depending on a form of weapon controlled solely by the USA. While the West Germans were forbidden by their treaty of membership in NATO from developing nuclear weapons themselves and, in any case, would have been unwilling to do so, they wished to have some say in their use.

The American plan for a multinational force, probably sea-going and equipped with **Polaris** submarines, was official policy briefly under President Kennedy, although nothing came of it. The ANF was a British plan, and it was widely thought at the time that it was introduced more to confuse the discussions inside NATO than with any real intent. The UK was still very sensitive about its own independent nuclear force, and saw the US enthusiasm for an MLF as a threat to the retention of the British Polaris squadron.

Atlanticist

Atlanticist is the description frequently given to those taking a particular position in the argument currently engaging the US defence establishment on the appropriate world strategy for America. The debate between the Atlanticists and those with a more 'Global' or 'Maritime' strategic view (see **Global Strategy** and **Maritime Strategy**) has always been present in the post-war period, but reached a new height under the Reagan administration, and is unlikely either to be resolved, or to fade away under the next administration. Atlanticists take US membership of NATO as the paramount obligation, see the defence of Western Europe as integral to the security of the USA itself, and advocate the concomitant force structures and military and diplomatic relations. The 'Maritime' theorists believe that the USA should be much more independent of Western Europe, indeed of any alliance, and are sometimes called 'Nationalist Unilateralists' as a result. The essence of this doctrine is that the USA must be able to protect its national interests anywhere in the world, and must be able to reply on its own power to do so. If an alliance, like that with Western Europe, is costing the USA more than it seems to be worth, because of failure by the other members to pay their share, then the US national interest requires it to withdraw, or at least to reduce its own contribution (see **Burden Sharing.**)

Among the major differences between the Atlanticists and the Globalists are their respective interpretations of where America's vital interests lie, and where they are most in danger. The Globalists tend to play down the extent both of US vital interests in keeping Western Europe free, and of the Soviet threat there. Instead they see the real threat to US interests as lying in insurgency movements and other conflicts in the Third World. For example, major concerns of this school of thought are the Middle East, the Persian Gulf, the Caribbean, and Central and South America. They are willing, and eager, to see the US intervene militarily in these areas, unilaterally if necessary, but believe that commitment to the European theatre has resulted in the wrong force structure, making US involvement elsewhere nearly impossible. What they want is a considerably enhanced naval strength, especially in amphibious warfare ships and **Sea-Lift** capacity. They argue that there is a need for additional **Light Divisions,** rather than the army units developed for the European theatre.

Atlanticists do not find it easy to argue against the thesis that the US has vital interests elsewhere, or that its current military force structure is ill-designed to meet them. However, they still stress the primacy of the European (and Korean) areas, and tend to want to rely more on their allies for help elsewhere. Thus, for example, even the most ardent Atlanticists complain bitterly at **European NATO's** refusal to take **Out of Theatre** operations seriously. As is inevitable in such debates, the positions taken up tend to reflect professional group interests. The submarine navy is entirely Atlanticist, because they have no role in the global strategy, while the carrier and most of the remainder of the surface-warfare navy see it as a chance to expand their influence. Similarly, the Marine Corps, who would like to be given the entire responsibility for intervention are dedicated

Globalists, having very little importance in the conventional NATO strategy. It is impossible to say which argument will prevail, but the Atlanticists have to fight against the persistent threat of **Isolationism** in US political thinking, which rather favours the Maritime strategy lobby. If any decision is, eventually, to be made, it will be by the US Congress, rather than the Presidency, but it is more likely that the tension will simply continue, ensuring that the USA is not fully prepared for *either* role.

Atomic Bomb

An atomic bomb is the type of nuclear weapon used twice by the USA in August 1945, on Hiroshima and Nagasaki. Until the early 1950s it was the only form of nuclear weapon, but has since been replaced for most uses by what is usually called, in lay discourse, either a **Hydrogen Bomb** or a **Thermonuclear** bomb. The atomic bomb is the product of **Nuclear Fission** rather than **Nuclear Fusion** power—the source of the hydrogen bomb explosion. The first atomic bombs were built in the USA by the **Manhattan Project,** but the scientific expertise was truly international. Apart from the UK and Canada, which had been interested in the project from close to its conception, many of the intellectual leaders were European refugees from Fascism. France and the UK, as well as Germany and Italy, had been the leaders in early experimentation and physical theorizing that had first suggested that the controlled release of the energy of the atomic nucleus was possible. A practical demonstration of this capability was not achieved until after the beginning of the project to build the atomic bomb, in a famous experiment in Chicago on 2 December 1942. Indeed, one of the decisive moves in persuading President Roosevelt to put in the enormous resources needed for the project was a letter from Einstein urging him to consider the appalling risk if Hitler were to develop such a weapon. (The Germans were actively researching a similar plan, but resources were side-tracked into other weapons.)

The initial atomic bombs were of very low **Yield** by modern standards, and were certainly of less than 20 kilotons. Nevertheless, as is well known, they were terrifyingly effective, even if it needs to be remembered that the initial death tolls from neither the Hiroshima nor Nagasaki raids were as high as those resulting from fire-storms after **Conventional** raids on Tokyo or Hamburg. A combination of the physics of nuclear interaction and design factors limit the yield of atomic bombs to only some tens of kilotons. Consequently the three earliest nuclear powers, the USA, the UK and the USSR, rapidly moved on from atomic bomb production in the late 1940s and early 1950s to the development of the hydrogen bomb. All three powers had finished testing and started production of these weapons, measured in **Megatons** of yield, by the end of the 1950s, less than 15 years after the first successful atomic bomb explosion. The technology of weapons design has now come full circle, with the basis of the hydrogen bomb, fusion power, used to produce intentionally low-yield weapons (although fusion weapons use an elemental fission explosion 'trigger' to generate the high temperatures needed for a fusion energy release).

It is interesting to note that the transition from the atomic to the hydrogen bomb is a good example of the dictum that what *can* be built, in weapons design, *will* be. It was not unknown at the beginning of the Manhattan Project that there was a more powerful form of nuclear energy release, but nor was it initially planned to move on to the development of such weapons. Indeed, there was a fierce debate inside the American atomic physics community as to whether or not the second generation of nuclear weapons should be produced. The final decision went in favour of those who wanted to build what were known as 'super bombs' largely because the USSR exploded its first test atomic device in 1949, years before the Americans had expected them to. There was no immediate military requirement for the weapon; indeed, in those early days of nuclear strategic planning the US **Strategic Air Command** largely saw nuclear war as a direct extension of **Tactics** such as the 1942-45 **Strategic Bombardment** of Germany, and not as a totally new form of warfare. In many ways the 'Atomic Age' arrived by accident, with the physicists producing destructive power in advance of strategic theory or military necessity. (See also **Enhanced Radiation Weapons.**)

Atomic Demolition Munition (ADM)

Atomic demolition munitions are nuclear mines buried in areas which Warsaw Pact troops would have to cross to invade NATO territory, or positioned to destroy vital road, rail and bridge communications. They are, however, likely to be removed, in keeping with the NATO plan to reduce considerably the stock of aged short-range **Battlefield Nuclear Weapons.** Few, if any, military tacticians regard this sort of weapon as of any use. As, by definition, one can only explode pre-deployed mines on one's own territory, the use of ADMs is understandably unpopular with the USA's European allies. (It is thought that only the USA has ADMs.) Their existence dates from the early days of NATO's **Massive Retaliation** strategy, when the USA developed many different sorts of small-scale and short-range nuclear weapons to offset Warsaw Pact conventional numerical superiority.

Atomic Diplomacy

Atomic diplomacy was a phrase used rather generally in the 1950s and 1960s to refer to the power politics of a nuclear-armed nation. There is no very precise meaning for this term, but any foreign policy posture which relies ultimately for its effect on even an implicit nuclear threat would be covered. There are some obvious examples, such as the overall US foreign policy of **Containment** in the 1950s, when America had a near monopoly of nuclear weapons. Other examples, such as the USSR's role in the Suez Crisis, when threats of atomic war in Europe may have been influential, are more open to question. (See also **Brinkmanship.**)

Atomic Energy Commission (AEC)

The Atomic Energy Commission was set up under the Atomic Energy Act of 1946 as the overall controller and co-ordinator of nuclear energy in the USA, whether for civilian or military purposes. It had become apparent that the entirely military-controlled process of nuclear construction that had operated during the Second World War could not be continued. The AEC was originally responsible for all nuclear energy programmes, it not being expected that there would be privately-owned nuclear electrical generating stations. At the same time it was responsible for acquiring and enriching material suitable for use in weapons design and construction. The original Act has subsequently been amended at various times, partly to accommodate the building of nuclear generating stations by private utilities companies. The AEC itself ceased to exist in 1974; control over civilian nuclear energy passed to the Nuclear Regulatory Commission, while the US Department of Energy is now responsible for the development of the nuclear weapons programme. The AEC and its successors have been subject to much more direct political control than its British counterpart (the **United Kingdom Atomic Energy Authority**), because the 1946 Act established a joint Congressional committee of the House of Representatives and Senate, the Joint Committee on Atomic Energy, with oversight powers. It was because of this Committee, for example, that for some time in the 1950s the AEC was forbidden to share its knowledge and expertise with any foreign institution, which temporarily set back Anglo-American co-operation, particularly in military nuclear matters.

Attrition Warfare

A major debate has been in progress within the US defence establishment for some years, with implications for the rest of NATO, about the whole conduct of land warfare. Part of the military reform debate, it is a contrast between two styles of warfare which have been characterized as **Manoeuvre Warfare** and attrition warfare. The proponents of manoeuvre warfare come mainly from inside the **Military Reform Movement,** led by civilian analysts with a Congressional staff background, such as Steven Canby, William Lynd and Jeffrey Record, and anti-establishment military theorists such as the highly influential US Air Force Colonel, William Boyd.

Attrition warfare involves amassing huge and heavily-armed land forces, which batter their way through opposition by sheer weight of numbers, artillery and tanks. It is the typical warfare of a highly-industrialized democratic nation which, because of domestic political pressures, is also very keen to spare its own soliders' lives. Thus, instead of taking risks and relying on brilliant generalship (itself not easily available in a country where a military career lacks respect and fails to attract the most able leaders), US armies depend on material and brute force in direct battlefield engagements. The defence reformers base their theory on a historical analysis of what they see as this characteristic American land battle tactic. They cite, for example, the US inability to win in Vietnam where its huge preponderance of equipment and **Firepower** was useless because of the refusal of the Vietcong and North Vietnamese armies to engage in pitched battles.

Whether or not attrition warfare *has* been characteristic of US, and indeed British, warfare is debatable. What cannot be denied to the reformers is that attrition warfare, depending on superiority in numbers of troops and heavy weapons, is simply no longer available as an option in the context of NATO. Beyond any doubt, the Warsaw Pact armies do, and can easily continue to, outnumber NATO forces through **Conscription,** and in all categories of heavy weapons, such as tanks, artillery, armoured helicopters and **Armoured Fighting Vehicles.** If it is true that attrition warfare is the predominant combat model for which NATO forces are trained, the outlook for them is rather gloomy.

The military reformers instead wish to rely on, train and equip forces for, and above all produce senior officers sympathetic towards, a different tactical approach, which they call manoeuvre warfare. This depends on lighter and very mobile forces capable of penetrating enemy battle lines, destroying communications, throwing the opposition's command into confusion and, literally, outmanoeuvring the bigger and more heavily-armed opponents. The model for the theory is, above all, the German Panzer unit of the Second World War, which practised the tactic of **Blitzkrieg.** Only such reliance on military skill, more suitable for the professional troops of the US **All Volunteer Forces** than the conscript armies of the Warsaw Pact, they believe, can compensate NATO for its numerical inferiority. The doctrine meets with mixed reactions from the professional military, many of whom dispute the right of such self-appointed experts to criticize professional army thinking (even Colonel Boyd was a fighter pilot, and is therefore regarded by them as suspect). However, the theory of manoeuvre warfare is more popular in the US Congress, where a lack of confidence in the **Pentagon** has been growing for some time.

B

Bacteriological Warfare (see Biological Warfare)

Balance of Power

The balance of power is arguably the oldest of all strategic concepts. It can be found as an idea in the speeches of Athenian statesmen. David Hume, the 18th century philosopher, used the phrase in an essay, and claimed that it had been known throughout the ages. The 18th and 19th centuries are often referred to by historians as the classic age of the balance of power, especially as established by the Concert of Europe in post-Napoleonic Europe. It remains both a common practical policy and a major theoretical concept in international relations.

The concept can be used in two different ways: it can be a description of how a stable international system functions, or a prescription for how a nation ought to handle its foreign and security policy. In either case its value is limited to certain international contexts. In brief, there must be at least three, but probably not more than five or six, states of roughly equal power in the system. The logic of the balance of power is that whenever a state or coalition appears to be reaching a position in which they may be able to dominate a weaker state or coalition, all other states will give their support to the weaker power. The most famous example is British foreign policy in the 18th and 19th centuries. Britain refrained from making alliances with European countries unless any one state threatened to be able to control the whole continent, and would then use its power to offset this potential 'superpower'. Those who believe in balance-of-power theory argue that the absence of a major war in Europe from 1815 until the Franco-Prussian war of 1870, or even to the First World War in 1914, was the result of a power balance both as a description of actual power distribution, and as a conscious foreign policy of European statesmen.

Even if the theory does work, the conditions necessary for it to apply are highly restrictive, and less easily found in the contemporary world. As an example, balance-of-power strategy assumes that nations have no ideological preference—that they will just as happily oppose or support any member of the international system. Secondly, it assumes that each nation thinks it is in its self-interest to maintain the existence of each other member state as well as its own. Why, otherwise, would a state not join the *stronger* force and share in the carve-up of the defeated weaker powers? Again, the model assumes that any potential new power is accepted into the balance-of-power system as an equal partner, although this inevitably diminishes the total combined influence of existing members of the system.

The main reason that a classic balance-of-power system cannot exist at the global level today is that the superpowers so dominate all other nations that no third or fourth party can significantly alter the relative influence, either of the USSR or USA, by joining or leaving a coalition with them. Another way of putting this is to say that a balance of power requires **Multipolarity;** the international system in operation today is one of **Bipolarity** (see also **Bi-Multipolarity**). One can, nevertheless, see remnants of balance-of-power politics as, for instance, in the US courtship of the People's Republic of China during the Nixon **Détente** period, when US attitudes shifted from seeing it as an unacceptable partner to trying to use its potential as a third superpower to alter the balance between the USA and USSR.

'Balance of power' can also be used in a purely descriptive way to describe the actual distribution of power in the world, without implying that an equal 'balanced' distribution exists.

Balance of Terror

The balance of terror is a term sometimes used to describe the contemporary version of peace through **Deterrence,** or **Mutual Assured Destruction,** rather than the traditional **Balance of Power.** Whereas the balance of power actually involved occasional recourse to war in order to restore an equilibrium, in which no nation dominated the international system, the balance of terror freezes that system by making war impossible because of the probability of utter destruction. 'Balance' is used with a different meaning here; with nuclear weapons two sides do not need to be equally powerful, as long as the weaker can still guarantee to devastate the stronger. A major theoretical problem is that while the classic balance-of-power system assumed **Multipolarity,** modern experience of the balance of terror is essentially one of **Bipolarity,** and major efforts have been made to prevent many countries owning nuclear weapons through the **Non-Proliferation**

Treaty. It is uncertain whether there could ever be a stable multipolar nuclear balance of terror.

Ballistic

The term 'ballistic', in modern strategic language, occurs usually in conjunction with 'missile', as in intercontinental ballistic missile **(ICBM),** submarine-launched ballistic missile **(SLBM)** or **Intermediate-Range Ballistic Missile (IRBM).** It means that the missile in question behaves essentially like an artillery shell or mortar bomb, launched by a powerful explosion and then coasting for most of its trajectory. A ballistic missile burns its engines very quickly for a relatively short time, giving it enough velocity to complete the remainder of its journey without further power (see **Boost Phase**). Most ballistic missiles, in fact, briefly leave the earth's atmosphere, and accumulate gravitational acceleration on re-entry. The entire track of the missile and its impact point are both determined by its initial position, thrust and angle, as with a bullet or shell, and unlike an aircraft or other continually-powered flying vehicle.

Because of the very high speed of ballistic missiles it is extremely difficult to intercept them. However, the essentially fixed character of their trajectory makes it relatively easy to plot their course and predict their impact. For this reason some weapons designers want to create a **Manoeuvrable Re-entry Vehicle (MARV),** with which the US **MX** would be equipped, and which would be able to alter course during the **Terminal Phase** rather than falling at a predetermined and predictable point. This would be an advance on the existing **MIRV** technology, which allows the independent targeting of multiple warheads at the point of release from the missile's **Bus.**

Ballistic missiles must not be confused with **Cruise Missiles,** which are literally 'flying bombs', operating at, and below, ordinary aircraft altitudes, at low speeds under constant power, and are correspondingly more vulnerable to defensive fire.

Ballistic Missile Defence (BMD)

Ballistic missile defence refers to any system or set of systems used to destroy **Ballistic** missiles or their **Re-entry Vehicles** (also known as warheads). Both superpowers engaged in research and deployment of **Anti-Ballistic Missiles (ABMs)** during the 1960s, but the ABM treaty which formed part of the **SALT I** agreement in 1972 severely restricted installation and testing of future systems. The USA dismantled its only ABM site later in the 1970s, and the Soviet Union never completed its permitted deployment of the **Galosh** system around Moscow.

In 1983 President Reagan launched the **Strategic Defense Initiative (SDI),** popularly known as Star Wars, and re-established BMD as a legitimate defence aim. SDI is more general than the anti-ballistic missile technique, because it covers a wide variety of possible destructive techniques. These include space- and land-based **Laser Weapons,** as well as **Kinetic Energy Weapons** (sometimes described as 'smart rocks') and possibly other methods such as X-ray lasers. Most BMD strategies envisage a **Layered Defence,** with successive attempts to destroy missiles early in their flight (the **Boost Phase**), during the **Mid-Course Phase,** at the stage when the re-entry vehicles separate and, finally, in the brief **Terminal Phase** as they re-enter the atmosphere before hitting their targets.

The problems of achieving any fully-effective BMD system are enormous, not only on the engineering side but, perhaps even more so, in the computing and software aspects. Some estimates of the amount of programming involved suggest that a full ballistic missile defence, such as is planned under the SDI, might run to 10 million lines of computer code, none of which could ever be tested in realistic conditions.

BMDs can be constructed to protect a large area, for example a city or region, or as a **Point Defence** protection around a missile silo. Naturally the more precise the target to be defended, the more possible the task. Even so many analysts fear that *any* BMD system can be beaten because it will always be cheaper for the opponent to build more and more offensive weapons to saturate the defence. There is a powerful and vocal school of thought, mainly but not entirely in Western Europe, which feels that even the attempt to build a BMD shield is destabilizing and inevitably leads to an **Arms Race.** Their argument is that international stability rests on **Mutual Assured Destruction;** anything that lets one nation feel less vulnerable automatically makes the other more fearful and behave less predictably.

Ballistic Missile Early Warning Systems (BMEWS)

The main ballistic missile early warning system for the USA, and indeed the UK, is that based on three enormous radar stations, two in Canada and one (Fylingdales) in Yorkshire, England; this system is known as the Distant Early Warning (DEW) Line. These radar stations can give between five (for the UK) and 20 (for the USA) minutes warning of an incoming missile attack over the polar route from the USSR. Several other subsidiary and interlinking chains of radar stations exist for the North American continent, including systems set up to spot missiles coming in from the west, for which the DEW Line is useless. Increasingly, however, these ground-based systems are of secondary importance to satellite observation. Given the ease with which such a system can be put out of action by the deliberate explosion of a large nuclear warhead at a high altitude, thus creating an **Electromagnetic Pulse,** reliance on them has declined somewhat. The development of **Phased Array Radar** has also made the construction of chains of stations at some distance from the territory they are protecting less important.

Ballistic Missile Submarine, Nuclear (see SSBN)

Basing Mode

Basing mode simply refers to how **ICBMs** are deployed or stored while on stand-by. The term has become familiar through the policy debates in the USA on the **MX** (Missile Experimental). The most familiar basing mode is the positioning of missiles in hardened underground silos (see **Hard Target**). These silos are spread around a large area so that a nuclear hit on one will do the minimum damage to any other. However, it is beginning to be impossible to build silos sufficiently strong to resist the force of a large nuclear warhead delivered with the accuracy possible with missiles such as the Soviet **SS-18** or the US MX.

In order to ensure the viability of land-based missiles, experiments are being made with other basing modes. One option, so far rejected for the MX, is the **Dense Pack** plan, which is dependent on the validity of **Fratricide** as a defensive option. Another is to have mobile missiles. One plan for MX, also rejected on grounds of cost, was to build a large number of concrete shelters at various points on a huge road network. The missiles would be shuttled round and round this network, so that at any one time the USSR could not tell which of the shelters housed real missiles, and which decoys. The advantage of mobile basing systems is that the enemy is forced to use up a large number of warheads on empty shelters. Thus the number of missiles which an enemy requires in order to eliminate one's own forces in a **First Strike** can be increased to an economically impossible level, while one's own investment remains much less. So far only the USSR has begun to deploy an **ICBM** in a mobile basing mode. This is the SS-24, mounted on specially-designed trucks and capable of being fired from anywhere on the Soviet road system. Both the USSR and the USA have had mobile **Theatre Nuclear Forces** for some time, and the **Scowcroft Commission** recommended that the USA should build a new generation of single warhead manoeuvrable ICBMs, known as **Midgetman.**

Battlefield

The 'battlefield' in **Conventional** warfare has been relatively small in area, and usually has been linear. That is, two lines of troops have fought over an area only a few miles long, with death and destruction restricted to a narrow strip, perhaps not much more than a mile wide. (It is true that the trench line in France during the First World War ran from the Swiss border to the English Channel, but the major battles were fought over much shorter segments of, at most, dozens of miles in length.) A projection from presumed Warsaw Pact and NATO **Strategy** and **Tactics** in the event of a **Central Front** conflict would suggest that the battlefield would be very different from the traditional model. Firstly, it is probable that the entire length of the central front would be assaulted at once, with Soviet **Operational Manoeuvre Groups** exploiting points of weakness. Secondly, NATO tactics, whether described as **Follow on Forces Attack** or under the American doctrine of **Deep Strike,** are designed to deepen the battlefield by attacking second echelon (see **Echeloned Attack**) Warsaw Pact forces as far as 500 kilometres behind the front line. Finally, new doctrines such as the Americans' **Manoeuvre Warfare** anticipate a much more fluid and non-linear battle, as well as **Defence in Depth,** rather than a firmly held, but thin, line (see **Battlefield Mobility**).

For obvious reasons the use of **Theatre Nuclear Forces** and **Battlefield Nuclear Weapons** will be planned to be as far away as possible from one's own troops, thus further deepening the area over which battle conditions apply. Consequently modern tacticians frequently refer to the 'battle area', as in the concept of the **Forward Edge of the Battle Area.**

Battlefield Mobility

New doctrines of how to fight a possible war on the **Central Front** in Europe, for example the US concept of **Manoeuvre Warfare,** have placed enormous stress on speed and mobility in battle. This has led, in particular, to a need for **Infantry** to be equipped with **Armoured Fighting Vehicles,** for sophisticated communications, and for extremely capable and long-range **Target Acquisition** artillery systems. Instead of the traditional pattern of **Attrition Warfare,** with troops in trenches and static emplacements, the emphasis is on surprise, on keeping the enemy off balance, and on rapidly exploiting weaknesses or gaps in the enemy's line. All of these tactics employ mobility on the battlefield and are ways of compensating for the numerical inferiority of NATO's armies against those of the Warsaw Pact. Mobility, however, involves a very different, and probably more demanding style of leadership and command than Western armies traditionally have been trained for. The closest to an equivalent is the **Blitzkrieg** warfare which Germany demonstrated so brilliantly in the early phases of the Second World War.

Battlefield Nuclear Weapons

Battlefield nuclear weapons and tactical nuclear weapons are, up to a point, interchangeable terms. Both refer to nuclear weapons used within the context of a battle, for direct and immediate tactical ends. **'Battlefield'** obviously suggests short-range weapons and, as such, they have to be weapons of low **Yield,** to prevent endangering the user's own troops. Typical of such weapons are nuclear shells for firing from field guns, with ranges approximately between 15 and 25 kilometres, or missiles like the Lance, deployed by several NATO armies and having a range of about 110 kilometres. Such weapons will have yields from as little as one kiloton (see **Enhanced Radiation Weapons**) to perhaps 10 kilotons. More modern designs of battlefield weapons in fact have what is called a 'dial-a-yield' capacity, so that the field commander can select the detonation power to suit the exact tactical needs.

The NATO stocks of battlefield nuclear weapons are almost all under direct, or **Dual-Key,** control by US officers, and are plentiful. There are more than 5,000 warheads in this category, the majority of them in the form of artillery shells. They are essentially a left-over from a much earlier NATO stance when, in the 1950s, the USA had a virtual monopoly on sophisticated nuclear design. At that stage it seemed that battlefield nuclear weapons were the obvious way to offset Soviet conventional superiority, and President Eisenhower equipped American forces in Europe with them in large numbers. There was even a plan to produce an ultra-small nuclear device, to be called the 'Davy Crockett', a sort of nuclear bazooka, which would have been issued at the **Infantry** company level. The impossibility of retaining control in the hands of senior officers and politicians, if nuclear weapons were distributed this widely, prevented the plan from being realized.

The Soviet Union did not follow suit, and has never developed short range low-yield warheads for artillery, preferring the longer-range missiles like the SS-4 and SS-5 which are more often placed in the **Theatre Nuclear Force** category. Increasingly this plethora of small nuclear weapons has come to be seen as having no military value at all, and NATO is in the process of cutting considerably the numbers deployed in Europe. One serious limitation is that the very short range meant that the warheads had to be issued to units very near the **Forward Edge of the Battle Area.** This, in turn, would require their very early use in wartime, because a successful rapid breakthrough by the Warsaw Pact forces would quickly overrun the positions. This, which came to be known as the **Use Them or Lose Them** problem, implied such a very low **Nuclear Threshold** as to be quite unacceptable within the doctrine of **Flexible Response.**

Obviously there is no clear dividing line between battlefield or tactical weapons on the one hand, and theatre weapons on the other. Recent arms control negotiations have tended to classify battlefield nuclear weapons as those having a range of less than 500 kilometres. To the extent that one *can* make a distinction, a nuclear phase in a **Central Front** war is much more likely to start straight at the theatre level, most probably as part of a **Deep Strike** attack. However, battlefield nuclear weapons retain their enthusiasts in the USA, and studies are proceeding on new ways of using them, and on new designs for such weapons. If they do again become a major element in war planning, very severe requirements would be placed on tactical training, because any form of concentration of troops, even at the battalion or company level, would become a target. The command and control problems involved in so dispersed a tactical plan are far beyond the current abilities of the officer corps in any NATO or Warsaw Pact army.

Bias

Bias has a specific technical meaning in reference to missile warfare. It is one of two logically very different measures of a missile's accuracy, the other being **Circular Error Probable (CEP).** CEP, the most frequently cited measure, refers to the size of the area within which it can be predicted that 50% of missile warheads will hit, and does not reflect the ordinary meaning of accuracy. Bias refers to the probable average distance a warhead will be from the actual aiming point, or **Designated Ground Zero (DGZ),** when it detonates. Clearly the degree of bias in a missile's flight path is crucial if the object is to destroy a **Hard Target,** but it is much more difficult to measure than CEP.

It is probably safe to say that no country has a very accurate idea of the bias of its missiles, because the deflection from the aiming point will be affected seriously by factors such as the earth's magnetic field, which varies in strength around the planet. Neither the Western nuclear powers nor the USSR have ever tested their missiles over the flight path which they would have to take, for the obvious reason that this would involve actually firing at a potential opponent. Consequently, some scientific critics of nuclear strategic doctrine are particularly sceptical of sophisticated war plans which call for high degrees of precision, on the grounds that there is no real reason to believe that weapons are capable of such accuracy.

Bi-Multipolarity

Bi-multipolarity is a term used by some modern international relations theorists in an attempt to keep alive the traditional **Balance of Power** theory of how to establish and maintain a stable world system. Balance of power theory requires the presence of several, perhaps five or more, roughly equal powers. Then, if one begins to grow so powerful that it threatens the others, or if two or three weaker nations form a potentially threatening coalition, the other powers will ally to offset the danger of destabilization. This can also be described as the existence of **Multipolarity.** However, the conditions of the post-war world, with two superpowers dominating the international system (see **Bipolarity**), undermine this theory, and the consequence should, in theory, be instability through rivalry and **Arms Races.**

Bi-multipolarity is an attempt to describe the reality better, by pointing out that there are still several powerful states in loose alliance around each of the 'poles', and that these act to constrain the superpowers. The superpowers themselves discipline the weaker allies so that they, in turn, live stably with their peer states. The main problem with this model is that the traditional multipolar balance of power worked because there were no ideological barriers to alliance formation, and whatever *ad hoc* coalitions were necessary could form. The second rank powers surrounding the superpowers are not, however, free to behave in this fashion, and cannot easily act to restrain an over-powerful 'polar' nation.

Biological Warfare

Biological (or bacteriological) warfare is a form of attack of peculiar horror to most people, carrying with it deep 'science fiction' connotations. There is no clear-cut evidence that it has ever been practised, and no power has openly claimed to have such a capacity. Nevertheless it is certain that research has been carried out in the field by several nations, including the UK. The Biological Weapons Convention, signed in 1972, provided for the destruction of all biological weapons by the end of 1975. However, because of the development of quantities of viruses with potential military applications for legitimate civil medical purposes, **Verification** problems familiar from the field of **Chemical Warfare** and, indeed, the production of nuclear material, are encountered.

The essence of biological warfare is simple. It involves taking a highly-infectious virus which occurs naturally and developing it so that its lethality is enormously increased. This would make it possible to infect large populations or large territorial areas with very small samples. Thus a war would be fought by deliberately infecting the enemy's armies or civilian population with deadly diseases, some of which, in their scientifically-enhanced form, could kill in minutes. The ideal biological agent is one with a very fast infection and death rate, but which also becomes inert very rapidly, allowing troops to occupy the no-longer defended areas. Although extremely virulent diseases can easily be bred in laboratories, and even deployed in weapons, the latter requirement, fortunately, seems harder to guarantee. So, for example, although British scientists experimented with anthrax as a biological weapon in the Second World War, the Scottish island (Gruinard) they bombarded has had to remain closed to the public ever since. Decontamination of the island was in process during 1987. A further danger inherent to the use of biological weapons would be the tremendous risk that airborne infected matter could be carried back into the user's own territory, or affect its own troops. (See also **Nuclear, Biological and Chemical Warfare.**)

Bipolarity

Bipolarity is a technical term from the study of the theory of **Balance of Power.** Developed by many thinkers and politicians from the 17th century onwards, balance-of-power theory has traditionally assumed that the international political system would consist of several states of roughly equal power. In its modern social science incarnation the theory has become much more technical, and even mathematical in certain hands. There is no agreement about exactly how many significant 'international actors' there have to be to ensure stability in a balance-of-power system, but estimates vary from five upwards (see **Multipolarity**). Perhaps even three would be sufficient, but no traditional version of the main theory accounts for international stability through a balance of power with only two actors.

Yet the commonest analysis of the post-war international system is that there are only two significant international actors—the two superpowers, the USA and the USSR. To accommodate this theorists have developed the concept of bipolarity. Some have taken the view that bipolarity can be a stable system given, for example, a new form of balance of power through **Mutual Assured Destruction** (see also **Balance of Terror**). Others have held that a bipolar international system is inherently unstable, bound to collapse through **Arms Races,** because there is no third party whose shifting alliances could cancel out any temporary advantage one of the bipolar powers may have gained over the other.

Whether the modern international system is truly bipolar is itself not entirely obvious. The only aspect on which bipolarity cannot be doubted is the sheer size of nuclear arsenals. On the other hand, a view that regards even a small nuclear force as capable of causing almost unimaginable damage would have to accept that some of the ingredients of a classic balance of power existed, with five acknowledged members of the **Nuclear Club,** and several others potentially so. Because of the rigidity of ideologically-defined alliances in the shape of NATO and the Warsaw Pact, however, the fluidity of coalition formation needed for the classic balance-of-power thesis is not present, even if we accept the effective power status of the small European nuclear powers. Hence bipolarity, or some similar concept, such as **Bi-Multipolarity,** seems a more supportable proposition.

Blitzkrieg

Blitzkrieg (translated literally as 'lightning war') was the tactical method used by the German army in the invasions of Poland, France and the USSR during the Second World War. It stresses speed and manoeuvrability, rather than the **Concentration of Force** in one spot. A blitzkrieg campaign involves a series of army columns searching for weak spots in the enemy line. By exploiting these weaknesses, units pass behind the enemy, destroying its lines of communication and disrupting its plans. The emphasis is on avoiding the opposition's strength rather than trying to meet it head on, and winning not so much by destroying the enemy's troops and material, but by paralysing them with multiple unexpected attacks that destroy military co-ordination. The goals the campaign is being fought for, the cities or industrial heartland beyond, are the prime targets, rather than the enemy's forces themselves.

If properly executed, a blitzkrieg tactic offers a weaker army the chance to overcome one it could not defeat in pitched battle. Although the German blitzkrieg in France is always associated with their much-feared Panzer tanks and Stuka fighter-bombers, it should be remembered that the British and French between them had far more tanks and aeroplanes than the Germans, and that the main French battle tank was indisputably superior to any German tank. The Allies operated a rigid emplaced front line, with tanks spread out thinly along it, so that the blitzkrieg technique allowed the Germans to have a local superiority of forces wherever they chose to attack.

The current importance of blitzkrieg is largely that it has become the model doctrine for those, especially the American **Military Reform Caucus,** who oppose existing NATO tactics and training for fighting a war on the **Central Front.** They argue that the local superiority which huge Warsaw Pact armies enjoy would make it impossible to mount an effective static defence, even with most of the NATO forces **Forward Based.** To do this would be to risk repeating the French experience of 1940 (see **Maginot Mentality**). However, it is not anticipated that the Warsaw Pact would employ blitzkrieg tactics (see **Operational Manoeuvre Group**). Instead, the reformers recommend what they call **Manoeuvre Warfare,** very similar to blitzkrieg except that it is defensive. They want the front line held only thinly, with the mass of NATO forces held-back to be employed in blitzkrieg-like counter-attacks which might outmanoeuvre the superior numbers of the Warsaw Pact forces. The great problem is that, whereas France should, perhaps, have employed such a technique of 'trading space for time', and even though the USSR did defeat the German armies this way in 1942, the modern West Germany cannot use the strategy. NATO *has* to be committed to a **Forward Defence** because the alternative would be to retreat deep into the Federal Republic (which is not very deep anyway) turning the whole country into a battlefield. Even if nuclear weapons were not used, the damage to West Germany could well be worse than any consequence of surrender. Therefore the plausibility of a blitzkrieg-like tactic in a defensive battle is, in any case, in doubt.

Blockade

A blockade is a military action in which one country attempts to prevent another from importing either some or all goods by use of force. Traditionally a blockade involves the exercise of sea power, where one navy patrols the coastline of another, stopping shipping from entering and preventing the enemy's merchant and naval ships from leaving harbour. The aim of a blockade is to starve the enemy into surrender by depriving it of food, raw materials and military supplies, and by impeding its international trade, thereby squeezing its financial resources. Classic examples include the blockade by the Federal navy of the coastline of the Confederacy during the American Civil War. In this case the blockade was especially effective because the South was lacking in heavy industry and needed to import weapons. It was also totally dependent financially on exporting its cotton crops to Europe, which was prevented by the northern fleet. Similarly, the British Royal Navy has always attempted to fight European wars by blockade, notably against Napoleonic France, although the huge length of the French coastline, as well as the intense activities by English smugglers and French 'blockade runners', made this less than totally effective.

Although a blockade is obviously an act of war, it is somewhat more constrained than many because there is a complex body of international law governing the behaviour of blockading fleets towards the shipping of neutral countries; this distinguishes a proper blockade from unrestrained naval activity against any shipping that might help the enemy. It was, for example, the German use of unrestrained submarine warfare against American shipping in both World Wars that was partly responsible for the USA taking up arms against Germany. (See also **Sea Lines of Communication.**)

Blue Steel

Blue Steel was the code-name for a **Stand-Off Bomb** built by the British in the late 1950s and early 1960s. It was developed to prolong the operating life of the **V-Bomber** force, which was originally designed for the delivery of nuclear **Gravity Bombs.** A stand-off missile is an aircraft-launched rocket that can be fired from a distance of some hundreds of miles from the target. It is neither a **Ballistic** missile nor a **Cruise Misssile,** and lacks all but rudimentary self-guidance systems (see also **Skybolt**). Effectively it is a rocket-powered bomb, the only point of which is to make the delivering aircraft slightly less vulnerable because they do not need to approach quite as close to the enemy's anti-aircraft defences as they would to deliver a traditional **Gravity Bomb.** Blue Steel was the last missile of any form built by the British, after the cancellation of the **Blue Streak** programme, and was deployed only during the last years of the UK airborne deterrent, from 1962 until the introduction of the Royal Navy's **Polaris** squadron in 1968.

Blue Streak

Successive British governments, from the end of the Second World War onwards, were determined to make the UK a nuclear power, and initially this was done by building the **V-Bomber** force for airborne delivery of nuclear **Gravity Bombs.** The move by both the USSR and the USA towards missile forces made the RAF's **Bomber Command** bases increasingly vulnerable; the maximum warning they would have received in order to get the aircraft off the ground before Soviet missiles could destroy them would have been about four minutes. Consequently the Conservative government of 1955 – 59 decided that Britain, too, should have a land-based missile deterrent, which was code-named Blue Streak. By 1957 the government was fully committed to a shift to a missile-based strategy, and development went ahead.

However, the UK had no adequate technological base or scientific experience from which to develop such a programme, and the Blue Streak project soon ran into technical problems and threatened to cost vastly more than budgeted. In addition, strategists became more and more doubtful that it made sense for the UK, a small and densely-populated land mass, to try to follow the route of the superpowers, who could place their missile silos deep in their heartlands and away from areas of dense population. The British missiles, even in concrete silos, would have been highly vulnerable, and would also have presented a threat to the UK population because of the targets they would become. Consequently the programme was abandoned in 1960, and Britain became dependent on the USA for its supply of strategic nuclear weapons. Initially the UK was offered the US **Skybolt** missile but, as this too was cancelled, in 1962, eventually had to make do with **Polaris.** (See also **Blue Steel.**)

Blue Water Navy

The term blue water navy has been in use since at least the late-19th century, and refers to a naval force with the ability to patrol and fight effectively anywhere in the world. The contrast is with the coastal protection forces which many countries maintain instead of investing in long-range naval capacity. (During the Vietnam War the phrase 'brown water navy' became fashionable in USA, and was applied to coastal and river patrol forces.) The main focus of interest in blue water navies since the 1960s has been the development of such a capacity by the USSR. Not since the Bolshevik Revolution had the USSR had, or felt it needed, a long-range oceanic navy, but it has built up a substantial fleet since 1960. This has led many analysts to the conclusion that the Soviet Union has developed a strategy of world-wide commitment, because it would not otherwise need a **Force Projection** capacity in this form.

There have not, in fact, been many true blue water navies in history. Today, the cost of major naval units has become so extreme that there may not be more than four such navies in the world. Even the UK, which probably has the third largest navy, can muster only between 50 and 60 surface combat ships. Even the much-vaunted US '600 ship navy', an ambition of the Reagan administration, will only be achieved by including non-combat ships in the total.

Blunting Mission

The blunting mission was the original plan for the use of nuclear weapons in a third world war, conceived by the US **Strategic Air Command (SAC)** in the early days of its monopoly of nuclear weapons. It was really a development of the traditional theory of **Strategic Bombardment,** made during a period when nuclear weapons were not numerous enough totally to determine a major war with the USSR. It meant that nuclear weapons would be used right at the beginning of the war against military targets to 'blunt' the enemy's capacity to fight on effectively. Later, some argued that the UK's nuclear force, which though smaller than that of the SAC was much nearer to the USSR, would attack first, carrying out a blunting mission against Soviet air defences, radars and so on to maximize the impact of the subsequent arrival of the US Air Force. In some ways the blunting mission doctrine was an early precursor, within the **Massive Retaliation** strategy, of the concept of **Limited Nuclear Options** that developed in the 1970s as a result of dissatisfaction with the **Assured Destruction** posture.

Bomber Command

Bomber Command was one of the original functional divisions of the Royal Air Force when it became a separate service in 1918. From then until the major restructuring of the RAF into a smaller number of divisions in the 1970s, when its separate identity ceased, it was the pride of the Air Force, with enormous political influence. Its political importance derived from the fact that it was the only instrument of air power that allowed the RAF to present itself as an equal, in terms of *offensive* warfare, to the British navy and army. Whereas the air forces of other powers in the inter-war years, especially Germany, were seen as essentially supports to naval and military power, with army co-operation as the vital element, the RAF wanted a totally independent role.

In the years of financial stringency after 1918 it was extremely difficult to win a large share of the small defence budget, and the two traditional armed forces, unconvinced that air power was very useful, wanted to abolish the independent air service, merge it with their own, and train air-crew for supporting roles to their own strategic missions. The early RAF leaders, above all the first Chief of the Air Staff, Lord Trenchard, adopted a theory of strategic warfare borrowed from the Italian military thinker **Douhet** to justify their service's existence. According to this theory a war could be won by **Strategic Bombardment** of the enemy's civilian population and industrial capacity, destroying both the technical capacity to wage war and, even more importantly, **Civilian Morale.** It was argued that a country could only be defeated by taking the popular support away from its government, and that it was, therefore, far more important to attack the home base than the armies in the field or the fleets at sea. Only a long-range bombing force could do this. Furthermore, this form of warfare was seen as essentially more humane, at least compared with the years of slaughter in the First World War trenches which had become the dominant model for warfare.

Bomber Command therefore got the bulk of air force funding, to the detriment of Fighter Command, which would otherwise have been in a better position to protect the British cities from Germany's own bombing campaign of the Second World War. To some extent this was politically inevitable: to justify Bomber Command involved the additional theory that 'the bomber would always get through'. To invest in Fighter Command would imply that bombing fleets could be stopped, and undermine the whole offensive strategy. At the same time the RAF paid very little heed to the army co-operation role, due to its unwillingness to become subservient to another branch of the armed forces. This left British ground forces seriously underprotected compared with the German army.

The subsequent conduct of air operations during the War, depending heavily on long-range strategic strikes into the German hinterland, has generally been considered a failure. German civilian morale did not crumble, nor did industrial productivity decline. Indeed, some post-war studies suggested that production *rose* in the most badly-bombed areas of the Ruhr. A very similar doctrine was held by the commanders of the heavy bomber sections of the US air forces, although they were still part of the army. After the war the USAF was established as a separate military service, and in its turn the equivalent of Bomber Command, the **Strategic Air Command,** became the dominant arm. The advent of nuclear weapons, originally existing only as **Gravity Bombs,** continued to provide Bomber Command with its supremacy because now the doctrine of strategic bombing, with the guaranteed capacity to destroy the industrial structure and civilian population seemed unquestionable.

Bomber Command could not retain this role for long, however, when it became apparent that long-range missiles, against which there was, at the time, no defence would supplant the piloted aircraft for strategic missions. When Britain's nuclear deterrent was transferred to the Royal Navy **Polaris** submarines in 1968, because of the vulnerability of ground-based bombing squadrons, there was no remaining independent role for Bomber Command. Indeed, some have argued that as there is no longer much of a role for air defence either, the entire mission of the RAF has become one of support for ground forces.

Although the USAF (and the French air force) has retained, and even increased investment in bombers as one leg of the nuclear **Triad,** this option was too expensive for the UK, and all combat aircraft in the RAF have been merged into one functional division, known as **Strike Command.**

Boost Phase

Boost phase refers to the first few minutes of flight of a **Ballistic** missile. This is the powered phase before the missile leaves the earth's atmosphere and starts to coast. The boost phase can last up to five minutes, with the missile accelerating until it reaches an altitude of some 200 kilometres and a velocity of about seven km per second. The missile is easily visible to orbiting satellites during this phase, not only by radar but by infra-red sensors, and would be highly vulnerable. The boost phase is the ideal time for a **Ballistic Missile Defence** system to operate against a missile. By destroying the missile at this stage, before the **Front End** or **Bus** has separated and the **Re-entry Vehicles** have been released, the problems of decoys and the multitude of targets a **MIRVed** missile ultimately presents are avoided. As the missile is still some distance and perhaps 15 minutes or more from its target at this stage, a miss can still allow other elements in a **Layered Defence** time to come into operation.

If the **Strategic Defense Initiative** is to work, and especially if it is to provide **Area Defence,** and possibly defence for Western Europe as well as the continental USA, a boost phase destruction capacity will be required. However, by the very fact of the distance a Soviet missile would still be from a US target, and the short time available to spot, track, aim and destroy, it is the greatest of all technical challenges. It is almost certain that boost phase destruction would not be possible by **Anti-Ballistic Missile** techniques, so this objective rests with the more novel directed energy methods, such as **Laser Weapons** and **Particle Beam Weapons.**

Bottleneck

The concept of a bottleneck is more familiar in economics than in strategic studies, but bottlenecks in industrial production are crucially important for targeting policies in strategic warfare. If the aim was to destroy a major portion of the industrial productive capacity of a nation, it would be necessary to strike a huge number of different factories making endless different capital and consumer goods. However, a simpler approach bringing the same result would be to identify a small number of products which are crucial to the manufacture of a much larger number of goods. If, in addition, these basic products were made in a small number of plants, a relatively small-scale attack could cripple production across wide areas of the economy for a long time.

The **Strategic Bombardment** of Germany by the British and Americans during the Second World War has often been criticized because it failed to have much effect on industrial productivity. But, for a long period during the War, there was just one plant capable of making a particular chemical additive required to refine oil into fuel for the high-performance engines of military aircraft. The reason this plant was not destroyed, thus creating a bottleneck that could have grounded the German *Luftwaffe,* was a lack of intelligence material, and the difficulty of carrying out effective precision bombing. A nuclear attack on the USA could create bottlenecks by using relatively freely-available information, and no great degree of precision bombing. For example it has been calculated that about 90% of aluminium production in the USA is carried out in only 19 plants, and that these could be destroyed in a carefully designed attack that would, at worst, lead to only 100,000 civilian casualties (see **Collateral Damage**). An attack like this could very well be carried out without leading to a massive reprisal by the USA but would, nevertheless, have very serious consequences for the US economy in general and its military production capacity in particular. Similar results could be obtained through a bottleneck attack on the oil refining and distributing industry. As far as the UK is concerned, it is quite obvious what the consequence would be of destroying the North Sea oil platforms—a very easy task with nuclear weapons—yet it would probably not be seen as justifying the use of Britain's nuclear weapons in retaliation.

Clearly bottleneck targeting only makes sense in the context of a prolonged war, but the mere fact that most NATO thinking assumes that a **Central Front** war could only last for a month does not, in fact, guarantee this. No major war has been fought in the last 150 years which the participants did not expect to be over in a few weeks. Bottlenecks are the strategic equivalent to **Choke Points** in tactical war planning of the kind NATO carries out under concepts such as **Deep Strike** or **Follow on Forces Attack.**

Brinkmanship

Brinkmanship is a term, somewhat like **Atomic Diplomacy,** which is used to describe the foreign policy of a government prepared to run the risk of war in taking an intransigent or uncompromising stand against the actions of another. The term was used in attacks on President Kennedy's policy during the **Cuban Missile Crisis,** when it was argued that he brought the world to the brink of nuclear war. It implies a policy of calling the opponent's bluff, and of not standing down or seeking a compromise position whatever the risks.

Obviously the phenomenon itself is not new. Hitler's repeated defiance of the European democratic powers in his demands and actions over Czechoslovakia, Austria and Poland were acts of brinkmanship, and only with Poland did he miscalculate and step over the brink. The risks of nuclear war, however, are thought by some to be so much greater that uncompromising adherence to a conception of the national interest is no longer an acceptable policy. There has probably not been a genuine example of brinkmanship in Soviet – American relations since the Cuban Missile Crisis.

British Army of the Rhine (BAOR)

The British Army of the Rhine is the main commitment of land forces which the UK makes to NATO. It has retained its name from the early post-war days when it was an army of occupation in West Germany. The BAOR covers a vital section of the **Central Front** some 100 kilometres long in northern West Germany, with a German army corps on its left flank and a Belgian corps on its right. Although there is no formal treaty requirement to keep any specific number of British troops in Germany, there is a very long-standing agreement which in practice ties the BAOR to the figure of 55,000 troops. It is organized formally as the British First Corps, although the internal structure varies from time to time as the **Ministry of Defence** tries to minimize administrative costs by keeping headquarters elements in the UK. The BAOR is regarded throughout NATO as among the best trained and led of the European contingents, in part because it is the only **All Volunteer Force,** apart from the US armies (see also **Conscription**). However, it is also less well-equipped than the American and West German armies. Only about one-half of the armoured regiments are equipped with a current generation tank, and the **Infantry** was only to be supplied with a modern **Armoured Fighting Vehicle** in the late 1980s. It is essentially a conventionally-armed force. What **Battlefield Nuclear Weapons** it possesses are under a **Dual-Key** arrangement with the USA.

While the British contingent of 55,000 is not large compared with Warsaw Pact forces, it is the second biggest non-German contribution to the central front defence, and represents by far the biggest task for the British army, which only totals a little over 160,000. As a result a major part in war would be played by reserves, the Territorial Army Volunteer Reserve, which would add about another 90,000 troops to the BAOR. The economic strain of the BAOR is very great because of the costs, especially in foreign currency, of keeping so large a part of the UK military strength **Forward Based.** Occasionally calls are made for the withdrawal of the British army from Germany. These come not only from those opposed to Britain's NATO membership, but also from those who regard a major land presence in Europe as an inefficient utilization of Britain's military capacity and tradition. A more moderate suggestion that has been made recently is to reduce the size of the BAOR, and to change its role from holding the front line to covering and defending the RAF Germany bases, on the grounds that the UK can make a more sensible and efficient contribution in such a way. Unless, as is possible, the USA withdraws its forward deployed troops and thus opens up the entire question of European defence arrangements, the political commitment both to forward deployment and to the figure of 55,000 is unlikely to be altered.

Burden Sharing

'Burden sharing' is a catch phrase for what may well be the most ominous cause of contention inside the **NATO** alliance. It usually refers to the relative share of total NATO defence expenditure paid by each member nation. Almost since the formation of the alliance there have been voices in the USA denouncing the European members for paying less than they should, thus forcing the USA to spend its own money to defend its allies. Originally this was certainly so, and was intentional. In NATO's early years the war-torn European members were in no position to defend themselves, and America shouldered the major share of the costs because it feared that a defenceless Western Europe would rapidly become as much a Soviet sphere of influence as Eastern Europe. Since then, however, the European states have more than recovered, and the collective GNP of Western Europe, as well as its total population, makes it at least as rich as the USA and notably richer than the USSR. It is therefore felt by many in America that there is no excuse for the USA still having to pay more than their 'fair share' of total NATO costs.

While the principle implied above does, indeed, seem beyond argument, it is extremely difficult to know how to assess the relative shares of the burden that each country 'ought' to pay. To take one specific example often used by American opponents of NATO budgets, the USA dedicates between 55% and 60% of its defence budget to NATO (depending on whose figures one accepts). The assumption is clear—more than half of all US defence expenditure goes on defending its European allies instead of itself. How can this be fair? The answer depends on just how dangerous, and how much of a threat a Soviet-controlled Western Europe would be to the USA. If Americans could live in perfect contentment with the whole of Europe under Soviet control, then they are indeed spending the money 'for' other countries. However, the position adopted by those Americans who defend US expenditure on NATO is that the defence of Western Europe is integral to the USA's own defence—that Western defence in general is a seamless web. In this case the question of whether 55% of the US defence budget is 'too much' or not is a subtle question of relative capacity, and involves some extremely sophisticated economic analysis.

Unfortunately the economic analysis engaged in is anything but subtle. The most common measure is to express defence expenditure as a percentage of each country's GNP. If this is the measure taken the resulting pattern is indeed varied, and seems to suggest some countries are not paying their way while others (notably the UK) are spending rather more than they can afford. The table below gives the figures for 1983.

Country	Defence Expenditure as % of GNP
Belgium	3.3
Canada	2.1
Denmark	2.4
France	4.2
Germany, Fed. Rep.	3.4
Italy	2.8
Netherlands	3.3
Norway	3.1
UK	5.5
USA	7.4

Source: *The Military Balance 1986/87,* IISS, London.

These figures naturally fluctuate from year to year, for example the 1985 US level is rather higher, because of President Reagan's military build up, but they represent a roughly continuous trend. Whether they do actually mean that some countries are taking advantage of the generosity of others is another question. Straightforward percentages of GNP do not, after all, say very much about relative internal costs. It is, for example, easier for a country as rich as West Germany to spend 3.4% of a very large GNP than for Italy to spend 2.8% of a smaller GNP, which is much more strained to deal with necessary social expenditure.

The real problem, which is slowly becoming appreciated, is that expenditure figures do not, in any case, measure what is really important, which is the amount of military capacity contributed to the alliance in any year. West Germany's army is recruited by **Conscription,** while the USA and the UK have very expensive **All Volunteer Forces,** so why should the Federal Republic, in a sense, be blamed for finding a way to reduce the internal cost of its contribution? Furthermore, West Germany allows huge and environmentally-costly exercises to take place on its territory every year, for which it gets no bookkeeping credit. Finally, it is difficult to claim that one country is paying too little if that country believes the total defence cost of NATO is overestimated. There is no absolute answer to the obvious question of just how much NATO should, in general, be spending, and neither can there be a value-free answer. For many American analysts the total NATO expenditure is much too low, because they would like NATO to enhance considerably its conventional capacity, in order to reduce the risks of nuclear war. To some Western Europeans, and to most Western European governments in the past, conventional rearmament was actually undesirable, because lowering the risk of nuclear war involved increasing the risk of a devastating conventional war (see **Nuclear Threshold**). The policy preferred by Western Europe for general strategic reasons simply did not need as much money, and any effort they made to match US defence expenditure was, in some ways, reducing their security.

However arbitrary the arithmetic, conflicts over burden sharing are endemic to NATO; in recent years they have become more bitter, and possibly more dangerous. Some critics of NATO in the US Senate are quite prepared to start withdrawing US troops from Europe in order to force European governments to increase defence expenditure (see **Isolationism**). Were this to happen it is most unlikely to have the impact desired, but could weaken alliance cohesion very considerably.

Bus

The bus on a **Ballistic** missile is the final stage of a multi-stage rocket. It is the bus which continues the trajectory path once the drive stages have burned out and been jettisoned (see **Boost Phase**). The bus contains the **Warheads,** more properly called **Re-entry Vehicles,** guidance computers, **Decoys** and other anti-**Ballistic Missile Defence (BMD)** technology. The bus and its contents are also referred to as the **Front End.** In itself inert, the bus may well be the most likely target for advanced BMD weapons, because it is the last stage at which the entire **Payload** can be destroyed with a single strike, and the post-boost or **Mid-Course Phase,** after the jettisoning of the engines, lasts longer than any other phase. To reduce its vulnerability, some weapons designers have hoped to give the bus itself motors and guidance systems so that it could manoeuvre during flight and present a much more elusive target. As far as it is known, no missile system currently deployed has this capacity in full, although the **Chevaline** front end for the British **Polaris** missile is thought, by some, to have a limited version of it.

C

C^3I

Command, control, communications and intelligence assets are usually referred to by the acronym C^3I (which is pronounced 'C cubed I'). They represent the combined capacity to deliver orders to military units, to continually monitor and control their presence, movements and status, to be well-informed of enemy movements and intentions, and to be able to relay and receive messages reliably, quickly and secretly. C^3I is more crucial to the successful conduct of warfare today than ever before. This is particularly so because the mass and detail of information involved is far more than any general, admiral or political leader can hope to take in and act on rapidly, and now has to be handled by computer. The speed of movement of military units, the ranges over which weapons can be fired and the size of the areas affected by them have all increased massively since the last time a major war was fought. At the same time the need for detailed and infallible control of interlocking operations has increased. The doctrine of controlled **Escalation,** for example, makes it imperative that orders for the use of different forms of weapon be obeyed with absolute precision. If the level of **Central Strategic Warfare** had been reached, the execution of a **Limited Nuclear Option** under the **Single Integrated Operations Plan** would require that exactly the prescribed missiles were launched at exactly the required time on precisely the pre-planned target co-ordinates. At the level of conventional warfare, tactical doctrines such as the **Air-Land Battle Concept** or **Follow on Forces Attack,** and the use of **Precision-Guided Munitions** call for what is known as **Real Time Information.** This means that the army commander and, for example, the artillery or missile commander, have to know the whereabouts and movements of enemy units immediately and instantaneously, rather than waiting a few hours for photographic reconnaissance to report where they *were.*

Clearly high quality C^3I is vital—and it is the weakest link in the military machines of both the Warsaw Pact and NATO alliances. It is no longer a matter of a few radios and field telephones linked to a camouflaged command truck. C^3I bases need to be 'hardened' (see **Hard Target**), with ever more sophisticated communications links, which makes them static, at risk both to attack, and to interception and jamming of communications as never before. Furthermore, the need for this form of control and intelligence exists right up to the highest political level, because so many of the decisions on escalation will have to be taken by national leaders, and only then after using secure communications with other national leaders. It is the vulnerability of such C^3I facilities which principally concerns those who doubt that escalation can be controlled and that sophisticated war plans can be implemented or relied on. Until recently, no NATO member has invested anything approaching the required amount in C^3I facilities, and there are grave doubts among professionals about the quality of what systems do exist. (See also **Electromagnetic Pulse, Electronic Intelligence, National Command Authorities, TACAMO** and **Ultra-Low Frequency.**)

Campaign For Nuclear Disarmament (see CND)

Carrier Battle Group (CVG)

The carrier battle group is at the heart of a modern navy, or, at least, of the US navy and, as far as they can manage, its imitators. A full carrier battle group consists of two aircraft carriers, such as the enormous USS Kennedy, nuclear-powered ships able to carry several squadrons of aircraft, and various smaller craft, such as **Destroyers** and **Frigates.** These carriers can launch massive air power against land or sea targets within several hundred miles. They are even equipped with nuclear bombs, as well as **Nuclear Depth Bombs,** and can therefore be regarded as strategic weapons. For the purpose of **Force Projection,** the CVG (the non-obvious acronym for a carrier battle group) is admirable. Their disadvantage is that they may be extremely vulnerable in a war zone. Carriers provide huge and slow-moving targets for submarines—nuclear **Hunter-Killer Submarines** can actually travel faster than them. If the carrier battle group disperses to make a less-inviting target to submarines, it then becomes much more vulnerable to aircraft attack because the main purpose of the destroyers and frigates making up the rest of the group is to serve as radar outposts and **Area Defence** ships. In fact, some calculations suggest that 50% of the air power of a carrier battle group is actually dedicated to self-protection.

If this arithmetic is right, the effective carrier battle group force, as projected by the **Department of Defense** into the 21st century, is not between 12 and 14 but, perhaps, seven. As one in five of the carriers will always be in dock for overhaul, the power of even the US navy in terms of aircraft carriers is seen to be decidedly limited. This combination of expense and vulnerability has prevented the USSR from trying to rival American naval air power in the same way, and has reduced the British Royal Navy to much smaller carriers, mainly intended for **Anti-Submarine Warfare** purposes.

Catalytic War

A catalytic war is a small nuclear engagement, or the use of nuclear weapons by a minor nuclear power, which triggers off the cataclysm of **Central Strategic Warfare** between the superpowers. During the early days of the development of nuclear strategic theory this was a topic of common concern. The main worry was the fear of such a major war starting by accident because of the fallibility of detection methods and the impossibility of controlling **Escalation.** Furthermore, **Scenarios** were created in which a minor nuclear power deliberately used its forces to trick one superpower into believing it had been attacked by the other—a more cynical version of the **Trigger Thesis** sometimes attributed to the French as a justification for their small **Force de Frappe.**

In general the fear of **Accidental War** has declined, in large part because the nuclear strategies of the USA have increasingly come to depend on the ability to control escalation in an *intentional* nuclear interchange between superpowers. Clearly if such controlled nuclear exchange is contemplated, one cannot afford to give credence to the idea that superpower nuclear confrontation could be triggered by the actions of a third party. It remains the case that fear of nuclear proliferation, sometimes called the **Nth Country Problem,** has as part of its grounds the possibility that a minor nuclear power might attack an ally of a major nuclear power. If this happened, the superpower might be forced to retaliate on behalf of its ally. If the minor nuclear power was supposed to benefit from the **Extended Deterrence** of the other superpower, a chain reaction could indeed lead to central strategic war.

CENCOM

CENCOM, which is an abbreviation of 'Central Command', is the current name of the US military organization set up by President Carter under the title Rapid Deployment Force (RDF). It is a response to America's potential need to intervene at some distance from the USA, in the Persian Gulf for example (see **Global Strategy**). The forces need not be heavily armed, but must be capable of fighting for at least a brief period before reinforcements and extra equipment can be delivered (see **Air-Lift** and **Sea-Lift**). CENCOM does not, in fact, consist of permanently prepared troops at all. It consists mainly of a headquarters and planning staff which would, in a time of emergency, be able to deploy combat forces that usually have a more general role, such as the 82nd Airborne Division and various Marine units.

It is generally thought that, at its conception, the RDF was meant to be a permanent force, stationed somewhere in the Mediterranean area—but no country was prepared to host the base. In any case, there are so many problems of force deployment involved that it is far from agreed within the US military that CENCOM could deliver sufficient forces to a distant area and provide them with adequate initial supplies. (See also **Force Projection** and **Light Division.**)

Center for Naval Analyses (CNA)

The Center for Naval Analyses is one of many US defence **Think-Tanks** that have come to be of profound importance in developing national security policy since 1945. Just as the **RAND** Corporation works entirely for the United States Air Force, CNA develops plans and technology for the navy. The ability of the services to call on civilian expertise in planning and design helps them to fight the endless internecine struggles for financial priority in Congress, and even against the political head of the whole armed services, the Secretary of Defense. Partly to offset this there is also a think-tank, the **Institute for Defense Analyses (IDA),** which works directly for the **Office of the Secretary of Defense.** Ironically, the CNA and IDA offices are situated facing each other on opposite sides of the same street in a Washington suburb.

Central Command (see CENCOM)

Central Front

The central front is the border between the two Germanies, along which are arrayed the **NATO** and **Warsaw Pact** armies. This is the area with the highest concentration of modern weapons and combat troops in the world, and traditionally is seen as the **Theatre of War** in which a third world war would be fought. Apart from the northern extension of the central front in Arctic Norway, it is the only place in the world where American and Soviet troops face each other, although, of course, one or other superpower faces forces deployed by its opponent's allies, and actually fights such allies, in several places.

The major troop concentrations on the central front for the Warsaw Pact are the East German army and the **Group of Soviet Forces, Germany (GSFG),** but NATO is represented by all its main partners. Apart from the whole of the West German army and air force, totalling about 450,000, there are large contingents of the US Seventh Army and US Air Force of about 300,000, and smaller contingents from the UK, Belgium, the Netherlands and, while not at the front and not technically under NATO command, France. The NATO contingents are, as far as possible, deployed near the border for fear of a surprise attack. They are concentrated to defend the two major avenues of possible invasion, the North German plains, where the British and the West Germans take the major burden, and, in the south, the Fulda Gap covered mainly by the American forces. The central front forces of both alliances are under the command of the respective superpower, with the Soviet commander of the GSFG controlling the Warsaw Pact troops and the NATO commander, known as **SACEUR,** always being an American general.

In recent years analysts have become more and more doubtful that confrontation on the central front will initiate a full-scale war, because although the NATO forces are outnumbered by the Warsaw Pact, the latter's superiority is not such as to give the USSR any great confidence that they could win an intentional war of aggression. However, the real fear is that superpower conflict elsewhere in the world is very likely to trigger fighting on the central front, if only because of the huge mass of high-technology forces assembled there.

There are enormous problems of defending the central front, which are aggravated by political pressures that force NATO into a linear or fixed defence along its entire length, and the unwillingness of the West German government to allow serious fortification of its border with East Germany for fear of the symbolic import of building a fortified line between two halves of what they still regard legally as one country.

Central Intelligence Agency (CIA)

The CIA was established as part of the 1947 reorganization of the whole structure of American defence, largely on the basis of the wartime Office of Strategic Services. The USA, almost alone among world powers, had not previously had a peacetime civilian intelligence service. It was heavily influenced by the British Secret Intelligence Service when the need arose to create an organization for wartime, and later peacetime, intelligence gathering.

At first the CIA was not expected to be a large or powerful organization. Its original function was to be more of a co-ordination and analysis centre for the disparate intelligence-gathering branches of the military services, the Department of State, the **Defense Intelligence Agency** and the **Electronic Intelligence** service, the **National Security Agency.** No one really intended the USA to operate serious 'covert' intelligence operations, and there has always been a tension inside the CIA itself between collection and analysis of open source information, and the running of espionage agents in foreign countries. Even less was there an intention to have the sort of 'dirty tricks' section carrying out revolutionary and destabilizing operations, and indulging in assassination and criminal activities worldwide. Those in charge of the CIA, however, worked for growth and for a dominant position in the US intelligence fraternity from the outset. There was soon a private slogan inside the Agency that ran 'Bigger than State by Forty-Eight', implying that it aimed to have more intelligence officers than the official foreign service within one year of its being created. It did.

The CIA has gone through periods of excessive power and blatantly illegal activities, always to be brought to heel by other forces in the US political and defence worlds. Its period of greatest power and excesses was, perhaps, in the 1960s and early 1970s. Subsequently Congress, as part of its general resurgence after the Nixon years, placed major constraints on the Agency. Its actions continue to be severely restricted by committees of both Houses of Congress, and even the President cannot easily order it to undertake foreign activities without Congressional control.

In general the valuable work the CIA does has never been affected by the rise and fall of its covert or illegal aspects. Most useful intelligence has always been gathered by open methods, and the analysis of the huge stream of available information by skilled and talented professionals is executed very efficiently, quite aside from any other activities. So sullied is the public perception of the CIA outside, and often inside, the USA that it is amazing to many to meet the hundreds of able and perfectly normal people in Washington who openly admit that they are, or have been, Agency officers.

Central Region of NATO (see Central Front)

Central Strategic Warfare

Central strategic warfare is the technical description of what many think of as the paradigm for a nuclear war—the exchange of long-range ballistic missiles between the USA and the USSR. Even central strategic war can come in a variety of forms, given the endless distinctions between **Counter-Force** and **Counter-Value** strikes, and the doctrine of **Limited Nuclear Options.** Nevertheless, it is widely accepted in strategic thought that the transition to central strategic war, whatever precedes it, would be the biggest single step in any **Escalation** ladder. To the extent that the only real purpose of American and Soviet possession of ballistic missile forces is to deter each other from attacking their respective homelands, often referred to as the theory of nuclear **Sanctuarization,** the onset of central strategic war would coincide with the breakdown of **Deterrence.** If central strategic war occurred, with all the attendant dangers of **Mutual Assured Destruction** and possible **Nuclear Winter,** only the more optimistic adherents of **War-**

Fighting theories believe that responses could be controlled so that one or other side could survive as, in some sense, 'the winner'.

The concept of central strategic warfare serves to distinguish it as a step in escalation from those which challenge less directly the very existence of the superpowers. Although usage of the term is normally confined to nuclear exchange between the superpowers, there are two other nuclear-armed powers in Western Europe, France and the UK, who are perhaps more likely to become involved in trading nuclear strikes with the Soviet Union. Clearly they would regard such a situation as very 'central', and deeply **'Strategic'.** Thus the use of **Theatre Nuclear Forces** against targets in Western or Eastern Europe (as long as they did not explode inside the borders of the USSR), the use of naval nuclear weapons against enemy ships, and even the nuclear bombardment of the UK by Soviet missiles would not quite step over the borderline into central strategic war.

Although, at one time, the concept was clearly defined, the development of versatile weapons means that it may no longer be of great analytic use. The single main problem, as the USSR itself continually points out, is that US missiles, such as the **Pershing II** and the **Ground-Launched Cruise Missile,** based in Western Europe are perfectly capable of hitting targets (including Moscow) in the western Soviet Union. There seems to be no reason why the USSR should regard the destruction of a site inside its own borders as 'different' just because the missile was fired from West Germany rather than from the American Midwest. (See also **Euromissile** and **Eurostrategic.**)

CEP (see Circular Error Probable)

Chemical Warfare

Chemical warfare is the modern way of referring to the use of gas as a weapon. The most infamous example is the use of poison gas in the First World War. Chemical warfare has a curious history, in as much as it may be the one clear example we have of the use of a **Weapons System** being deterred by two alliances possessing an equal capacity. This, at least, is the explanation for the non-use of gas or chemical weapons in the Second World War, despite the widespread fear of the 1930s that such weapons would be used in air attacks on civilian populations.

A wide variety of chemical weapons has been developed by both NATO and non-NATO forces, although NATO had, until the mid-1980s, done very little to modernize and develop its chemical warfare stocks. There are, perhaps, two main categories of such weapons, differentiated according to whether their main impact is on the respiratory system or on surface skin. The prinicpal objective of chemical attack is to incapacitate rather than necessarily to kill. Thus the use of various blistering agents, such as mustard gas, which burn and irritate the skin, even though they are not usually lethal, is as effective as the traditional 'poison', or nerve gas, which asphyxiated.

Modern chemical weapons, in the form of artillery shells or bombs, are of a design known as binary rounds. In such a design the poisonous effect is caused by combining two different chemicals, each of which is innocuous in isolation. The rounds are manufactured so that it is only on impact that the two materials mix, making the storage and use of them much safer. NATO stocks of chemical weapons are currently entirely 'unitary' rounds, and their age has begun to present a serious problem. Consequently, in July 1984 the US Congress authorized the development of a new generation of binary chemical rounds. However, there is some doubt as to whether these will in fact be built or, if built, deployed, because chemical warfare arouses if anything even stronger moral opposition from the public than do nuclear weapons.

Nevertheless, chemical warfare is one area in which it is widely agreed that the Warsaw Pact has a very considerable superiority over the West; the UK, for example, unilaterally discontinued making and researching such weapons in the 1950s. This imbalance is seen as especially important because the threat of the use of chemical weapons, by forcing the enemy to take precautions like wearing protective suiting, seriously impedes their efficiency. An air-base under threat of a chemical attack, for example, may have the number of missions which it can launch reduced by over 50% by this factor alone.

Chemical Warfare Convention (CWC)

One of the less well known, but possibly more potentially succesful among **Arms Control** arenas, is the **Forty Nations Conference on Disarmament,** which has been meeting since 1983 in Geneva. The Conference is linked to the United Nations, but is generally autonomous of all other international bodies or talks, such as NATO or the **Conference on Security and Co-operation in Europe (CSCE).** The 40 members include all admitted nuclear powers (see **Nuclear Club**). Its principal task is to draft a convention that will ban all chemical warfare and all preparations for it. By mid-1987 it seemed that there was a good chance of achieving a verifiable and total ban on the development of such weapons. Most Western commentators think that the USSR has been noticeably less obstructionist here than in some other areas of arms control. The conclusion of a CWC has been made a priority, in particular by the British Foreign Office.

The major remaining area to be negotiated was, as usual, one of **Verification.** A formula acceptable to most members of the conference, possibly including the USSR, does exist, but the USA has refused to agree to it. The problems of verification for this convention are probably greater even than for nuclear weapons, as there seems absolutely no alternative to on-site inspection. The voices raised against such inspection come from a much wider field than usual. Many chemical industries, in Western as much as in Eastern countries, are extremely worried that on-site inspection will damage their industrial security, because the overlap between chemical research intended for peaceful use and for military applications is very great. For the same reason there is an enormous problem of definition involved, far exceeding the problems of establishing the difference between peaceful nuclear energy and military nuclear installations.

Chevaline

Chevaline is the code-name for the programme to modernize the **Bus** (or **Front End**) of Britain's **Polaris** missiles. This programme was carried out during the 1970s and early 1980s at a cost of at least £1,000 million (in 1980 prices). It was seen as so politically controversial that Parliament was not informed of it until 1980, and it is believed that many members of the Conservative and Labour cabinets, under Prime Ministers Heath, Wilson and Callaghan, were also kept in ignorance.

The purpose of the modernization was to maintain the ability of Britain's Polaris missiles to penetrate the **Anti-Ballistic Missile** shield around Moscow (see **Galosh**), thus retaining Britain's prime war object, the so-called **Moscow Option.** The new front end is designed to achieve this by carrying a large number of **Decoys.** Although it is now believed to carry only three, or possibly even only two, warheads, these have been hardened to withstand a near miss by a nuclear interceptor rocket. The most revolutionary aspect with which the new design has been credited is the capacity of the whole front end to manoeuvre just before releasing the warheads, thus giving further confusion to Soviet missile defences. Other unconfirmed suggestions include the possibility that some of the missiles on each submarine are intended to explode before reaching Moscow, ahead of the rest of the attack, to put the Soviet defence radars out of operation with a huge **Electromagnetic Pulse.** Whatever the true details of Chevaline are, it is generally accepted that it confirms the importance of the Moscow Option to British planning, and may well have the effect of limiting the usefulness of Polaris to this one task.

Chiefs of Staff

The Chiefs of Staff of a country's armed services are the professional heads of each of the branches recognized as separate services. Thus, for example, Britain has a Chief of the General Staff (Army), Chief of the Air Staff (RAF), and a Chief of the Naval Staff, while the USA also recognizes the head of the Marine Corps. The real political importance of these officers, however, is not as head of their own services, as this function is in practice carried out by officers deputizing for them, but as a sort of collective military leadership. Thus in both the USA and the UK there is another officer, drawn in turn from the different services, who serves as the chairman of the Chiefs of Staff in their collective role. (In the USA the post is actually entitled Chairman of the **Joint Chiefs of Staff,** while in the UK it is known as the Chief of the Defence Staff.)

These small groups of military officials are responsible for formulating overall defence policy and advising the political leadership in times of crisis and war. This system of command by committee replaced the older idea of having a single commander-in-chief, or the traditional German idea of a single integrated general staff with one actual commander. At the same time the system allows for more or less unified advice, rather than having each service giving separate, and probably conflicting, advice to its government. A further consideration may be that such a committee system avoids the danger of the military producing a commanding figure who could challenge the legitimate government.

Since the early 1980s defence analysts and politicians in both the USA and the UK have had serious doubts about the merits of the system. As it has worked in practice, the system has tended to detract from full inter-service planning and co-operation, and has failed to give firm and clear advice to the political leadership. A major problem is that each individual service chief fights for the interests of the corresponding service. As a result the deliberations of the Chiefs of Staff committees culminate in compromise and group bargaining, rather than representing a true overall non-partisan policy, whether it be on funding and procurement or on military strategy.

The problem has been compounded because the chairman has not been allowed any real command authority, and has been required to represent faithfully the compromises reached by the service heads rather than being able to present the best professional advice from an all-service perspective. Reforms to give the chairman much more authority and to force genuine inter-service planning have been undertaken in both countries. However, it is improbable that such administrative reforms could do much to solve the long-running problem of **Inter-Service Rivalry.**

Choke Points

Choke points are places where transport of military material and personnel is concentrated and slowed down. A typical choke point would be one of a small number of bridges across a major river that could bear the weight of tanks. Another example would be a major railway junction through which supplies for the front line had to pass. The importance of choke points is that they provide targets for **Interdiction** strikes by an enemy, who can hope to cause serious delay to troop movements and reinforcements by a small number of attacks. In this way a military power can even win a battle in which it would be outnumbered and face certain defeat were the opponents able to bring their full force forwards. Most probably the vulnerable power would not have the aircraft or the time needed actually to destroy the opposing armoured divisions, but preventing their use by attacking a choke point would have an equivalent effect.

The existence of such choke points is important to NATO's current strategy of **Follow On Forces Attack (FOFA).** In order to defeat the numerically superior Warsaw Pact forces, from which an **Echeloned Attack** is anticipated, it would be necessary to destroy or hold up the second echelon at a point deep inside Warsaw Pact territory, by the use of air power, while the NATO armies defeated the first echelon troops. However, NATO's dependence on US reinforcement from across the Atlantic provides many similar choke point targets for Warsaw Pact air power. These include the English Channel ports and the small number of locations in West Germany where the US divisions have to collect equipment stored under the **POMCUS** arrangements. Some analysts also argue that interdiction of choke points in Eastern Europe will not be very effective, given, for example, the bridge-laying capacity of the Soviet army.

CINCHAN (see SACEUR)

Circular Error Probable (CEP)

This concept, which is usually abbreviated to CEP, is the most commonly cited measure of the accuracy of a missile or other long-range weapon. It has a simple and precise definition. The CEP of a missile is the radius of a circle within which 50% of all warheads fired at the same target will fall. Thus, if 10 single warhead rockets, each with a CEP of 100 metres are fired at the same point, at least half of them should fall within a cluster where they are not more than 200 m apart from each other. (CEP is measured by the *radius* not the *diameter* of the hypothetical circle.) This measure is used to gauge the suitability of a missile for a particular task. Against a city, for example, high CEPs can be accepted because the force of a nuclear explosion will cause enough damage, even if the missile lands some way from its actual target.

Early generations of missiles, particularly **SLBMs,** had very high CEPs: the Titan II, the earliest of the US **ICBMs** still in operation in the mid-1980s, had a CEP of over 1,300 metres, and the Polaris submarines used by the British navy carry missiles with a CEP of about 900 m. **Hard Targets,** such as missile silos or command posts built from super-reinforced concrete and buried underground, are likely to be destroyed only by a direct hit or by a very near miss. This requires a missile with a typical US warhead power of 300 kilotons to have a very low CEP, perhaps as little as 60 m. The larger the force of the explosion, the further one can afford to miss by, naturally, which is why the high CEPs estimated for most Soviet missiles matter less than they would for the USA, as Soviet warheads are much more powerful, often in the multi-**Megaton** range. Except for the latest generation of US **Minuteman** III missiles, the CEP/explosive force combination is too weak for reliable use against very well-hardened targets, and no existing SLBM, American or Soviet, can be used in such a role. This is partly to be remedied for the USA (and, should it be deployed, the UK) by the introduction of the **Trident** II (or D5) submarine missile, which will have a CEP as low as any ICBM.

Increasingly, technically competent critics have been calling into question the validity of CEP as a measure of accuracy. The trouble with CEP is that it does not really measure accuracy in a conventional sense at all, because arranging for warheads to fall in a tightly packed cluster is compatible with *all* of them being a long way from their target. Those familiar with rifle shooting will be aware of the competition known as 'grouping', in which points are scored not for how many bullets are put into the centre of a target, but for the consistency of aim in getting them grouped tightly together. A group of five shots in which two are in the bull's-eye and three spread widely round the outer circle of the target will score less than a group of five, all on the outer circle, which can be covered by a penny. It is this latter achievement that is measured by CEP. An alternative measure has been created, called **Bias,** which more closely measures the ordinary sense of accuracy—the likelihood that a single missile will actually hit its target. In fact it is extremely hard to measure accuracy in this sense reliably, because of the problems of testing missiles. Many of the factors that will affect the accuracy of intercontinental missiles fired from the USA towards the USSR, or vice versa, cannot be tested at all. A major source of inaccuracy, for example, is the variation in the earth's magnetic and gravitational fields over the route such missiles would have to travel. Given that US missiles will have to travel, more or less, on a west to east trajectory, it is notable that the US missile ranges all point east to west! A considerable amount of caution is, therefore, required in accepting claims to accuracy, especially when measured by CEP.

Civil Defence

Civil defence refers to all arrangements for reducing death and damage to non-combatants in the home territory of a nation during wartime. The distribution of gas masks to all UK citizens in 1939 and the building of bomb shelters were classic civil defence measures, as was, in a different way, the evacuation of children from London to safer rural areas. In a modern context the civil organization for controlling and distributing food and medical supplies after a nuclear attack qualifies as civil defence just as much as planning for the evacuation of major cities. There are several schools of thought on the efficacy, and even the desirability of organizing civil defence measures in the face of nuclear war. In some left-wing circles in Western Europe civil defence is thought to be not only useless because of the inescapable horrors of a nuclear war, but actually dangerous. Belief that adequate civil defence measures had been taken could encourage governments to take nuclear war risks, or persuade the population to support nuclear armament policies.

There is, in the 1980s, little general emphasis placed on civil defence in the West, even by conservative governments, because most available evidence suggests that the investment would, indeed, be useless. For example, evacuation of major cities could well lead to more deaths, not only through panic and possible riot, but because it could mean more people being caught in the open during an attack when many of them would have been indoors, and marginally better protected, were they to be following an ordinary daily routine. Furthermore, some of the comparative statistics of injuries compared with health facilities are so stark as to strengthen the case for taking no action. A few 200-kiloton warheads exploding over the centre of London or New York would produce more serious burn injuries than there are beds in burns units in the whole of the respective countries—probably by a factor of 10. Thus the provision of nuclear air-raid shelters in the USA has been allowed to lapse and no Western European government has had a policy of providing sophisticated shelters for the general populace; civil defence provisions *have* been made, to a certain extent, in Switzerland and Finland, but even these do not approach the standard of Soviet preparations.

A large part of the difficulty in making civil defence policy is that very little is known either about the actual results of a nuclear attack, or the nature of the attack that is to be prepared for. One of the great popular fears of a nuclear strike has always been the idea that it would result in a huge amount of **Fall-Out,** so that protection against the immediate blast damage would be irrelevant; death through radioactivity would occur in a few weeks or months even among those unharmed by the original attack. This, however, is by no means inevitable because **Air Burst** nuclear explosions, which are more efficient against most kinds of target, may cause very little fall-out compared with **Ground Bursts.** Even such calculations leave an enormous amount to chance. Analyses of the effect of a Soviet attack on the Midwestern missile fields in the USA range from only two million human casualties to 20 million, depending on the time of year and the direction of the wind.

In general, civil defence precautions almost certainly would reduce casualties in at least some **Scenarios,** and it must be noted that both superpowers probably have targeting plans designed to minimize **Collateral Damage.** This is not to say that Western governments are necessarily wrong not to invest more in civil defence, because it is obviously a complex matter of balancing the relative costs, but it is absurd to treat civil defence as *ipso facto* pointless. Perhaps the best argument in favour of developing civil defence policies with shelters and evacuation plans is that the USSR *does* invest heavily in this area. Again, the wilder estimates by some American writers, which suggest that the USSR could escape relatively unscathed from a major nuclear war because of their civil defence precautions, are simply wrong, and are usually based on obvious methodological errors. Nevertheless there is a curious imbalance in approach between the two superpowers which, given the paucity of high quality research in the public domain on the possible effects of civil defence measures, makes it difficult to be sure that the current vulnerability of Western civilians really is inevitable.

Civilian Morale

Civilian morale denotes the willingness of the population to support its government during wartime, to work for the 'war effort' and to accept the sacrifices, scarcities and miseries that full-scale **Total War** involves. Since the First World War, which has often been seen as the first total war in which the active support of the whole civilian population was vital, there has been a theory that destroying this intangible but vital asset was the real path to victory. This was the doctrine of **Douhet,** the Italian airman whose ideas served to support strategists in the RAF's **Bomber Command** and many other air forces (see also **Strategic Bombardment**).

It should be noted that it is not merely a matter of starving a population into surrender—the history of naval **Blockades** and, indeed, the whole history of warfare has rested on such a strategy. What Douhet and others were arguing for was not necessarily a particularly bloody destruction of civilians, but a psychological onslaught by bringing the fear of death to those many miles from the fighting at the front, and otherwise safe. The objects were to break the political will of the civilian population and to cause massive disobedience to the government. Such a doctrine touches on much older theories of the nature of war, and is particularly suitable in a modern context where, even if governments are not directly dependent on formal democratic support, they at least need the acquiescence of the mass population. The 18th-century philosopher Immanuel Kant, for example, argued in his famous essay *Perpetual Peace* that war would be much rarer between democratic political systems when those who bore the cost of war were ultimately responsible for policy. Neither is the strategy of terrorizing the civilian population to break their morale a 20th-century invention. The famous Union general, William T. Sherman, when marching his northern army through the Confederate South, deliberately urged his troops to devastate a corridor 60 miles wide along the route of the march in order to terrify the Southern civilians, and to make them realize, in his own words from a speech made in 1880, that 'war . . . is all hell'.

What is not clear is whether the strategy of breaking civilian morale actually works. Germany suffered incessant bombing raids during the Second World War, in the daytime from the American and at night from the British air forces. Yet no real impact was made either on morale in general, or on industrial morale in particular. The bombing offensives did not have a major impact on the German war effort until close to the end of the War, when the raids were switched away from ordinary civilian targets to specific industrial targets of immediate **Bottleneck** importance to vital war supplies. The available evidence tends to point in the other direction; the inhabitants of both German and British industrial cities were strengthened in their resolve, and the more easily convinced of the cruelty of the enemy when their families and homes were destroyed. The one previously incontrovertible example, that the atomic destruction of Hiroshima and Nagasaki ended the war against Japan, has even had doubt cast on it recently.

Classical Strategists

Obviously the study of warfare, and the attempt to derive theories about how best to achieve war aims, is as old as any intellectual endeavour. Although the conditions of modern nuclear conflict are dramatically different from any past experience, reference is still made to the tradition of strategic theorizing. The great figures in this tradition are often referred to as the 'classical strategists', the implication being that they offer rules and maxims good for all ages and all conditions of war. As with any putative list of 'the greats', there are few totally agreed entries, but strategic thought has not thrown up so many commanding figures that an indication of the range is impossible.

Without any doubt the first entry in most people's list would be the 19th-century Prussian soldier-scholar Karl von **Clausewitz.** The only figure from an earlier period very commonly cited today is the classical Chinese thinker, Sun Tzu, although his inclusion may be partly the result of a fashion set by influential American analysts in the last few years. Others would include Jomini and Mahon (particularly influential in naval circles), who were both Americans and both influenced by Clausewitz. In the 20th century Basil Liddell Hart and his contemporary and fellow Englishman, Fuller, drew out the strategic lessons of the First World War, although they were probably followed more seriously by German generals, such as Rommel, than by those in their own country. While there have been several influential thinkers since 1945, such as the Frenchman, Raymond Aron, and several Americans, these belong to the 'post-classical' world.

There are few, if any, principles common to these thinkers and, indeed, they often contradict each other. Sun Tzu, for example, preaches a doctrine that has influenced modern American advocates of **Manoeuvre Warfare.** It stresses the importance of destroying the enemy's co-ordination and confusing it rather than fighting it, while Clausewitz is firm in his emphasis on the impossibility of avoiding the decisive battle. Mahon stresses the need for **Concentration of Forces;** Liddell Hart advocates many rapid thrusts avoiding entrenched strength in favour of destroying communications, and so the conflicting theories mount up. What the best of classical strategists have in common is summed up by Clausewitz's dictum that war is 'nothing but the continuation of politics by other means'. This emphasis on the need always to keep the political reason for war in mind has endeared him to Marxist thinkers, and is the main contribution that tradition can make to the flights of fancy of modern strategic theorists.

Clausewitz

Karl von Clausewitz (1780 – 1831) is beyond any doubt the most famous and influential military thinker in Western history. He was a Prussian general who fought in the Napoleonic wars and later dedicated himself to writing a massive study, simply entitled *On War,* which was not published until the year after his death. His thought cannot be summarized here and, indeed, much of it is obscure and concerns highly technical details of staff work. The most significant part is probably the idea that war is a 'continuation of politics by other means'. By this he meant to stress that all **Strategy,** all military activity, must be subordinated to clear political motivation and aims. In a sense he was preaching the same **Doctrine** that became famous much later in phrases such as 'war is too important to be left to the generals'. A second very famous concept of Clausewitz concerns what he called 'the friction of war'. He argued, effectively, that everything that can go wrong will, and that all plans, however elegant or simple, will not quite work out properly. (See also **Classical Strategists.**)

CND

The Campaign for Nuclear Disarmament, very often referred to just as CND, is probably the oldest of the West's anti-nuclear protest movements. It was formed in the UK in 1958 and has had two periods of extreme activity and considerable political success. The first was from its foundation until the early 1960s, when its demonstrations, especially the traditional Easter march from Aldermaston (the major nuclear weapons development centre) to London, vividly caught the public imagination. During this period CND became extremely influential in part of the trade union movement, to such an extent that a unilateral disarmament motion was passed by the Labour Party at its annual conference in 1960 although, when in government, Labour has always veered away from unilateralism.

From the election of the Labour government in 1964, with their acceptance of the plans to equip Britain with **Polaris** submarines, until the mid-1970s, CND and the rest of the anti-nuclear movement became quiescent. Enthusiasm for the cause was reawoken in 1977 with the advent of the **Euromissile** issue, and although CND has had to share with other groups in this second phase of activism, it has remained the most successful movement. Once again it has had political success by convincing large sections of the Labour Party of the need unilaterally to abandon nuclear armaments, which has again been official Labour policy since 1983. (See also **European Nuclear Disarmament** and **Freeze Movement.**)

Coalition Defence

Coalition defence is how the USA defines its basic long-term security strategy. It means that it recognizes its inability to defend all areas of the world which affect its vital interests without assistance. Given this, there are two possible reactions. One is to redefine, more narrowly, the USA's national interests to bring them within its defensive capacity. In a sense, the old perspective of **Isolationism,** as enshrined in the Monroe Doctrine, was one in which national interest and national capacity coincided. The alternative, if the interests are to remain world-wide, is to rely on coalition defence, that is to find community of interests with others. NATO is the lynchpin, though not the whole, of the US coalition defence strategy. It is assumed that Western European powers are not able to defend their own interests independently, that the USA has an interest in a free Western Europe, and that the USA could not afford to guarantee this interest by itself either, but that in combination both European and American interests in Western European security can be defended. The other major alliances to which the USA belongs, such as **ANZUS,** are justified in a similar way.

The problem with coalition defence is that, while two or more nations may have generally compatible and overlapping interests in defending a certain area, the interests are not likely to be identical, and hence strain will occur. Furthermore there is a natural tendency for weaker powers to rely rather more than is in some sense 'fair' on the fact that the USA will, for its own reasons, defend them anyway, thus giving rise to arguments of **Burden Sharing.** The USA, for its part, naturally expects to be the leader in these coalitions, and to define the specific targets and methods of coalition policy. The increasing discontent with NATO, which has led to argument between the **Atlanticist** and **Global Strategy** schools, is typical of the strains of a coalition defence policy, as are disputes over **Out of Theatre** commitments.

Cold War

The phrase cold war began to be used in the late 1940s to describe the state of tension and international competition between the USSR and the West, particularly the USA, which arose soon after the end of the Second World War. The tension arose initially because of a mismatch between Soviet and American expectations for the post-war world, especially concerning the partitioning of Europe. Whether plausibly or not, the USA had expected Eastern Europe to be independent of superpower control, and hoped that the Soviet army would be demobilized as quickly as the armies of the Western Allies. A series of Soviet actions, notably the Berlin blockade and a Communist *coup d'état* in Czechoslovakia, made it clear that the USSR had no intention of relaxing its control over eastern Germany and the rest of Eastern Europe. Furthermore, the influence of the Communist Party of the Soviet Union over Western communist parties, particularly in France, seemed to threaten the stability of those European nations that were in the American sphere of influence.

The US intervention in Europe's development through the huge and generous financial aid grants of the Marshall Plan was interpreted by the USSR as essentially the use of economic strength to compete with Soviet military strength. The creation, at American and British instigation, of **NATO** (which pre-dated the **Warsaw Pact**) and the nearly simultaneous North Korean invasion of South Korea, apparently at Soviet instigation, set the basic terms of superpower relations for much of the rest of the century. This condition of diplomatic and economic competition, combined with active military support in conflicts which are sometimes referred to as 'wars by proxies', earned the situation the title 'war'. However, as there has never been a single case of Soviet and American troops engaging in combat the modifier 'cold' is added to distinguish it from a real, or 'hot', war.

Setting dates for and measuring the intensity of the cold war is open to a large degree of imagination. It is often suggested that the era of **Détente** in the 1970s was a temporary truce in the cold war, which some think began again at the end of that decade with the Soviet invasion of Afghanistan and the breakdown of **Arms Control** negotiations after **SALT II.** However, such distinctions only highlight the highly subjective nature of the concept itself. The obvious question to ask of those who adhere to the cold war theory is just what their perception of real peace would be. For most of recorded European history major powers have vied with each other and used every tool available to manoeuvre for relative advantage between wars: yet these conditions, when no actual combat was taking place, have been seen as the normal conditions of peace. Only some very idealistic conception of what relations could and should be like between two nations, each of over 200 million people, committed to thoroughly antithetical ideologies, can serve to describe a more genuinely 'peaceful' situation than has existed between the superpowers since 1945.

Collateral Damage

Collateral damage is simply damage, or casualties, which are caused by an attack over and above the destruction intended or required. It has become an important concept in strategic theory because of the desire in some nuclear targeting **Scenarios** to minimize civilian casualties. Since the abandonment by the USA, and probably by the USSR, of a strategy deliberately based on maximizing urban destruction and human casualties, considerable attention has been paid to using nuclear weapons against very specific targets. The US conception of **Limited Nuclear Options,** for example, requires a capacity to hit targets such as missile **Silos,** command and control bunkers (see **C³I**), political leadership shelters or very carefully chosen industrial assets without inflicting mass destruction. Any such strike would defeat its own purpose if the collateral damage to people, or to industrial structures in general, was to be great. Similarly the American fear of Soviet **Strategic Superiority** was based in part on a perception that the USSR would be able to destroy US land-based missiles in a **First Strike** without causing so many civilian deaths that the President would be forced to retaliate with a major strike on Soviet cities.

The problem which faces the strategy of minimizing collateral damage is that it is probably impossible to achieve. Even if such damage limitation was possible in theory, it would still be extremely difficult to predict the practical effects of a nuclear detonation. So little is really known about the effect of nuclear explosions, and so variable are the factors that will determine the outcome, that no one could be sure that the outcome of their strike would fit within acceptable casualty limits. Studies and computer modelling of a USSR strike on American missile bases have produced estimated human casualty rates varying between two and 20 million, depending on factors such as wind direction and other weather conditions. One other reason for the importance of collateral damage limitation is the growing sense of the need to produce a moral justification, preferably along the lines of orthodox **Just War Theory,** where a distinction has always been drawn between legitimate and illegitimate targets.

Collective Security

Collective security was introduced as a practical idea in international relations at the end of the First World War, although it had been a theoretical concept much earlier. It was intended as an alternative to the **Balance of Power** system for ensuring international stability. Under the balance of power peace, or at least the continued existence of the states in the system, depended on shifting alliances. The self-interest of individual states was supposed to ensure that whenever one major country or coalition increased in strength to the extent that it might be able to conquer another, the other major powers would join with the threatened nation to protect the status quo. One of the many disadvantages of this mechanism is that it ultimately depended on conflict, and on the pursuit of national self-interest.

Because it was thought that the First World War stemmed from a breakdown of the balance system, the victorious powers of 1918 sought an alternative guarantee of stability. The negotiators at the Versailles peace conference were heavily influenced by American diplomacy, particularly the ideas of President Woodrow Wilson, which were generally impatient with the traditional European powers, the Allies almost as much as the defeated Austro-Hungarians and Germans. Instead of a system based on international rivalry, it was proposed that all nations enter a binding agreement to respect the new national borders and to refrain from force as a method of international politics. So that no country would feel itself at risk by taking up such a self-denying position, it was argued that all nations should share a duty to come to the aid of any country attacked by another. Instead, therefore, of having to rely on its own allies, and facing an enemy which might have more allies of its own, each nation would know that an aggressor would have to face the combined might of the law-abiding nations.

The original institutional form of this collective security was the League of Nations, although there were a series of treaties and pacts signed during the inter-war period which supplemented it. Manifestly the League, and therefore the whole idea of collective security, did not work. Although members of its successor, the United Nations, are officially bound to the same doctrine of collective security, the international system now rests once again on a form of power balance in the **Bipolarity** of the superpower system, sometimes known as the **Balance of Terror.** The main problem with collective security under the League of Nations was that it could only have worked given two conditions. Firstly, with the disparity in national power, the small number of really powerful nations (the USA, the UK and France) would have had to shoulder the actual burden of military assistance to any country suffering aggression. Secondly, for there to be a general acceptance of the new frontiers there should have been no nations, certainly no potentially powerful ones, that felt aggrieved. Neither condition held. The USA went into a period of intense **Isolationism** and would not even join the League. Britain and France consistently failed to back League mandates. Germany could not accept that the war settlement was just, and the Nazi Party rose to power partly on a mood of revanchism. Whether collective security could successfully replace power balances between alliances may never be known because the advent of nuclear weapons has changed the whole concept of relative power. Certainly the UN has been no more successful in preventing wars outside the main power system than the League was within it.

Combined Arms

Combined arms refers to a military activity, whether it be planning, training or warfare, in which the separate functional sections of a military force operate in close integration under a single plan and commander. In modern armies the assumption is that all operations will be of a combined arms nature, but it took the Second World War to make it clear that this was necessary. Initially **Divisions** would consist purely of one type of, for example, **Infantry,** and an armoured division would have only tanks. A modern army differentiates between divisions according to the balance of each type of troops, as they each contain all necessary units.

Combined Operations

A combined operation is one in which two or more of the different armed services, navy, air force and army, collaborate. A typical example of a combined operation, and reputedly the most difficult military operation to carry out successfully, is the invasion of a hostile country by seaborne landing of army units (see **Sea-Lift**). Here at least the navy and army, and almost certainly the air force as well, have to integrate their planning completely and work to a minutely-detailed timetable. Combined operations of this type are notoriously difficult for a host of technical reasons. Different communications methods, probably different and possibly incompatible communications machinery, rival traditions, ambiguity about relative rank structures and authority relations all interact to maximize the chance of failure.

Apart from technical difficulties, serious tensions can easily arise between the commanders of the different services because of contradictory priorities. The naval commander knows that ships are most vulnerable while near to land, and wishes to get them away, while the army commander wants time to ensure that the troops are safely on the beachhead and that all equipment has been transferred. The air force commander is required to keep aircraft flying patrols over the beachhead rather than seeking targets deeper inland. Most countries make some effort to prepare for these problems by having joint planning staffs to accustom officers from the different services to working together, but these do not usually amount to very much. Part of the difficulty is that **Inter-Service Rivalry** often means that the route to top positions is via experience with a single service, and time spent on combined operations staff training is time taken away from the pursuit of career advancement.

Command, Control, Communications and Intelligence (see C^3I)

Comprehensive Test Ban Treaty (CTBT)

A comprehensive test ban treaty has been sought after in **Arms Control** negotiations ever since the early 1960s. In 1963 the UK, the USA and the USSR signed the Partial Test Ban Treaty (PTBT), which confined nuclear testing to underground explosion sites. The PTBT was more a reaction to fears of atmospheric **Fall-Out** than for purposes of arms limitation. This treaty has since been signed by over 100 states but, significantly, neither France nor the People's Republic of China has added their name to it. The Threshold Test Ban Treaty (TTBT) of 1974 confined underground explosions by the USA and the USSR to a maximum limit of 150 kilotons. Although the TTBT was never ratified by the US Senate, both parties have continued to observe its terms.

The impetus towards a CTBT has come, firstly, from professional arms controllers. To such professionals testing is crucial—if the development of the next generation of weapons can be prevented by a test ban it should be much easier to control existing weapons. Secondly, the idea of a total ban on all nuclear testing has a great moral appeal. At the same time the superpowers have tended to manoeuvre on the issue. For example, the Soviet Union observed a moratorium on testing between August 1985 and February 1987. If one country has finished a series of tests which will allow them to field a **Weapons System** which the other side has not yet perfected, to call for a CTBT is obviously strategically advantageous. Various devices are used to deny the need for a comprehensive ban, such as that used in the mid-1980s by the UK, which needs to continue testing for its new **Trident** warheads, and that such a treaty would, in any case, defy **Verification** measures.

Concentration of Force

Concentration of force was one of the traditional 'principles of warfare' taught in staff colleges throughout the world. It underlay the skills of the most famous generals throughout history, and has often allowed numerically inferior forces to defeat their larger enemies. The idea was that all troops, ships, or other units were gathered together and used to strike a decisive blow at one point against an enemy whose forces were distributed over a large area. In this way it could be hoped to have local superiority in numbers, despite being outnumbered overall. The opposite tactic, to divide an army or navy into a set of small units guarding particular points, or carrying out independent operations, ran the risk of having them 'defeated in detail' by a concentrated enemy.

Concentration of force generally puts an attacker at an advantage over a genuinely peacefully-inclined enemy. The defender has to cover a whole border, or at least all of its vulnerable points, while the attacker could choose the target area and force a way through the defender's lines. When warfare became static, as with the trench war of 1914 – 18, the generals on both sides would try to amass forces at one point in their lines to break through a thinly-defended single section. However, such concentration of forces depends on surprise, for if the enemy knows where the attack is coming it too can, at least temporarily, concentrate its forces. The lengthy build-up to the set piece battles on the western front, taking months of supply organization, and with the battles preceded by days of artillery bombardment, prevented surprise.

The advent of nuclear weapons has caused the abandonment of this doctrine. **Battlefield Nuclear Weapons** are essentially designed to destroy large battlefield areas, and concentrations of forces are ideal targets. Should a **Central Front** war involve mass destruction weapons (chemical as well as nuclear), no army will be able to risk concentrating the bulk of its forces against one part of the opponent's front. This is the main reason for the Soviet doctrine of **Echeloned Attack;** it would make possible the use of numerical superiority to force through NATO's lines without the risk of massing troops. Instead, the Warsaw Pact forces would attack in waves, along virtually the whole front line. NATO might well defeat the first wave, but it would be so weakened in the process that the second, or even third wave should be able to break through. But at no point will the mass of the Warsaw Pact armies be concentrated to present ideal nuclear targets. Whether this strategy would work, and indeed whether it is still the main **Soviet Doctrine,** it is impossible to be sure. In recent years the Soviet doctrine has been amended to include the idea of **Operational Manoeuvre Groups (OMGs).** This takes into account the continued need to have concentrated forces to break through points of weakness, and would enable the Warsaw Pact to seek a rapid victory by taking advantage of NATO's lack of **Defence in Depth,** and to operate behind NATO's front line. Such OMGs may present targets for nuclear weapons, but much would depend on speed, and therefore above all on Soviet and Western **C³I** and **Target Acquisition** capacities. (See also **Blitzkrieg** and **Manoeuvre Warfare.**)

Conference on Disarmament (see Forty Nations Conference on Disarmament)

Conference on Disarmament in Europe (see Conference on Security and Confidence-Building Measures and Disarmament in Europe)

Conference on Security and Confidence-Building Measures and Disarmament in Europe (CDE)

The CDE talks began in Stockholm in January 1984, as a result of a decision made at the Madrid follow-up meeting to the **Conference on Security and Co-operation in Europe (CSCE).** The CSCE talks involve all European States (except Albania), the USA and Canada; it was at the suggestion of the NATO representatives in Madrid that the CDE was convened. The Stockholm Conference can be seen as taking its inspiration from the agreement to exchange military exercise information that was contained in the Helsinki Final Act, which was signed at the conclusion of the CSCE's first round of talks. The main objective at Stockholm has been to seek an agreement which would limit the size of military exercises that can take place in Europe without the participating countries informing the other members of the CDE. The purpose is to remove the fear that traditionally attends major exercises, which is that they are really a cover for troop mobilization prior to war. To this end the CDE talks have had to face up to the problem of **Verification.** In a proposal made in September 1986, signed by all members, a plan was accepted to allow real inspection by teams of observers, as long as the aircraft are flown by pilots from the countries carrying out the exercises. Although this is only a modest achievement, it is a valuable first step, especially as the provision for a form of on-site inspection is the first time that the USSR has agreed to any such 'intrusive' inspections.

Conference on Security and Co-operation in Europe (CSCE)

No one could deny that the plethora of **Arms Control** and disarmament discussions going on in, and sometimes mainly about, Europe is confusing. Senior British diplomats have been known to despair of ever getting their ministers to grasp the difference between, for example, the CSCE and the CDE, the **Forty Nations Conference on Disarmament (CD),** the Geneva discussions on the CWC, or the CTBT. One way of understanding two of these, the Conference on Security and Co-operation in Europe (CSCE) and the **Conference on Security and Confidence-Building Measures and Disarmament in Europe (CDE),** is to appreciate that they both arise from the process which is usually referred to in journalistic terms as the Helsinki talks.

The first phase of the Helsinki talks ran from July 1973 to August 1975 in Helsinki, with the meetings also taking place in Geneva. The initiative came from the Eastern bloc, and all European states (except Albania) and the USA and Canada took part. In August 1975 the conference produced a document, known as the Helsinki Final Act, which was signed by all participating states. The Final Act is a mixture of humanitarian ideals, agreements for economic co-operation, and **Confidence-Building Measures** relating to exchanges of information about military activities. Two follow-up meetings were held, in Belgrade between October 1977 and March 1978, and in Madrid from November 1980 to September 1983. The Belgrade and Madrid meetings were concerned largely with assessing progress on the implementation of the Helsinki Final Act, although the arms control aspect became much more important in the second meeting. The concluding document of the Madrid meeting included provision for a new round of talks, the Conference on Security and Confidence-Building Measures and Disarmament in Europe—the acronym for which is truncated, confusingly, as CDE. The proposal for this conference was put forward by NATO and the meetings, which began in Stockholm in January 1984, are seen by many Western diplomats as the best realistic hope for serious arms control progress, especially in comparison with the deadlocked **Mutual and Balanced Force Reduction (MBFR)** talks. The CSCE process itself goes on; a further follow-up meeting opened in Vienna in 1986, principally to review the first stage of the CDE talks, which have, in fact, produced a modest confidence-building agreement.

Confidence and Security-Building Measures (see Confidence-Building Measures)

Confidence-Building Measures (CBMs)

Confidence-building measures are also referred to as confidence and security-building measures (CSBMs). They are a form of **Arms Control,** and one with which it is probably easier to make progress than with numerical limitations on troops or weapons. Their objective is to reduce the fear and risk of surprise attack and, indeed, to make it hard to mass for such an attack. Thus international crises are less likely to develop into war because one side feels that they need to make a **Pre-emptive Strike.**

As yet CBMs are limited to the **Conventional** arms control arena, and involve undertakings to give notice of large-scale exercises.Traditionally there has been a fear that major exercises were, in fact, masking troop concentrations intended to launch an attack. By letting a potential enemy know that an exercise is planned, and inviting them to send observers to check that it really is an exercise, this fear is alleviated. As with all arms control agreements there is a major problem of **Verification,** and the openness of any military organization to surprise inspection by officers of a putative enemy is never easy to arrange. However, real progress has been made in developing CBMs in this area at the series of negotiations held in Stockholm and elsewhere since 1982 under the auspices of the **Conference on Security and Co-operation in Europe.** This should have a general effect on **Détente** as well as reducing the atmosphere of threat and tension between NATO and the Warsaw Pact. The real problem is that it is very hard to see how the CBM approach can be extended to cover nuclear arms. The form of inspection which would be required—checking, for example, that nuclear weapons had not been moved into a forbidden area, or put on board ships—would be far too intrusive to be accepted.

Conscription

Conscription, the system in which young men, and sometimes women, are required to serve a period in the nation's armed forces, is still the basic method of military recruitment for most of the world's armies. In NATO, for example, the only major countries not to rely on conscription are the UK, which abolished **National Service** in the early 1960s, and the USA, which replaced 'the draft' with an **All Volunteer Force** after the Vietnam War. These two countries were already unusual in not having a long historical tradition of conscription in peacetime. For different reasons, neither had felt that they required a large army for its defence and, apart from the latter stages of the First World War, had enforced conscription only from the Second World War onwards. The European powers, in contrast, had based their defence on very large land forces, consisting of both active service personnel and rapidly-deployable reserve forces. Germany, France and Russia, for example, could each field armies of several million in 1914, after only a month to mobilize their reserves. Such forces could only be achieved by conscription, in which nearly every member of the male population would be drafted into the army when he reached a certain age. The main purpose of conscription was training. After a period of two or three years the age 'cohort' would be released to civilian life, but with an obligation to undergo periodic refresher training and inspection so that, in a crisis, several 'years' of men could be mobilized to reinforce the current conscript army.

Various political systems found the use of conscription justifiable. Sometimes there was a simple notion of the duty of the citizen to the state. In Republican France the argument was that it was safer not to have a standing army, but to rely on the 'citizen soldier' whose loyalty to the regime was not suspect, unlike that of the traditionally monarchist professional army. (The fear of political interference was a major reason for both Britain and the USA not having large standing armies.)

Conscription has remained in force in Western Europe for rather different reasons. While it might be difficult to provide enough soldiers for a large standing army without conscription, there is much less need for the large reserve forces, given NATO planning and assumptions for a future war. Furthermore, there is no doubt that short-service, inevitably less well-trained troops are neither as reliable nor as efficient as long-service regulars, especially in the highly technical battle conditions of NATO's **Central Front.** However, conscript armies are very much cheaper than professional ones. A conscript soldier is paid little more than pocket money, accrues no expensive pension and family support entitlements, and is housed cheaply in barracks. In addition the traditional idea of duty to the state has evolved into a belief that it is only fair that military service, just like other unpleasant burdens, such as taxation, should be fairly distributed in society, rather than reliant on economic conditions to force the poor or unemployed into the services.

From a military point of view much depends on the terms of service. The efficiency of the West German army, with its relatively lengthy service and intense post-service training, is beyond doubt, as is that of, for example, Israel, the vast bulk of which is a reserve army called up very rapidly in times of crisis. There is more doubt, however, about the reliability of the Belgian forces, which have less than a year of conscript training. Similarly many-analysts in the USA attribute their difficulties in fighting the Vietnam War, in part, to the use of unwilling and disaffected conscripts who never had to serve more than one year outside the United States. The ease with which the small British contingent sent to the Falkland Islands won the land battle, compared with their near-defeat in the air and at sea, says more about the relative efficiencies of Argentina's conscript army and professional air force than about anything else.

The Warsaw Pact forces rely almost entirely on conscription, to the extent that many of their **Divisions** on paper are only at cadre (skeleton) strength, and could not be deployed until supplemented with reserves. Some Western commentators believe that this seriously reduces the real threat from these forces, partly because of their low level of training, especially in the more technical services such as the submarine navy, and partly because of an assumed lack of enthusiasm and loyalty. In major wars of the past, however, most conscript units, and certainly reserve units, have suffered from low **Morale** no more than regular troops.

Containment

Containment was the official US foreign policy towards the Soviet Union from 1947 onwards, developed by President Truman as a response to the apparent expansionist intentions of Stalin. In many ways it represented a moderate approach, because at the time there was considerable support in the USA for actively pushing Soviet influence out of Eastern Europe, the policy sometimes known as 'roll-back'. The immediate consequences of the containment doctrine were the commitment to Western Europe, out of which NATO developed, and financial and military support to countries, such as Greece and Turkey, which were under Soviet threat. The Korean War, although technically a United Nations action, was perhaps the first test of American resolve in holding the USSR to its post-war boundaries.

While containment is not in itself an aggressive doctrine, it led to the Eisenhower doctrine of **Massive Retaliation,** and is the precursor of US foreign and security policy aimed at preventing the development of *any* form of radical socialist or communist government in the Third World. Although the subsequent policy of **Détente** relaxed somewhat the tensions that arose from containment, the policy has never been explicitly renounced. From the Soviet point of view the doctrine has a serious negative symbolic meaning. Part of the motivation of Soviet foreign policy is the wish to be treated as an equal to the USA as a superpower. The USSR argues, publicly as well as internally, that the traditional notion of 'spheres of influence' entitles it to exert pressure and to have an overseeing role in its surrounding areas, just as the Monroe Doctrine makes the USA in some ways the 'supervising power' over the American continent. Containment is thus seen as an arrogant refusal by the USA to accept parity of superpower status.

Continental Strategy

In British security policy there has always been a debate about the wisdom of what has traditionally been known as the 'continental' strategy. It usually means a firm commitment to help a continental European ally fight a land war in Europe by the dispatch of a British expeditionary force, as happened at the beginning of both world wars. Under the NATO alliance the continental strategy is even more entrenched in British security policy, as it involves keeping a major part of the British army and air force permanently **Forward Based** in Europe.

Those opposed to a continental strategy in this sense do not necessarily disagree that the UK should tie itself into a mutual support alliance with continental European nations, but oppose the strategy of contributing to a land campaign in Europe. They argue that Britain's traditional role and expertise is as a naval power, and that the UK should contribute to NATO principally by patrolling and controlling the North Sea, the English Channel and the eastern Atlantic, and perhaps by providing UK-based air support for the armies of its allies on the European mainland. The debate has become more pressing and heated in recent years because the UK's defence posture is seriously underfunded; there is serious doubt whether the country can afford, in the long term, to maintain all its existing defence commitments. Consequently the idea of saving perhaps as much as 50% of long-term defence costs by removing the **British Army of the Rhine** and RAF Germany, and putting the resources into enhanced naval power and UK air defence, is attractive to many.

However, the historical argument that Britain's contribution to alliance obligations has always been made more through the Royal Navy has little relevance to the late 20th-century world. The traditional role of the British army was the defence of the Empire and, in particular, India. Now that the UK's colonial commitments are minimal, and it has joined Europe in so many ways, the only arguments about the continental strategy that make sense are those based on both military expedience *and* political realism. A major reduction of army and air force deployment in West Germany would almost certainly be interpreted as a general weakening of Britain's commitment to NATO, rather than as a more efficient policy for carrying them out, and would have serious diplomatic and political consequences.

Conventional

In the language of strategy or defence analysis the adjective 'conventional' has, of course, come to mean weapons, or strategies for their use, that do not involve atomic or nuclear explosions. It is a less clear cut concept than is often assumed, as there is still ample room for confusion. What, for example, are **Chemical Weapons?** They are not nuclear, and are not even based on technology that originated after the First World War. But they are not 'conventional' if we mean by that the idea that the world has extensive experience of fighting with them. These are not idle problems of hair-splitting, because how weapons and strategies are classified can often affect how and by whom policies are developed. In the UK, for example, **Arms Control** policy is partly determined by two government departments, the Foreign Office and the **Ministry of Defence.** In the Foreign Office the Arms Control and Disarmament Research Unit handles both nuclear and chemical weapons, and a completely different department deals with conventional arms control. Just 300 yards along Whitehall, the MOD's Defence Arms Control Unit divides policy into two areas, nuclear and non-nuclear. This often results in considerably less coherence in policy than might otherwise be the case.

Modern conventional weapons are being designed to do the same job as, and have the same destructive force as **Battlefield Nuclear Weapons.** These, often described as **Emerging Technology** systems, use **Precision-Guided Munitions** to give them the capacity, for example, to destroy an entire Warsaw Pact tank squadron 200 kilometres inside Eastern Europe. As this is precisely what NATO had originally planned to use tactical, or battlefield nuclear weapons for, it seems odd to place these new conventional weapons in a different category from nuclear weapons. To do so seems to suggest that they are on a level with the familiar limited-capacity high explosive artillery rounds.

Apart from the problems for arms control agreements when munitions with the impact of small-scale nuclear weapons cannot be counted as such, there may be serious consequences for the current doctrine of **Escalation Control.** Will the Warsaw Pact commanders really feel obliged to respond to damage of nuclear proportions with the same restraint that we expect them to use when hurt by traditional conventional weapons? It may well be time for a radical rethinking of the categories of tactical and arms control debate. If for no other reason, the increasing deployment of **Dual-Capable Systems** may require this. The best example is the **Cruise Missile,** because exactly the same missile design can be equipped with either a nuclear warhead, with a **Yield** of as much as 200 kilotons, or with a conventional warhead. There would be no way in which an opponent could know what sort of missile was approaching, but the application of a military worst-case analysis could produce disastrous responses.

Conventional Defence Initiative (CDI)

The Conventional Defence Initiative is just one in a long line of attempts by NATO to get member states to co-operate on increasing defence expenditure to improve the conventional arms balance between NATO and the Warsaw Pact. As with all such programmes it started as an American initiative, in this instance in 1985, with moves by the US Congress to fund at least some degree of defence procurement co-operation with

Western Europe. In its first two years the attempt was relatively successful, mainly because it concentrated on more modest targets than its predecessor, the **Long-Term Defence Programme.** Furthermore, Western European politicians had been quite badly frightened when, in 1984, a motion sponsored by Senator Sam Nunn calling for US troops to be withdrawn from Europe unless **European NATO** members spent more on defence failed to pass the Senate by only one vote (see **Burden Sharing**).

The CDI is capable of adaption to either modest or ambitious goals. In its early stages the emphasis was on familiar NATO concerns such as improvements to sustainability, **Infrastructure** developments, and further attempts to achieve a degree of **Interoperability.** There has been some suggestion of the possibility of investing in **Force Multiplier** equipment employing **Emerging Technology,** and the US army has been given 'pump-priming' money for this purpose. Estimates of the cost of such an investment programme, however, make it very unlikely that any European government will follow this example. Such a programme would probably cost, in the UK's case, the equivalent of increasing defence expenditure by more than 3% a year, for between five and 10 years, in a period when defence expenditure is planned to decline.

Correlation of Forces

The correlation of forces is a term of great significance in Soviet military, and also political, theory. It has been noted so often by analysts of the USSR's huge body of military theory that it is even beginning to be used unconsciously by some Western strategic thinkers, although whether they use it in quite the same sense, and with the full implications of its Marxist theoretical genesis, is doubtful. It is remarkably difficult to give a clear definition, as there does not really exist an English, or a non-Marxist, equivalent.

It is easiest to approach the concept at its broadest, the way it would be used by a strategist or someone concerned with the whole spectrum of national security policy. In such a context the correlation of forces refers to the Soviet Union's comparative capacity to compete with the West in all important ways—the combined economic, military, diplomatic and ideological strength of one country against another. Military force is seen much more as just one component of an essentially political conflict between Soviet communism and the capitalist world, whereas Western attitudes tend more to separate the various forms of power and influence a nation has. It is often noted by Western analysts that all Soviet theorists since Lenin himself have recorded their agreement with **Clausewitz's** dictum that war is 'nothing but the continuation of politics by other means'. To the Soviet military thinker this is axiomatic, and therefore any particular piece of military hardware is also considered as, in part, an ideological and diplomatic weapon. An extension of this argument can be seen in the example of small strategic nuclear forces, like those of Britain or France. To many American analysts there seems no rational explanation for the UK spending so much to remain in the **Nuclear Club,** and they dismiss the British policy as being 'merely' a matter of international status. Soviet analysts might also see the UK policy as a status matter, but would not regard it as 'mere', because international status is itself a source of power; they would accept and understand this motivation. To them, Britain's correlation of forces would be enhanced by the country having impressive nuclear weapons more than investing the same money in, for example, **Anti-Submarine Warfare** fleets. The Americans, however, would think the latter a more rational policy because it would increase Britain's technical and usable military power.

In its more narrowly military sense, the correlation of forces is a way of expressing the power capacity, or the probability of achieving a *politically set* goal in, perhaps, a **Theatre of War.** It would, of course, primarily be a matter of troop numbers and equipment, but would also include confidence in the relevant doctrine, the likely response of both one's own and the enemy's civilian populations, the possibilities of tactical or strategic surprise, and any factor that might influence the final outcome. Any competent Western analyst will, in principle, consider all of these matters, but there is much more of a tendency to think about them separately, and therefore often to miss connections between different aspects of force, power or influence. Thus it is easy to miss the implications that, for example, offensive doctrine and diplomatic posture have on each other.

Counter-Force

In nuclear targeting theory there is a basic distinction between two types of strike: counter-force and **Counter-Value.** A counter-force target is, broadly, a military target, or, more narrowly, a target which is part of the enemy's own nuclear arms system. In its more narrow use counter-force refers to enemy missiles, **C³I** facilities, and other heavily-protected or vital operating bases, such as aerodromes and nuclear-missile carrying submarine ports. Counter-force can be taken to include **Other Military Targets** for example, troop concentrations, arms factories and **Logistic** support bases. Considerable effort is taken to reduce the **Collateral Damage** to the civilian economic and social structure in planning counter-force strikes.

Counter-force targets are often divided into two further categories, according to whether or not they are **Hard Targets** or **Time Urgent Targets** or, indeed, both. A hard target is an installation, usually underground, such as a missile silo or control bunker, which is protected by super-hardened concrete walls. Such targets can only be destroyed by missiles with a combination of very high accuracy (see **Bias** and **Circular Error Probable**) and high explosive power, while some counter-force targets, such as air-bases, although very important, are 'soft', capable of being destroyed by an **Air Burst** from a much less accurate missile. Time urgent targets are usually those which are part of the enemy's retaliatory or defensive systems. These must be destroyed at once if a **First Strike,** or even a retaliatory **Second Strike** against the remainder of an enemy's nuclear force, is to succeed. Thus an unused missile in its silo, or a nuclear command bunker will be both time urgent and 'hard', while strategic bomber fields will be equally time urgent, but only 'soft'. The major wartime political control centres of an enemy state, where the **National Command Authorities** are housed, will be hard target bunkers, but may not be time urgent, because many **Scenarios** call for leaving the political leaders safe, at least at first, to facilitate **Intra-War Bargaining.**

The counter-force/counter-value distinction is crucial in strategic theory because the two concepts are, to many, morally different (the American Catholic Bishops have given a degree of support to strategic weapons as long as they are intended only for second strike counter-force targets), and have contrasting **Escalation** risks and values. So, for example, counter-force targets are at the centre of strategies involving **Limited Nuclear Options,** as well as being the main concern for those who fear **Strategic Superiority.**

Countervailing Strategy

Countervailing strategy, or the search for 'countervailing power', was another name for the **New Look** in strategic posture initiated in 1974 by the US Secretary of Defense, James Schlesinger. It indicated that the USA would ensure that it always had a capacity to match any Soviet threat at a particular level, so that there could be no advantage hoped for (see **Essential Equivalence**). It was an attempt to produce a more subtle alternative to **Assured Destruction,** but in a world where **Strategic Superiority** could not be obtained. The language was chosen in such a way as to avoid getting into an argument about whether a nuclear war could be 'won', but the idea of the USA 'countervailing' certainly implies that it will not lose. The doctrine involved what came to be known as **Limited Nuclear Options,** ensuring that the USA would never be in the position of having to choose between doing nothing or responding too massively were the USSR, for example, to launch a restricted strike against a few **Counter-Force** targets. In many ways it is an application of the policies of **Escalation Domination** and **Flexible Response,** which characterize NATO's theatre strategy. The doctrine later became enshrined in President Carter's **PD-59,** and is perhaps a precursor of the **War-Fighting** theories that became popular during the Reagan years.

Counter-Value

Along with **Counter-Force,** counter-value is the major categorization of targets in nuclear strategy. A counter-value target is anything which does not have a narrowly military use to a society. In most targeting theories, counter-value targets are economic, political and industrial structures—factories, power plants, mines, transport systems, agricultural storage depots, government offices, police headquarters and so on. It is rare, nowadays, for city centres or residential areas to be listed as targets, although in practice the death rate and destruction of hospitals, schools and housing would be appallingly high even where targeting plans were not deliberately intended to maximize human casualties (see **Collateral Damage**).

The idea behind most counter-value targeting is to destroy a major part, often put at around two-thirds, of the industrial capacity of a society. This is seen as being an adequate level to ensure **Assured Destruction,** as defined in most **Deterrence** theories. Increasingly more-refined target lists are being developed, and the latest version of the US **Single Integrated Operations Plan** defines the prime counter-value targets as being industrial and economic **Recovery Capacity,** rather than industrial capacity in general. Similarly the official French target plan has changed from being an overtly 'anti-demographic' plan, to aiming at what are called 'vital centres', which seem to be those economic and political targets the loss of which would most hamper the functioning of a modern system. Some theories specifically call for targeting the Soviet *state,* rather than the Soviet society, although it is unclear both how this can be done, and whether there is any real distinction. However, all such distinctions have only abstract value. The secondary death rate from nuclear **Fall-Out,** starvation, and the breakdown of services, quite apart from the primary death rate from blast, would be enormous. Nearly all of the counter-value targets are in urban locations. As a 150-kiloton warhead can produce a lethal overpressure of five psi (that is, five pounds per square inch above normal atmospheric pressure) over an area of 45 square kilometres, it will make no difference to most cities whether the ground zero (see **Designated Ground Zero**) was a factory, government building or a housing estate.

Counter-value strikes remain the essence of deterrence, and require notably less sophisticated missiles, in terms of power and accuracy, than counter-force strikes. However, the latter cannot be guaranteed to cause sufficiently low levels of collateral damage to make them in practice as different as strategic planners maintain.

Credibility

Credibility is at the heart of all theories of **Deterrence** because a threat that lacks it cannot be relied upon to prevent an opponent taking the first aggressive step. There are two main elements to credibility: technical capacity and political will. Clearly a threat that cannot be carried out technically will lack credibility, and thus the value of a country's deterrent force is linked to its opponent's arms programme. Even the development of defensive weapons, such as the American **Strategic Defense Initiative** programme which could render the Soviet **ICBM** force unable to penetrate to US cities, would remove the credibility of the Soviet deterrent. However, even if a deterrent threat is technically feasible, its credibility is further dependent upon the political will to use it.

What is really crucial is the ability to convince a potential enemy that under certain conditions a threatened action will, in fact, be carried out. This equation has many components, all of them difficult to calculate or even to express precisely. The central problem is that of forcing an opponent to believe that the risks of carrying out a threat are acceptable. For example, British nuclear strategy is based on the ability of **Polaris** submarines to deliver an **Assured Destruction** strike on Moscow (see **Moscow Option**) should the USSR do whatever, presumably a nuclear attack on the UK, they are there to deter. There is little doubt that the fleet could do this, nor that it would be a terrible blow to the USSR. But such an attack would do nothing to reduce the huge third strike that the USSR could deliver to the UK. That strike, instead of killing about 8 million people (the population of Moscow) could easily kill 80% of the British population. In which sense can the British threat be seen as credible? How could the British Prime Minister take such a risk?

It is not only minor powers that run the risk of making a non-credible threat. In general a deterrent threat is only credible to the extent that the damage it threatens (and the retaliatory damage it invites) is in proportion to the seriousness of the enemy action it is intended to deter. So the US threat totally to destroy the USSR if that country destroys the USA may well be credible. However, a threat to use nuclear weapons on the Soviet Union to deter any lesser action is correspondingly less credible. For this reason the doctrine of **Extended Deterrence,** by which the US Central Strategic Forces are supposed to be the ultimate step in NATO's **Flexible Response** strategy, lacks much credibility. Is it really believable that a US President would risk the destruction of New York by threatening Leningrad, in order to prevent the Warsaw Pact armies from capturing Paris?

Large-scale threats may be non-credible even when the enemy lacks the capacity to respond to them, if they are used to deter relatively minor offensives. Under the Eisenhower presidency the official US strategy was one of **Massive Realiation,** according to which any incursion by the USSR on the territories of America's allies would provoke a paralysing nuclear blow on the Soviet homeland. The US **Strategic Air Command** certainly had the capacity to do this and, at that time, the USSR could have done little by way of nuclear retaliation. But no responsible political leader could be thought to be quite so amoral as to destroy millions of civilians in response to a conventional invasion of at least limited scale, and the deterrent effect was probably very limited. The real problem with credibility is not even logically soluble. The credibility of a threat depends on how it is perceived by the enemy. The more rational and moral a power appears, the less the opponent is likely to believe any threat of destruction, or apparent preparedness to accept destruction in reply to an attack.

Deterrence theorists have long recognized this problem, and sometimes seek what is called 'automaticity', some way of guaranteeing that a threat really will be carried out, regardless of the consequences. There is no practical way of achieving this facility, which would be akin to the mythical **Doomsday Machine** of the film **Dr Strangelove** by which nuclear weapons are triggered by tamper-proof computers if they register a nuclear explosion anywhere. Instead there is much concern about factors that might limit the likelihood of a nuclear response. Western Europeans have always been less enthusiastic than Americans, for example, that extensive conventional defences should be provided for Western Europe, because the more *non-*nuclear options NATO has, the less credible it is that it will resort to **Central Strategic Warfare.** Similarly, French theorists have always stressed the need for individual national ownership of nuclear weapons, because they do not believe that a threat by an alliance is as credible as one by a 'unitary actor', that is, a single nation state actually facing destruction. The French are a case of some interest in this context because their nuclear threat, unlike that of the UK, is intended to deter conventional invasion of their country as well as nuclear attack. Because of this they have always sought to increase the credibility of the threat by stressing that their nuclear capacity would allow them to opt out of a general European war (the doctrine of **Sanctuarization**), and have deliberately not built up conventional forces that would offer them any possibility of fighting off such an invasion by non-nuclear means.

Considering how vital a role credibility plays in deterrence theory, little progress has been made in developing a convincing doctrine. Doubts about the credibility of the deterrent threat have tended to be the driving force in strategic weapons development. Many of the design developments in the US forces since the 1960s have been devoted to increasing the possibility of **Limited Nuclear Options** and **Counter-Force** targeting, as with President Carter's famous plan **(PD-59).** These developments arose from earlier Presidents, especially Richard Nixon, doubting publicly the credibility of a nuclear threat that seemed to be an 'all or nothing' option.

Crisis Management

Crisis management refers to the problem of reacting in an unstable situation in such a way as to maximize one's own interests while not pushing the opponent too hard, and not forcing it into an **Action – Reaction** cycle. In many ways crisis management is simply good diplomacy, but the concept has come to be increasingly important since the 1970s. The contrast with a crisis management approach to international tension is perhaps best seen in the idea of **Brinkmanship.** Whereas the latter approach is essentially one of taking a tough line and calling the opponent's bluff, a good crisis manager will avoid raising the temperature in the situation as much as possible.

Tough or threatening behaviour may be necessary, to give a clear signal of resolve, as when President Nixon brought all US forces, including nuclear, to a high readiness state at one stage during Middle Eastern tension, but it should only be used with great care. The stress is on absolute clarity of international expression and to avoid causing the opponent to panic, or allowing it in one's own staff. In order to facilitate crisis management the famous **Hot Line,** allowing leader-to-leader communication, may well be relied upon. Some US Senators have urged the establishment of a crisis management centre, perhaps in Switzerland, jointly staffed by US and Soviet officers, to ensure that each superpower knows exactly what the other is thinking, and that no mistakes are made in interpreting, for example, the implication of troop movements. Although most professional diplomats reject this as being superfluous, because such a centre would be bypassed in a real crisis, the idea demonstrates the concern which exists to avoid the risks inherent in brinkmanship, and a new understanding of the whole problem of international communication.

Crisis Stability

Crisis stability is clearly a most desirable quality in the international strategic system. Its basic meaning is that **Parity,** or an overall balance of military strength, exists so that there is no reason for one power which could potentially come into conflict with another to fear that the opponent will think it advantageous to launch a major strike without warning. If there is no such fear, then the first nation will not be tempted into a **Pre-emptive Strike** to avoid such an attack. There are no magic guaranteed means for creating crisis stability, but some of the ingredients are fairly obvious. First comes the possession by both relevant actors of secure **Second Strike** capacity. As long as that exists neither side can hope to benefit by escalating the crisis, and therefore neither need be led by fear to pre-emption. Secondly, it might be argued that this **Deterrence** needs to operate at each stage of a possible **Escalation** ladder, so that neither side can think it has **Escalation Domination** and thus be tempted to solve the crisis by resorting to the use of a higher level of force.

An alternative approach to averting catastrophe resulting from a crisis situation is to concentrate on **Crisis Management** possibilities. Here the crucial element is clear and unambiguous information and signalling. For example, troop movements that might seem like mobilization need to be avoided, and very precise signals need to be sent about exactly what is to take place. The more the international system is equipped to transmit information in this way, the less a danger exists of escalation by accident into some form of **Catalytic War** or **Spasm War.**

Cruise Missile

There are three basic types of cruise missiles in the US armoury. The one that European political attention has focused on is the **Ground-Launched Cruise Missile (GLCM),** fired from mobile launchers and intended to be used, most probably with a nuclear warhead, on theatre targets, or conceivably at strategic targets in the western USSR. The other two forms, the **Air-Launched Cruise Missile (ALCM)** and the **Sea-Launched Cruise Missile (SLCM),** are far more numerous and have a diversity of roles.

Essentially a cruise missile is similar to the famous 'flying bomb' or V1 rocket used against Britain by Germany in the later part of the Second World War. They are pilotless jet aircraft which fly, by a combination of radar guidance and pre-set computer control, relatively slowly over ranges that can vary from less than a hundred to several thousand miles. They can be armed with either nuclear or conventional warheads, and used in either a strategic or tactical role. They are capable of very great accuracy, but currently are relatively vulnerable to ordinary anti-aircraft defences. A second generation of supersonic cruise missiles has been developed by the USA, and is to be deployed in at least some roles.

The USSR has, in fact, had cruise missiles of a rather unsophisticated design in their inventory for longer than the USA, but is not thought to put as much reliance on them, except perhaps for the Soviet navy, which includes naval aviation squadrons.

Cuban Missile Crisis

The Cuban Missile Crisis occurred in 1962 when the US government discovered that the USSR was in the process of deploying medium-range nuclear missiles in Cuba, capable of attacking much of the American eastern seaboard. At this early stage in the strategic nuclear **Arms Race** the **ICBM** forces of both superpowers were small, and the USSR was at a considerable disadvantage. The USA had medium-range missiles and bombers based in the territories of their allies, especially Turkey, near the USSR borders, while all Soviet strike capacity had to come from Europe. To redress this, and as a more general expression of power (see **Correlation of Forces**), the Soviet leader, Khrushchev, decided to place missiles in Cuba, the USSR's only ally in the Americas.

Once President Kennedy was fully convinced of the danger of these preparations from photo-reconnaissance evidence, he publicly demanded that the USSR withdraw all their rocket forces. At the beginning of the crisis the rockets themselves were not actually in Cuba, but were on board ships *en route.* Kennedy announced that he could not allow the ships to be unloaded, and ordered the US navy to blockade Cuba and intercept the Soviet merchant fleet that was delivering the missiles. He made it clear that he would, if necessary, order the sinking of the ships.

For a period of a little over a week there was intense fear that neither side would give way, and full-scale nuclear war was seen as imminent. Finally Khrushchev backed down, and the Soviet fleet turned around before any serious incident could occur with the US navy. The situation was particularly dangerous because the confrontational attitudes of both sides could easily have transmitted themselves to relatively junior officers, who could have exceeded their orders, engaged in conflict and escalated the crisis whether or not the political leaders had decided on war. It is known that some US naval officers in **Anti-Submarine Warfare** ships did, indeed, exceed their orders to shadow Soviet submarines, and instead forced them to surface.

For some time it was publicly believed that nuclear war had been a real alternative for the USSR, and Kennedy's stance, called **Brinkmanship,** has been judged, favourably or otherwise, in that light. However, it has subsequently become known that the available ICBM force at the command of the Soviet leadership was still extremely small at that period, a fact almost certainly known to the US administration: nuclear warfare was *not* an option for the Soviet Union. Nor, indeed, does the US success in forcing the USSR to change its plans stand as evidence for the utility of nuclear weapons at all. The real advantage the USA had, as subsequently pointed out by analysts, was a huge preponderance in locally-available *conventional* forces. Finally, the USSR's action was almost entirely political symbolism, and had been strongly opposed by the Soviet military leadership, who knew that they could not succeed were the Americans to act forcefully.

D

Damage Limitation

Damage limitation is a wide-ranging concept in nuclear strategy, with many disparate applications. Generally it refers to a nuclear strategy that reduces the amount of damage one's own country will suffer from an enemy, particularly during **Central Strategic Warfare.** Although it could refer to damage of any type, the usual meaning is the limitation of **Counter-Value** damage, that is, to prevent strikes against industrial and population centres.

The term covers both active steps a country can take itself, and passive steps intended to persuade the enemy to act in such a way that the damage suffered will be limited. For example, both an extensive **Civil Defence** system, and the deployment of **Ballistic Missile Defences** around major cities would clearly be active steps to limit damage. Equally, an offensive strategy can be designed in such a way as to dissuade the enemy, whether engaging in a **First Strike** or a retaliatory strike, from targeting cities and industrial bases. There are a variety of possible targeting strategies which might accomplish this, but they have one common theme: not to attack the enemy's cities, but to hold in reserve a secure **Second Strike** capacity which could do this. In this way every incentive is given to the enemy to avoid attracting the sort of destruction one is trying to limit on one's own territory.

It was this theoretical argument which produced the **No Cities Doctrine** in the USA, and has led both superpowers to seek to avoid **Collateral Damage** in their targeting policies. Like so many concepts in nuclear strategy it has an air of unreality about it, both because it implies such cool rationality in a crisis and because it places enormous demands on **C³I** capacity. It also tends to ignore the fact that collateral damage, if only from **Fall-Out,** is likely to be extremely hard to restrict.

Declaratory Policy

A declaratory policy, in strategic language, is the publicy-announced doctrine about when, and how, a country will use its nuclear forces. For example, Eisenhower's **Massive Retaliation,** McNamara's **Mutual Assured Destruction** and Schlesinger's **Limited Nuclear Options** were declaratory policies. To the extent that it was intended for publication, Carter's Presidential Directive 59 **(PD-59)** was also a declaratory policy. The French policy of using tactical nuclear weapons just before the invasion of France to signal a **Fire-Break,** and then following this with an all out anti-city strike, is their declaratory policy. Whether the UK really has a declaratory policy is more debatable. Officially, British governments insist that they will never declare their policy on the use of nuclear weapons because uncertainty increases **Deterrence.** On the other hand, the **Moscow Option** is so widely acknowledged that it amounts to a declaration of targeting policy, even if it does not detail the occasions of use.

The connection between declaratory policy and employment policy is complex. It is not generally suggested that declaratory policy is bogus—that it is merely propaganda or bluff. Nevertheless, it usually does not coincide directly with what is known about detailed military plans. The best evidence for this is that the American **Single Integrated Operations Plan,** on which employment policy depends, is known to have been remarkably constant since 1960, exhibiting merely incremental development. Yet this same target plan and set of strike options has served several apparently radical changes in declaratory policy. There is every indication that an analogous contrast, although in reverse, is about to happen in the UK. It is most unlikely that official statements will alter from a reliance on the simple and limited Moscow Option in the future. Yet it is equally obvious that the deployment of the **Trident** squadron would increase the UK's strike power far beyond the requirements for destroying Moscow. The military planners are not going to fail to develop an employment policy utilizing Trident to the full, but this will not be presented as a new declaratory policy.

Declaratory policy does not, therefore, give a reliable account of what might be expected to happen in the event of nuclear war. Rather it highlights the political aims which nuclear weapons will be used to protect or achieve, and presents a justification for their use.

De-Coupling

Ever since NATO first accepted, in the mid-1950s, that it was not going to be able to match the conventional force levels of the Warsaw Pact, it has relied in one way or another on American nuclear strength. This reliance has taken the form of an explicit, though never official guarantee that the USA will take any and every necessary and appropriate step, up to and including **Central Strategic Warfare,** to prevent the defeat and occupation of Western Europe by the Warsaw Pact. In this way America's entire nuclear arsenal is 'coupled' to the defence of Europe. The Warsaw Pact, including the USSR itself, must anticipate nuclear destruction should they seem likely to win a European conventional war.

The commitment to couple US nuclear forces to Europe's defence was originally made as part of the Eisenhower doctrine of **Massive Retaliation,** itself a development of the 1947 **Truman Doctrine** of **Containment.** At the time, the USA's **Nuclear Supremacy** was so great that there was no reason whatsoever to doubt both its capacity and resolve to carry out that commitment. As the policy allowed Western European countries to invest in economic recovery rather than building up the conventional force levels otherwise needed to defend themselves against the USSR, the US policy, which came to be known as the **Nuclear Umbrella,** was understandably popular.

Not everyone was so confident even in these early days of the guarantee. France officially used its doubts of US fidelity in these extreme conditions as a justification for building its own independent nuclear **Force de Frappe;** although the UK has been careful never to admit to the same doubts, it has acted as though it shares them. As the nuclear capacity of the USSR increased rapidly towards the **Parity** levels of **SALT II** more and more voices, mainly in Europe but also in America, were heard doubting that the nuclear guarantee would hold in a crisis. This idea, that the US central strategic force would *not* be used against the USSR to protect or avenge Western Europe is what is meant by 'de-coupling'. The problem is often put succinctly as follows: what American President, if the USA has not been directly attacked, is going to risk the destruction of Detroit in response to the destruction of Hamburg?

Much effort has been put into ensuring that both Western Europeans and members of the Warsaw Pact really do believe that the coupling still holds. Military deployment policies that seem to have relatively slight practical justification often result from a fear that a belief in de-coupling will either encourage the USSR into adventurism, or weaken Western European loyalty to NATO and the USA. The best example is the deployment of **Ground-Launched Cruise Missiles** and **Pershing II** missiles in Europe, which are often called **Euromissiles,** as part of the modernization of **Theatre Nuclear Forces.** (It is often forgotten that the original call for this programme, in 1977, came from Europeans, and especially the West German government, and was *not* pushed on Europe by the USA.) The main idea behind the deployment was that these intermediate-range nuclear missiles would further couple the American strategic forces by forming a linkage between their conventional commitment to Europe (they have 330,000 troops in Western Europe) and the home-based nuclear force.

In the end the question of whether or not US (or, for that matter, British and French) strategic forces are coupled to the conventional defence of Europe, as opposed to serving merely for **Sanctuarization** of the homeland, is a psychological one. Worse still, it is the Soviet, not the Western psyche, on which coupling depends. Not only can the USA not guarantee that the leadership of the USSR will believe in the coupling, they could also not force them to believe in de-coupling were that ever to become official policy.

Decoy

In modern strategic language a decoy is something intended to defeat **Anti-Ballistic Missile (ABM)** or other **Ballistic Missile Defence (BMD)** systems, and is carried as part of the **Payload** of a ballistic missile. Decoys are part of the larger category of penetration aids, which also includes more basic devices such as radar-reflecting metallic strips (chaff). The simplest, but potentially very effective, type of decoy is a metallic-coated balloon. Packed inside the **Front End** of a ballistic missile, along with the **Re-entry Vehicles (RVs)** or **Warheads,** the decoys would be ejected from the missile during the **Mid-Course Phase** at the same time as the RVs. Inflated by compressed air, they would give the same **Radar Signature** as a real warhead, and travel at the same speed because of zero gravity, presenting any BMD system with the problem of distinguishing between real targets and fakes. It would be perfectly possible to pack missiles so that the ratio of decoys to RVs was at least 10 to one. Thus a strike by 10 US **MX** missiles, which could carry a combined total of 100 RVs, would present Soviet BMD radar with perhaps over 1,000 targets to deal with. As the current capacity of the USSR's ABM shield around Moscow (see **Galosh**) is at most 100 interceptor rockets, it is clear that decoys could be extremely effective against current generation BMD.

Such simple decoys will not defeat future developments in BMD, of course. For example, the type of decoy described above works only because the current generation of Soviet BMD intercepts RVs outside the atmosphere, during the relatively long mid-course phase. As soon as the decoys and warheads re-enter the atmosphere the decoys, if they do not just burn up, will be detectably different from the real targets and will travel at much slower speeds because of air-drag. Other characteristics can be tested for, in principle,—infra-red signatures would be different, for example, even if radar signatures were not. Nevertheless, it is highly likely that advances in design of decoys will keep pace with detection methods sufficiently to ensure that any BMD shield will have to be capable of handling much larger numbers of objects than the opponent needs build actual warheads.

Deep Strike

Deep strike is one of a series of US doctrines for fighting a war on NATO's **Central Front,** and is best thought of as the air force equivalent to the US army's official doctrine of **Air-Land Battle Concept.** As with all current theories of countering the Warsaw Pact's conventional numerical superiority and planned **Echeloned Attack** strategy, it involves **Interdiction** of Warsaw Pact armies as far from the battlefield as possible. In particular, deep strike calls for rapid action to make Warsaw Pact airfields unusable, and for establishing **Air Superiority** in order to make the land battle winnable. As such it requires a new generation of weapons, especially bombs for cratering runways, and **Precision-Guided Munitions** to destroy the Soviet tank armies before they can arrive at the front line.

Defence

The need to defend a nation and its citizens is apparently innocent enough, but there is, nevertheless, a moral and ideological debate over aspects of the term 'defence'. This probably accounts for the decline of 'Departments of War' and their replacement with Departments of Defence (or Defense) in many countries. Eisenhower's **Massive Retaliation** doctrine drew a clear distinction between defence and **Deterrence.** This doctrine contrasted these two ways in which harm from an enemy might be avoided—by using weight of troops to fight a possibly lengthy defensive war, or by threatening a massive nuclear attack as retaliation for any damage inflicted on an ally—and opted uncompromisingly for the latter. A slightly earlier example would be the RAF's two major combat commands in the Second World War. Fighter command existed to shoot down German bombers, and in that way to defend the UK from harm. **Bomber Command,** on the other hand, could (at least in theory) do so much damage to Germany in retaliation for an attack on Britain that no such attack would take place.

Thus, although in military contexts 'defence' is often merely regarded as preventing the enemy from inflicting harm or defeat on one's country or alliance, whatever the means, there are obvious reasons why a moral evaluation might favour, say, defence to deterrence. The traditional military maxim that 'the best defence is a good offence' is at odds with calls from, among others, the British Labour Party for a defence policy based on 'genuine defence'. By this they mean a set of weapons, and presumably **Tactics,** which would be incapable of being used in an attack on the Warsaw Pact. One of the policies which this doctrine is meant to counter is the current US policy of **Deep Strike,** which specifically involves attacking deep into Warsaw Pact territory within the context of a *defensive* battle. In an attempt to minimize the problems of their obviously 'defending' troops the Allied air forces would try to destroy the Soviet second echelon of troops (see **Echeloned Attack**). There is a genuine political debate as to whether such a tactic would be regarded as 'defensive' compared, for example, with troops firing static weapons from fortified positions. The answer to this question depends on motivation and political insight, and defies objective analysis.

Defence in Depth

Defence in depth has been recognized as a desirable tactic since the middle of the First World War, when it became official German doctrine. It contrasts with the fixed linear **Forward Defence** system currently employed by NATO. Defence in depth treats the front line that separates one's own forces from the enemy merely as the forward border of an area, the whole of which will be defended. Several lines of prepared positions will be set up, with tactical strongholds positioned throughout, and with one's forces distributed over the whole area. Although this means that the enemy may find it easier to penetrate through the first line of defence, it will not then be free to move at will behind that line. Instead any attempt to exploit the breakthrough will continually come across hindrances as well-prepared positions attack the enemy forces in the flank. As with all such tactical doctrines, actual practice is a matter of compromise between the ideal and the availability of troop numbers. However, many analysts think that the current NATO doctrine of forward defence makes inadequate allowance for defence in depth. (See also **Forward Edge of the Battle Area.**)

Defense Intelligence Agency (DIA)

The Defense Intelligence Agency is the much less well-known counterpart to the **Central Intelligence Agency (CIA)** which, along with the **National Security Agency,** make up the US 'intelligence community'. The DIA is essentially an analysis and research organization, lacking the 'covert operations' brief of the CIA. Its concerns are not, however, confined to purely military matters, unlike the intelligence branches of the individual services which investigate matters such as the enemy **Order of Battle.** The DIA's major source of intelligence information comes from the service attachés in US embassies. In addition it employs a large staff of mainly civilian analysts in Washington, who report on all aspects of foreign countries (including US allies) which might affect their military capacity. An example of the range of its concern can be seen in the long-running and highly-publicized clash between the CIA and the DIA on the correct estimate of the size of the Soviet defence budget, in which the **National Security Council** and Congressional committees were regularly given radically different figures by the two agencies.

Delivery Vehicles

Delivery vehicle is a catch-all concept for ways of 'delivering' nuclear **Warheads** on an enemy. They can be **ICBMs,** theatre or **Intermediate Nuclear Forces,** bombers, **Cruise Missiles** or whatever is capable of doing the job. The need for the general concept is largely in **Arms Control** negotiations, where an agreement may be possible on how many warheads and how many ways of getting them to the target a country can have, but where national differences in weapons systems make it much harder to get specific limits on, for example, bombers and ICBMs. The USA keeps a far higher proportion of its warheads on **SLBMs** than does the USSR, and also retains a much larger long-range bomber fleet, while the USSR has greater numbers of, and more powerful, ICBMs. Thus agreement in **SALT II** on equal numbers of each weapons system would have been much harder to achieve than agreement on the total number of delivery systems of all types which the two superpowers were to be allowed.

Dense Pack

Dense pack was one of the **Basing Modes** suggested by the US administration for the new **MX** missile, in order to ensure that it is less vulnerable to a Soviet **First Strike** than the existing generation of **Minuteman** missiles, which are contained in concrete **Silos** at several bases in the American Midwest. The original plan for deploying the MX missiles was to put them on movable launchers which would travel continually around a complex of purpose-built roads with hundreds of shelters. At any one time only a small number of the shelters would, in fact, contain missiles, but the USSR would never be able to know which. Consequently, if they hoped to destroy the US MX missile force in a surprise first strike attack, they would have to target every shelter. Because of possible targeting errors, misfires and other problems it would, in practice, be necessary to target each shelter with several warheads. The result was supposed to be an impossibly high requirement in warhead numbers for the USSR.

This plan would also, however, have been impossibly expensive for the USA, and a variety of other basing modes were canvassed. Dense pack was possibly the most original of them all, because its logic involved a reversal of all previous assumptions. Instead of trying to emplace the missiles as far apart as possible to make it impossible for one Soviet warhead to damage more than one silo, the plan sought to do the opposite; the MX missiles should be placed in silos very close together—hence 'dense pack'. The theory behind this otherwise apparently lunatic plan is known as **Fratricide.** A nuclear warhead close to a nuclear explosion, and not protected by a hardened silo, is likely to be to defused by the burning out of the complex ignition system. That apart, a series of nuclear explosions would inevitably throw the aiming and steering mechanisms of the other incoming missiles into chaos, through the **Electromagnetic Pulse** phenomenon. As the silos would be super-hardened, near misses or premature explosions high in the atmosphere, caused by the initial detonations, would not matter.

The US Congress declined to vote funds for the dense pack basing mode, presumably on the grounds that it was an unreliable policy to rely on an essentially untestable theory. Instead, the huge and expensive MX missiles, originally justified by the **Vulnerability** of the less powerful, but more numerous, Minuteman series of **ICBMs,** have been placed in the original Minuteman silos.

Department of Defense (DOD)

The United States Department of Defense was created in 1947 as a result of wartime experience. Before then there had been, as in Britain earlier, a separate Department of War for the army, and a Department of the Navy, with the air force role split between them. Problems of co-operation and defence planning had proved very difficult during the Second World War and, as the USA faced up to the need for a permanent large-scale military establishment during the **Cold War,** a major reorganization was felt necessary.

The DOD was created as an umbrella administrative structure, containing within it, and responsible to the Secretary of Defense, separate departments of the army, navy and air force. The US Air Force was, by then, a separate service, although the navy retained its own naval aviation units, as did the Marine Corps—also under general navy jurisdiction. Each service has its own Secretary and civilian administrators, and these are quite separate from the Secretary of Defense's civil service. Indeed, when people loosely talk about the DOD, or the **Pentagon,** they most usually mean the administrative unit known as the **Office of the Secretary of Defense (OSD),** without realizing how little control this often has over the three separate services.

The authority that the DOD and its Secretary wields over the subordinate service departments varies, to some extent depending on the personality and ideology of the Secretary of Defense. In its early days the OSD was relatively weak, but President Kennedy's famous appointee as Secretary, Robert **McNamara,** introduced powerful budgetary and planning controls, along with the whole concept of systems analysis; since then the departments have been under a tighter control—should the Secretary of the time so wish. President Reagan's appointee, Caspar Weinberger, to some extent deliberately gave the departments more autonomy. This was possible only because of an across-the-board increase in defence expenditure since 1980, which reduced the need for central DOD arbitration between competing service bids.

Criticism of the DOD is more rife today than ever before, above all in Congress, and tends to concentrate on two aspects. The first is really a complaint about the military culture of the USA, with a weak **Chiefs of Staff** system and virtually uncontrolled **Inter-Service Rivalry.** The complaint that more specifically concerns the civil service and secretariat is that financial planning and control of procurement is appallingly lax, leading to huge cost overruns, duplication of **Weapons Systems,** and a general failure properly to match service planning and consumption to actual real defence needs. Just as the most commonly-preferred cure for problems in the military culture is to increase the power of the Chairman of the **Joint Chiefs of Staff,** so the remedy suggested for the DOD as a whole is to strengthen the role of the Secretary of Defense and the OSD, possibly even abolishing the secretarial-level appointments for the individual service departments. In both cases this would be an emulation of reorganizations made in the British **Ministry of Defence** since the mid-1970s.

Deployment

Deployment refers to the placing of troops and equipment. It is possible, for example, to talk of 'forward deployment' (see **Forward Based**). This is a concept often encountered in American strategic thinking, referring in particular to the USA's need to keep a major part of its army stationed in Europe, it being thus 'deployed'. Similarly a basic **Infantry** tactic of dispersing troops widely to present a difficult target is sometimes called 'loose deployment' (although the British army slang of 'thinning out' is a more graphic definition). 'Deployment capability', the ability actually to deliver troops to some distant part of the world in fighting order (see **Force Projection**), is often seen as the basic weakness in US military capacity. (See also **POMCUS.**)

Designated Ground Zero (DGZ)

Ground zero is jargon for the centre of the circle of destruction from a nuclear warhead detonation. It may not literally be a point on the ground, because most nuclear weapons are likely to be primed for **Air Burst,** in order to minimize **Fall-Out** and maximize blast damage. However, whether **Ground Burst** or air burst, it represents the theoretical point of impact. Designated ground zero means the aiming point, or the spot where the theoretical centre would be if the weapon was perfectly accurate. In practice the uncertainty attendant on both the **Circular Error Probable** and **Bias** of strategic missiles means that actual ground zero for **Re-entry Vehicles** is likely to be some hundreds of metres from the designated ground zero for at least 50% of warheads. Such discrepancies between designated and actual ground zero would be of little practical importance unless the intention was to destroy a **Hard Target.**

Destroyer

Anyone familiar with 20th-century naval history will think of destroyers as small, dashing ships, perhaps the maritime equivalent of fighter aircraft. Most navies still have classes of ships called destroyers, or, more likely now, Guided Missile Destroyers. (The acronym for this is, somewhat misleadingly, DDG: the 'D' for Destroyer is doubled, imitating 'SS' for steamship, and the 'G' stands for Guided.) However, developments in technology, and attendant changes in naval tactics and roles, have generally reduced the number of actual different classes of ships. Ship-to-ship combat between surface craft within gun-range of each other is unlikely ever again to be important, and, whatever the nomenclature, vessels serve one of three roles today. One is **Anti-Submarine Warfare (ASW),** another is air-power and the third is air defence of the ships involved in the first two functions. The usual title for ships involved in ASW is **'Frigate',** although older classes of anti-submarine ships still in service are often called destroyer. Air-power is obviously the function of carriers of one sort or another. Air defence for the fleet depends on specially-equipped ships capable of **Area Defence.** These again may be called destroyers but, because large craft are needed to carry the necessary radar and missile capacity, new classes tend to be designated as 'cruisers'. An example of this is the newest generation of US navy air defence ships, the Aegis class of cruisers. Frigates, which had previously been smaller and less powerful than destroyers, are now often bigger than a Second World War destroyer, and are the basic fleet ship. Britain's naval plans, for example, centre on the idea of a 50-frigate fleet, and there are no plans for new ships of any greater size.

Détente

Détente usually describes the period of calmer and less antagonistic relations between the USSR and the USA inaugurated by President Nixon's famous visit to Msocow in 1972, and given substance by the **SALT I** treaties. If détente is over, as some commentators would argue, the Soviet invasion of Afghanistan, the **Arms Control** tension caused by the **Euromissile** problem, and the harder anti-communist rhetoric of the Reagan administration would mark its end.

In many ways both the birth and supposed death of détente are exaggerations, or at least highly subjective and partly media-created interpretations. It is more that the extreme tension of the **Cold War** period, from 1948 to the early 1960s, tends to be used as a bench-mark, against which anything would look like détente. In reality there was no really significant reduction in superpower conflict during the 1970s, nor has there been any dangerous clash between them in the 1980s. Détente is precious to those, mainly in Western Europe, who share the ideology of neither superpower and fear both more or less equally as potential causers of war. The countries in which détente has probably meant something practical would be the two Germanies, where the détente between the superpowers who influence East and West Germany allowed their respective leaders to establish much stronger economic and diplomatic relations with each other than had previously been encouraged.

To the extent that détente is real, rather than a style of rhetoric, it is based on factors such as mutual economic interdependence (for example, the USSR needs to import grain and the USA needs to reduce its surplus) or a shared desire for arms control. However much US Presidents might denounce Soviet heavy-handedness in Eastern Europe or Afghanistan, they will not allow an irrevocable deterioration in relations with the Soviet Union. Where interests do conflict, as in Latin America, the Soviet Union will not let its protests about American breaches of the spirit of détente prevent it backing radical governments. SALT II, indeed, is the perfect example of détente in action: a treaty limiting US missiles more than Soviet ones, never actually ratified by the US Senate, has been observed, with only minor technical breaches, by the most rhetorically anti-communist US administration for 20 years.

Deterrence

In some ways deterrence has always been the role of all military forces except those specifically intended for wars of conquest. In its broadest use, deterrence means any strategy, force position or policy which is intended to persuade a potential enemy not to attack. In this sense it is becoming common to blur the distinction between deterrence and **Defence** by talking about NATO's conventional defensive armies as 'deterring' the Warsaw Pact from starting a war by making it impossible for them to win. This alternative usage is known as **Deterrence by Denial,** but it serves little purpose except to confuse an otherwise useful distinction.

That deterrence and defence are normally used differently is shown, for example, by the way the Eisenhower administration's **Massive Retaliation** doctrine was described precisely as substituting nuclear *deterrence* for conventional *defence* as a cheaper means of protecting Western interests. Given this, deterrence can be seen as a policy of preventing aggression by making the direct cost to the aggressor's society greater than any benefit that might be gained by military success. While this is the logic of all deterrence policies, it is at its clearest in the French doctrine of **Proportional Deterrence.** This is the belief that a weak country can deter attack by a stronger one as long as it can impose damage in proportion to its own value as a prize. Thus France, with a population of 55 million need only be able to kill 55 million Soviet citizens to deter the USSR, while a more populous or economically stronger nation would need to threaten more. Similarly, the British doctrine of **Graduated Deterrence,** a predecessor to **Flexible Response** developed in the 1950s because massive retaliation was thought to lack **Credibility,** involved relating the size of the deterrent strike to the size and nature of any Soviet aggression.

The major difficulty with theories of deterrence, as opposed to defensive **Strategy,** is that they run into problems of credibility, with widespread doubts that a nation attacked in a particular way would necessarily carry out its earlier threat. Defence in the conventional sense does not require credibility. No one seriously doubts that the NATO armoured divisions would fight if attacked by the Warsaw Pact armies, but many question whether the US **Strategic** missiles would indeed be fired at the USSR if those tank divisions were to be defeated.

Will deterrence succeed? It is clear that little thought has gone into devising a rational war plan for the event of its failure. The most that can be quite confidently asserted about weapons which are deployed for their deterrent value is that they effectively deter the use of a similar weapon. In this way the possession of **ICBM** and **SLBM** forces by the USA and USSR probably would successfully deter either side from using those precise weapons against each other to devastate the opposing homeland, much as it is argued that the possession of chemical weapons (see **Chemical Warfare**) by the Allies deterred Germany from using them in the Second World War. The only way that strategic nuclear weapons can be said to deter war in general is through fear that any form of conflict will escalate (see **Escalation**) into **Central Strategic Warfare.** It is largely because of this residual hope that the two small nuclear powers, France and the UK, deliberately refrain from any statement about the conditions under which they would use their deterrent forces (see **Declaratory Policy**). Governments of both countries expressly say that uncertainty about nuclear employment policy increases deterrence. In contrast the USA has tended to publish rather detailed declaratory policies for potential nuclear use.

So what would happen if deterrence failed? Suppose the USSR did launch a **First Strike** which obliterated US society, and destroyed all but the safe submarine-based US 'deterrent'. As the event which the force was built to prevent has happened, what should be done with it? There might be nothing to lose by launching the retaliatory strike, but what would be gained by doing so? Unless one holds to a very primitive retributive theory of justice of the 'an eye for an eye' variety, this is not at all an easy question to answer. As a consequence, employment policy tends to deny the total finality of the **Assured Destruction** which the deterrent threat emphasizes. Targeting plans are aimed at **Recovery Capacity,** and seek to destroy the enemy state rather than society. In other words employment theory views nuclear war as fightable and winnable in ordinary terms, as indeed it has to do if it is to answer the 'what happens after deterrence failure?' question. In so doing, of course, it is partly inconsistent with the *deterrent* emphasis of the supreme threat.

Deterrence by Denial

Deterrence by denial is not, as it might seem, really a branch of general **Deterrence** theory. Rather it is a somewhat jargonistic way of talking about the traditional role of military defence. The argument is that one can prevent war by having such strong defences that a potential enemy knows that an aggressive war cannot be won. Thus the enemy is being 'denied' victory, or the territory which might be gained, and is in some general sense being 'deterred' from trying. However, there is no good reason to complicate the otherwise useful distinction between **Defence** and deterrence, and the concept is of little use.

Disarmament

Disarmament has been a goal of peace movements and individual peace leaders, as well as out-and-out pacifists, since the mid-19th century or even earlier. The idea of stopping war by destroying military weapons could only make sense in a context with a fairly high degree of military technology, because it requires specific single-function objects that can clearly be identified as weapons. Preventing warfare in the Middle Ages, to take an obvious example, by abolishing swords and spears would not have been a supportable proposition. Any sharp-edged piece of steel could serve as a sword and any pointed stick as a lance, and, in any case, legitimate reasons for having such weapons for hunting would abound.

In fact, seeing the manufacture and sale of weapons as a cause of war, rather than an inevitable consequence of man's warlike nature, really began to be popular with the rise of powerful arms merchants in the late 19th and early 20th centuries, as demonstrated, for example, in George Bernard Shaw's play, *Major Barbara.* It is the uniqueness of modern weapons—a jet bomber really *does* have only one purpose—as well as their destructiveness that has made disarmament seem a way of controlling warfare. There are two separate arguments within this principle. Firstly, even if conflict is in man's nature, a war fought without modern weaponry is clearly less dangerous for the human race. Secondly, the possession of rival complex armouries, and the possibilities of consequent **Arms Races,** can be seen as themselves actually causing wars, particularly **Pre-emptive Wars,** which would not naturally happen otherwise.

The first important disarmament movements occurred during the period between the two World Wars; they were not necessarily politically radical, and tended to appear in the most developed countries. It was at this time that the division in disarmament theories first occurred, between what might be called 'weapon specific' disarmament and what is now usually known as 'general and complete' disarmament. The weapon specific schools concentrate on some particular form of armaments that are thought to be especially horrific or dangerous. Obviously the modern examples are the multifarious nuclear disarmament movements, such as the British **CND** (Campaign for Nuclear Disarmament), the **European Nuclear Disarmament (END)** movement, or the even more limited American **Freeze Movement.** But there were important precursors to these, including the attempt, taken seriously by many European politicians, to abolish the military aircraft that emerged in the early 1920s or, more obviously, the attempts to outlaw all forms of **Chemical Warfare** enshrined in the Geneva protocols.

General and complete disarmament has never made as much progress, although the United Nations does have it as an official policy goal. Clearly it runs into tremendous definitional problems, as well as being simply utopian in its impracticality. No disarmament movement has ever been known to succeed, giving rise to the common doctrine that nothing can be 'uninvented', and that if scientists discover how to make a weapon, someone somewhere *will* make it.

Division

Divisions are usually defined as the smallest unit in an army which can operate independently. Although a division can be predominantly of one arm, **Infantry,** for example, they are usually mixed. Thus the difference between an infantry division and an armoured division is *not* that one has only foot soldiers and the other only tanks, but rather a matter of proportion (see **Light Division**). An infantry division will have some tank squadrons attached as support, and an armoured division will certainly have several infantry battalions. Both will have their own artillery, signals, engineers, and **Logistics** services integrated into a single command under a general. As a consequence, they are the 'building bricks' of armies, and assessments of one's own and the enemy's forces are often presented in terms of the number of divisions in the **Order of Battle.** Ever since the founding of NATO there has been a tendency to assume that the Warsaw Pact conventional forces are much stronger than NATO's because of its greater number of divisions. By 1962, for example, the Soviet Union was seen to have 175 divisions while the US army could field barely one-tenth as many.

One of the major achievements of the systems analysts brought into the **Pentagon** by Robert **McNamara** was to point out how inadequate such a measure was. To a very large extent the number of divisions an army has reflects organizational doctrine, rather than either personnel levels or **Firepower,** and the US army at the time had almost twice as many soldiers in each of its divisions. Furthermore, as many as half of the Soviet army divisions were cadre (skeleton) divisions, maintained at only about 30% of their full strength. Such divisions could only be used after extensive call up of reserves, and would be far less well trained and integrated than a combat ready division. Despite such obvious arithmetic, the number of divisions is still very frequently used as a comparative measure, often with seriously misleading results.

Doctrine

Doctrine is a word much used by the military professionals of some countries, including the USA, the Soviet Union and West Germany, and almost completely unused by others, notoriously the UK. To equate doctrine with 'theory', as some do, is perhaps misleading, because a theory, whether in politics or physics, is a body of principles from which one can derive solutions to new, previously unconsidered problems. A military doctrine, in contrast, tends to be somewhat a-theoretic, and to consist of 'correct' answers, of ready-constructed solutions to problems which can be anticipated. An instance of this is the latest version of the US army's official operations manual, held to be the repository of 'doctrine' and known variously by its designation (FM 100-5) or its more catchy title **Air-Land Battle Concept;** it is the first ever produced which deliberately leaves out any statement of the supposed permanently valid 'principles of war'.

An army's doctrine is the general set of rules intended at least to guide, and possibly to control, combat decisions at all unit levels from platoon leader to the generals commanding army corps. Although often set out in lengthy documents, and subjected to endless debate in professional military journals, doctrines tend to be based on a few simple ideas. The doctrine of the French army in 1914, assiduously taught in staff colleges for the preceding 30 years, was that 'attack' was the only mode of fighting suitable, because it matched the *'élan vital'* held to characterize the French military tradition. The current emphasis in US military doctrine is on rapidity of action, on taking the initiative and on mobility. The doctrine it replaced, written in about 1968, has usually been characterized as stressing attrition—digging in strongly, massing troops, equipment and munitions, and not becoming committed to combat until overwhelming superiority had been built up.

In practice doctrine is both necessary (otherwise battles would rapidly disintegrate into anarchy), and largely irrelevant in detail. Senior US army officers are prone to point out that, within their own service careers of perhaps 20 years, there have been three radically different official doctrine publications. This could lead to quite different battle decisions from officers at junior, middle and senior ranks if it were not that most combat responses are conditioned by much longer-term general aspects of a country's military culture. For example, the entire political and authority system of the Soviet Union is perfectly in keeping with the official military doctrine, which allows hardly any initiative to junior officers. But even were some reforming general to rewrite the text book, and demand flexibility and initiative, these officers, accompanied by the Political Commissars and drilled in social orthodoxy from childhood, would carry out orders to, and not beyond, the letter. Similarly the US army doctrine can disparage **Attrition Warfare** as much as it likes, but the power of the US Congress will still ensure that American generals take more care to reduce casualties to their troops than the generals of most armies. The British faith in their soldiers in defensive positions, and the tradition of losing all but the last battle, guarantees a more static and positional campaign by the **British Army of the Rhine** than, for example, the **Blitzkrieg**-oriented West German army, even though the politically-imposed doctrine for that force is linear **Forward Defence.**

To a large extent 'doctrine' is a question of politics; it is either politically imposed, or developed by the military to please the civilians who control budgets. Just as the Soviet **General Staff** *must* enunciate their war plans in Marxist terms, the US army *has* to bow to Congressional pressure through, for example, the **Military Reform Group.** Current US naval doctrine, for example, calls for the use of its powerful **Carrier Battle Groups** to 'bring the war to the enemy' by sailing north of Norway and attacking the USSR's prime naval targets around the Kola peninsula. This plan was much-loved by President Reagan's Secretary of the Navy (until 1987), John Lehman. *He* was much-loved by the admirals because he fought hard and successfully for navy appropriations. It is not, however, easy to find an admiral who would, in practice, risk the highly-vulnerable carriers anywhere near such a concentration of Soviet naval air power.

There is a vital use for doctrine, for military theory, but it is at the highest strategic levels; without it a country could not define its general war aims or equip itself for possible war. Doctrine tends to be much less important at the **Operational Level.** The one doctrinal point agreed on by competent soldiers of all countries at all times could be summed up as, 'No plan survives the first contact with the enemy'.

Domino Theory

The 'domino theory' was the title given, both by those who accepted and opposed it, to a significant **Cold War** definition of international communism. The theory was that communism spread from country to country almost as if it were a medical epidemic. Thus if one Third World country fell to a communist revolution, the countries bordering it would in due course follow suit, and then the countries bordering them, and so on. Explicitly or otherwise, this fear of radicalism by contagion lay behind the US intervention in Vietnam. Successive US administrations from Presidents Eisenhower to Ford felt it necessary to shore up the anti-communist regimes in South Vietnam to prevent the whole of South-east Asia from ultimately falling under the rule of communist parties. More recently, the right wing in the USA has analysed the dangers presented by the left-wing government in Nicaragua in the same way: if Nicaragua stays communist, the rest of Central America will follow, and eventually Mexico as well, leaving the USA with an enemy at its southern border.

These theories are not as manifestly absurd as critics claim. Communist governments do, indeed, tend to infiltrate neighbours and give aid to rebel movements nearby. Socialism *is* an internationalist doctrine, and subscribes to the theory of international class war. The countries surrounding Vietnam *are* now under radical left-wing influence. Cuba most certainly tried to export its revolution elsewhere in Latin America during the 1970s, and still supports revolutionary movements, not only in the American continent but also in Africa. What the theory ignores is that the similarity of politics is caused by a similarity of socio-economic conditions as much as by a deliberate proselytizing or political emulation. More importantly, the theory only makes sense in the context of the much-outdated notion that there is some general world-wide school of government called 'communism'. In fact rivalries between national styles in radical socialism, and competition for influence between the two communist superpowers, the People's Republic of China and the USSR, show this idea of a unitary spreading political creed to be too simplistic. Vietnam's neighbours are indeed communist, but relations between Vietnam and Kampuchea, or Vietnam and China, could hardly be worse.

Doomsday Machine

A doomsday machine is an imaginary concept in strategic theory which captured the public imagination through novels and films like **Dr Strangelove.** The idea is simple, if frightening. It supposes that one nation built an enormously powerful nuclear device which was capable of annihilating the world's population. It further supposes that this machine was automatically programmed so that it would explode if the country that built it was the subject of a nuclear attack. Finally, it supposes that this automatic programme could not be turned off, even by the builders. Would not this be the perfect deterrent against nuclear war? Could any aggressor ever risk a nuclear strike under such a threat?

The strength of the threat does not come so much from the power of the doomsday machine, because nuclear warhead stocks reached theoretical **Overkill** years ago; it comes from the automaticity. There is no problem of **Credibility,** no question of an aggressor deciding to take the risk that the opponent will fire back. It is for reasons such as these that some schools of strategic thought would seek to make nuclear response as automatic as possible. An example that is sometimes debated in the USA is the question of what orders should be given to captains of nuclear missile submarines **(SSBNs).** If a major nuclear strike on the USA killed all the **National Command Authorities (NCAs),** and prevented all communication with submarines on patrol, all that the captains would know was that they had lost contact with Washington, and the whole of the USA. What orders should they be provided with for such a contingency? As far as is known, nuclear attack commands work on a positive release system, that is, no one has the authority to launch missiles unless specifically ordered so to do, and absence of an order does not entitle one to act. The system is sometimes described as **Fail-Safe** for this reason. But would a potential aggressor not be better deterred if the system were reversed? Why not have, for example, a radio station continuously beaming a 'Do Not Fire' order, which, if interrupted for any length of time would mean that missile launch should take place? Under such a system it would be inviting **Assured Destruction** to carry out a **Pre-emptive Strike** on the NCAs. The SSBN fleet would have been turned into a doomsday machine. Obviously the risks of malfunction, as well as the moral doubts attendant on automaticity of nuclear retaliation, make the system impossible to adopt, but the logic of **Deterrence** does indeed point in this direction. There are those who would argue that the danger to the whole world inherent in the theory of **Nuclear Winter** may, in effect, represent a doomsday machine.

Double Zero Option

After the Reykjavik Summit meeting between President Reagan and Secretary-General Gorbachev in October 1986, **Arms Control** negotiations at all levels, but particularly **Intermediate Nuclear Forces (INF)** were pursued with greater energy than for years. The breakthrough towards fruitful negotiations occurred after the Soviet Union accepted that INF negotiations could go ahead independently of any US concessions on its **Strategic Defense Initiative.** The 1981 **Zero Option** proposed by President Reagan was revived, in a modified form, under which all weapons in the range of 1,000 – 5,500 kilometres would be withdrawn; this would affect principally the **Pershing II** and **Ground-Launched Cruise Missiles** in American hands, and the Soviet Union's **SS-20s.** Each side would be allowed to keep 100 such missiles, but not in the European theatre. This clause was effectively to allow the USSR to retain a missile stock for its Asian border with China, and the USA was to be allowed an equal number to be kept in America as security against the USSR redeploying its remaining SS-20s back to Europe in a crisis. In July 1987 the Soviet Union offered a concession on this clause, raising the possibility of a global INF ban.

While in general welcoming such a policy, **European NATO** members immediately pointed out that it would leave NATO seriously exposed in another area, which has come to be known as Short-Range Intermediate Nuclear Forces (SRINF), missiles with a range between 500 and 1,000 km. In this category the USSR held an enormous superiority, probably in the ratio of 10:1. It was clear that European NATO would not agree to the zero option unless something could be done about this disparity. Hence the policy which came to be known as the 'double zero option' (or, sometimes, 'zero-zero') was invented, under which all missiles in the SRINF and INF categories, that is, any missile with a range between 500 and 5,500 km, would be removed from Europe. This would still leave the considerable stocks of **Battlefield Nuclear Weapons,** and would not control the deployment to the European area of ballistic missile submarines (see **SSBN**), but would, at least, remove an entire category of nuclear weapons. At one stage the West German Chancellor, Helmut Kohl, raised the possibility of a 'triple zero option', which would also remove battlefield nuclear weapons from Europe. However, this idea was never entirely accepted even within his own government, and would present seemingly insuperable **Verification** problems. An issue of contention which threatened to block a superpower agreement on INFs in Europe was the 72 Pershing IA missiles held by the West German air force, which have a range of 750 km. NATO maintained that these missiles, which are, in any case, seriously outdated, should not be considered in an agreement between the USA and the Soviet Union as they represent a 'third country' force. However, the USSR insisted that, as the warheads for the missiles were in American hands, they could not be separated from the negotiations.

Double-Key System (see Dual-Key System)

Douhet

The Italian, Giulo Douhet (1869 – 1930), is sometimes seen as the founder of much modern strategic theory. He developed the early doctrine that a war could be won by direct **Strategic Bombardment** of the civilian population from the air. This doctrine of strategic bombardment influenced the British Royal Air Force from its inception, as well as the US army air forces in both the European and Asian theatres during the Second World War. Douhet drew less on experience (there had been very little which even approached strategic bombardment during the First World War) than on logic. It seemed evident to him and to many leading airmen of his time that terror could destroy the **Civilian Morale** of a combatant nation, and that this would bring down its government and thus end the war. This was simply an adaptation to modern circumstances of a tried and tested technique of shortening a war. General William Sherman's famous march through Georgia, for example, during the American Civil War was an intentional infliction of terror on civilians to bring about the overthrow of the Confederate government. The difference was that Douhet saw a way to deliver this terror without having to defeat the enemy's armies first. As with all such theories, however, the establishment of **Air Superiority** is essential to guarantee that the bombers can get through. Therefore the first task of an air force in a war is, in fact, to fight a battle against the enemy's air force, rather than to deliver the supposed war-shortening strike against the civilian population.

Dr Strangelove

Dr Strangelove was the eponymous hero of a film, with Peter Sellers in the starring role, which satirized the nuclear strategy profession and the entire strategic force structure. Dr Strangelove himself is a composite characterization of a small group of thinkers, mainly mathematicians and political scientists, who largely created the doctrines of nuclear strategy in the 1950s and early 1960s while serving as consultants to the **Pentagon** and various **Think-Tanks.** Some of them, such as Herman Kahn, whose work is aptly characterized by the title of one of his early books, *Thinking the Unthinkable,* have become famous public figures. They broke many taboos and forced intellectual clarity into the emotionally repulsive world where 'megadeaths' have to be faced up to, and where even the horror of **Mutual Assured Destruction** is preferable to the anarchic over-use of weapons that led Kahn to describe early **Strategic Air Command** war plans as 'not war plans, but war orgasms'. Nevertheless, the very coldness of the theories, the stress on esoteric tools like **Game Theory,** and the intellectual fascination with Armageddon made them splendid targets for satire. (See also **Doomsday Machine.**)

Dropshot

Dropshot was the code-name for the earliest thorough US nuclear war plan, drawn up in 1949. Although there had been outline plans developed from 1946, the weapons stock was too small for them to have any real importance. Fleetwood, for example, was a plan drawn up in 1948 requiring 133 **Atomic Bombs** against 70 cities, yet at that time the **Strategic Air Command** had only 50 bombs, none of them assembled, and only 30 planes capable of using them. Even Dropshot assumed that the war it envisaged would not happen before 1957. The basic nuclear planning in the **Single Integrated Operations Plan (SIOP)** has not changed greatly since Dropshot. However, the target list has expanded drastically: Dropshot called for 300 atomic bombs on about 100 urban – industrial areas, identifying some 700 individual targets, whereas modern versions of SIOP included perhaps 10,000 targets from a list of over 40,000.

Dropshot had four main categories of target: stockpiles and production facilities of Soviet atomic and **Chemical Weapons;** governmental and control centres (see **C³I**); military supply and troop concentrations; and industry. The scale of damage was very much lower than modern expectations because planning assumed the use only of atomic bombs, using **Nuclear Fission,** the much more powerful **Nuclear Fusion,** or **Hydrogen Bomb,** not then having been developed. The committee which drafted Dropshot estimated a destruction of less than 40% of Soviet industrial capacity, and less than seven million human casualties.

Because destructive potential was limited there was no real grasp of the radical difference between nuclear and conventional war. At this stage atomic bombs were still seen as an extension of the conventional **Strategic Bombardment** of the Second World War, hence the concentration on essentially military targets. Indeed, Dropshot and Fleetwood both assumed a lengthy conventional bombing campaign of some 20,000 tons of high explosive in addition to the atomic component, and the major object, known as the **Blunting Mission,** was still to use air-power to restrict the Soviet's atomic and conventional military advance, rather than to win the war outright. Although nuclear capacity was rapidly increased and NATO strategy came to be based on **Massive Destruction,** no serious revision of strategic war planning was attempted until the SIOP concept emerged in 1959. Dropshot lingers on in one particular way. As is well known, the US President is accompanied everywhere by a military aide carrying the nuclear 'Go Codes'. The briefcase carried is ofen referred to as the 'football'—because Dropshot was the original content.

Dual-Capable System

Dual-capable systems are **Weapons Systems** which can be used in either a conventional or nuclear role. A typical example is the US F-111 fighter-bomber, which can carry either nuclear or conventional bombs, as can the European Tornado ground attack aircraft. Another example is the standard 155mm artillery cannon used by most NATO armies, which is also capable of firing nuclear shells. Dual-capable weapons pose particular problems in **Arms Control** negotiations because it is virtually impossible to find a way of verifying that such a system is not, in fact, being readied for a nuclear role. In this context the most dangerous weapons development has almost certainly been the deployment of **Cruise Missiles.** These have the range to be regarded as quasi-**Strategic,** and those which are the subject of intense political controversy (see **Euromissile** and **Ground-Launched Cruise Missile**) *are* intended for nuclear use. But exactly the same missile can carry a conventional warhead. The US navy has plans to arm almost all of its ships and submarines with their Tomahawk **Sea-Launched Cruise Missile (SLCM).** Although the intention is probably to use them in a conventional role, there is no means by which the USSR could tell that an incoming raid on their ports or air-bases was not the beginning of a nuclear strike. The problem also applies in reverse, of course. Both SLCMs and **Air-Launched Cruise Missiles** are also important features of Soviet naval planning.

Dual-Key

A dual-key system is an arrangement by which the USA shares control of a nuclear weapons system with the country in which it is based, or the national military service primarily responsible for operating it. An early example was the deployment of **Thor** (an **Intermediate-Range Ballistic Missile**) in the UK during the late 1950s and early 1960s. These were provided by the USA, but maintained and operated by the RAF. However, they could not be launched without the consent of both governments. To ensure this officers from both services were required to carry out part of the launch sequence together, in this case literally by turning keys.

Many of NATO's **Battlefield Nuclear Weapons** are held under a version of this system. The **British Army of the Rhine,** for example, has some Lance short-range nuclear missiles in its artillery, and several regiments have 155mm guns capable of firing nuclear shells. The warheads for these weapons are held under US control in separate locations, and would only be released to the British army by order of the NATO commander, **SACEUR,** who is always an American. Here the 'duality' is ensured because while only an American can order release of the warheads, the British artillery commanders would only fire them on the authority of the British Prime Minister.

The politics (and economics) of the dual-key arrangement tends to depend on the USA's perception of its interests. In the case of Thor the USA very much wanted the **Forward Based** missiles, because it was before the era of reliable **ICBMs.** They were prepared to provide Thor free, and have it staffed at British cost, regarding the sharing of firing authority as an acceptable sacrifice. Much more recently, when the **Euromissile** issue became politically sensitive in the UK, it was suggested in some circles that the British government should only allow **Cruise Missiles** to be based in the UK if given a dual-key share in control. The US response was very clear—the UK could have dual-key authority, but only if it paid for the missiles. In this case the US government's position was that as it was carrying out the **Intermediate Nuclear Force** modernization entirely at the request of Western European governments, it saw no reason why it should also give up control of an asset it would not otherwise have deployed. In fact dual-key arrangements are not common because most of **European NATO,** positively wish not to be associated with the responsibility of deciding to 'go nuclear'.

Dual-Track (see Twin-Track)

E

Echeloned Attack

Echelon is a technical term in military science that has come to have a vital meaning in discussion of a future confrontation between NATO and the Warsaw Pact on the **Central Front.** All echelon means is a line or formation of troops on a battlefield. A traditional military use is in the phrase 'real echelon' which meant, for example, the headquarters and **Logistics** services attached to fighting units, to distinguish them from the troops in the front line. Since the mid-1960s **Soviet Doctrine** has been based on the idea of attacking in several waves or echelons (see **Concentration of Force**). In this way they could avoid having too many troops at one place on the front line, which would present an easy target for NATO's **Battlefield Nuclear Weapons.**

The first echelon would attack along the whole length of NATO's front line. Even if the attack was repulsed, the weakened NATO forces would then face renewed attack from the second echelon, arriving fresh and at full strength perhaps 48 to 72 hours after the first echelon. By then gaps or weak spots would be expected to have opened up in NATO's defences, through which forces held in reserve, called **Operational Manoeuvre Groups,** could infiltrate to attack deep into NATO's rear area in a **Blitzkrieg** fashion. The entire doctrine is to take advantage of the Warsaw Pact's conventional superiority in troops and weapons, without prompting NATO **Escalation** to the use of battlefield or **Theatre Nuclear Forces.** This tactical approach is also dependent on NATO's commitment to a fixed **Forward Defence,** therefore presenting a thin linear battle array against which the successive echelons can attack. If NATO was able to make concentrations of forces to repulse selected parts of the first echelon, as well as holding the whole line, the Soviet strategy would be of much less value.

It is to counter echeloned attack that NATO has adopted **Deep Strike** and **Follow on Forces Attack** as its own doctrine, in which the second echelon will be held up and damaged deep inside Warsaw Pact territory. As a consequence it is common for NATO commanders to talk of fighting two battles at once—the immediate confrontation between the NATO troops and the first echelon being directed at one level of command, while the higher command 'fights the battle two days away'.

Effective Parity

Effective parity is one of the large number of concepts involved in the general consideration of **Parity** in weapons. It cannot, by definition, conform to a specific technical equation; it is a measure of the overall balance of destructive capability. One country might, for example, have superiority in launchers and **Megatons,** another in warheads and accuracy. If both were thus able to carry out their respective war plans, this situation could represent effective parity. Thus **SALT I** certainly, and **SALT II** probably, was an example of effective parity, although on any single measure of military strength one or other side might have held a clear superiority. **Essential Equivalence,** or nuclear sufficiency, is a further refinement of effective parity in that, rather than looking at the general question of whether two countries can destroy each other, it calls for parity of destructive capability in each of the legs of the nuclear **Triad.**

Eisenhower Doctrine (see Massive Retaliation)

Electromagnetic Pulse (EMP)

Electromagnetic pulse is a phenomenon of nuclear explosions. It is an electromagnetic wave, similar to a radio wave, caused by the secondary reactions occurring when gamma radiation is absorbed into the air or ground. Its effects are very much like those of an intensely-magnified radio wave, but the magnification is so great that it destroys or temporarily disables the same aerials and circuits that receive radio transmissions. There are two main differences between the two kinds of electromagnetic radiation. Firstly, a radio wave produces an electrical field of perhaps only one-thousandth of a volt in a receiving antenna, but an EMP can cause a field running into several thousands of volts. Secondly, an EMP is very much faster than other electromagnetic waves such as a lightning flash, because the entire energy is in a pulse that disappears in a fraction of a second. Thus equipment meant to protect antennae and internal circuits from lightning cannot cope with the overload, and the delicate electronics of radars, radios and short-wave telephone communications can be literally burned out.

Clearly EMP poses a major threat to **C^{3}I** facilities in a nuclear war. A nuclear explosion in the one **Megaton** range at a height of 2,500 metres will produce an EMP in the order of 10,000 volts per metre over the area, some 75 square kilometres, where it produces an overpressure of 10 psi (that is, 10 lbs per square inch above normal atmospheric pressure), and as much as a few thousand volts per metre at much greater distances. The real danger would come from the deliberate detonation of a few missiles at heights of about 35 km above the surface, where the EMP effect is extremely strong. A really large explosion of around 50 megatons at this height could cause extensive EMP damage across the whole of the USA.

Electronic Counter-Measures (ECMs)

Modern warfare has come to depend very heavily on electronic means, principally radar, for detecting targets at great distances. It is seldom realized that even a single combat between two fighter aircraft may now take place with the two planes more than 100 miles apart, because of the range and speed of air-to-air missiles. Moving army units may be detected and fired on by artillery and missiles at ranges of several hundred miles, while at short ranges anti-aircraft and anti-missile defences, at sea and on land, rely on radar totally because of the reaction speed needed.

Consequently considerable research effort has gone into what used to be called 'jamming'—methods of blinding radars and of blanking out radio frequencies to prevent **Target Acquisition** and interrupt communications. These procedures go under the general name of electronic counter-measures and represent their own **Arms Race,** every bit as vital as the competition to build faster or more destructive weapons. Even ECMs come in two modes, active and passive. So powerful is modern radar that it can itself be detected, with the result that techniques for warning an aircraft or ship that it is the subject of a radar search are as important as the searching technology itself.

Inevitably, the age of 'counter-ECM' has been reached, that is, technology to 'burn through' an ECM jamming screen. As these technologies develop an increasing proportion of equipment and personnel is taken up in mounting and countering these defences, rather than actually carrying and using offensive ordnance. One result is that there are critics in the USA increasingly stressing that human capacity is not keeping up with technological developments, and, for example, pilots are tending to depend much more on simply looking for an enemy target, and engaging it at more traditional short ranges. The moral drawn by these critics is that a larger amount of simpler, and therefore more robust and cheaper, war material may be preferable to buying continually more sophisticated electronically-equipped aircraft and ships.

Electronic Intelligence (ELINT)

Electronic intelligence covers all forms of information gathering based on radio and radar, but the principal source is the monitoring of the opponent's own radio and radar emissions. Ever since the First World War the process of listening in to the enemy's radio traffic has been crucial as a way of discovering its position, intentions and movements. The famous ENIGMA code-breaking operation run in the UK during the Second World War was, in many ways, the forerunner of the very extensive modern electronic intelligence gathering systems.

The general ELINT process can usefully be broken down into two activities. One aspect, directly following from ENGIMA in style, is the intercepting and recording of communications between enemy units and, indeed, inside the opponent's governing and administrative structures, in peacetime as much as in wartime. Often referred to as **Signals Intelligence (SIGINT),** this is, for example, the main activity of the Government Communications Headquarters (GCHQ) in Britain, and of the **National Security Agency (NSA)** in the USA. The amount of information that can be picked up is enormous: recently the British became concerned that calls on the internal telephone system in the **Ministry of Defence** itself could be overheard by suitable receivers elsewhere in London. Even when the communications are coded the sheer volume of radio traffic, as well as its source and destination, can often provide crucial intelligence in wartime.

The second general area of ELINT involves gathering information on an opponent's radio and radar procedures, wavelengths and electronic characteristics. This way it is possible to identify particular **Radar Signatures,** for example, which tell one exactly what sort of ship, aircraft, or other transmitter is operating at what range, for what purpose and so on. All major powers carry out intelligence gathering operations, often involving taking ships or submarines close to the opponent's shores, even sometimes deliberately triggering an alert in order to record the signal characteristics and operating procedures. Similar practices can be conducted with **Sonar** in **Anti-Submarine Warfare** operations, where each craft will have a characteristic 'sonar signature'.

As military procedures become more and more dependent on long-range electronic observation, and on increasingly complex and rapid communications (see **C³I**), the sheer level and variety of electronic 'noise' available for monitoring and analysis make them, in many ways, less and less secret than in more technologically primitive times. Already the vast bulk of useful intelligence comes from agencies such as GCHQ and NSA with their listening posts world-wide, rather than from the traditional espionage agencies using 'human intelligence'.

Electronic Warfare

Electronic warfare is a general description of the techniques by which the extremely high-technology weaponry of modern military services can be countered. A good example is that of a strategic bomber squadron in a modern air force, such as the **Strategic Air Command,** which contains several aircraft carrying no weapons at all because they are so full of **Electronic Counter-Measure (ECM)** equipment. The ECMs are intended to blind the enemy's radar, misdirect attacking missiles, mislead hunting fighter aircraft and so on. Probably the most important aspect of electronic warfare today is not, however, these defensive measures, but offensive long-range 'jamming' of radio and radar to facilitate an attack. All modern military services invest in electronic warfare capabilities of both offensive and defensive types, but they are among the most secret of all military procurement areas, and very little knowledge of them is in the public domain. One serious suggestion is that the electronic warfare capacities of NATO and the Warsaw Pact are so developed that they will mutually paralyse the command and control facilities (see **C³I**) of both sides in a future war.

Emerging Technology (ET)

Emerging technology is a shorthand way of referring to a number of **Weapons Systems** and **C³I** links which are being developed in the USA. It is advocated by many as a way of solving NATO's conventional inferiority problem. The hopes are based on a combination of two technological developments: **Precision-Guided Munitions** and **Real Time Information** remote **Target Acquisition.** Put at its simplest, it is believed to be possible to pursue the development of **Smart Bombs** so that a missile could carry dozens of warheads which, when dropped over a large target such as a squadron of tanks, would steer themselves individually to the tanks, with a near-perfect probability of hitting them. Furthermore, this is supposed capable of accomplishment at launch ranges of several hundred kilometres.

Even if the weapons can be developed, they are obviously useless without the capacity to detect these very distant and mobile targets; this is where the target acquisition technology is essential. Airborne radar and aeroplanes carrying infra-red search facilities, and possibly pilotless drones, are to be designed capable of detecting such targets and fixing their positions very precisely at great ranges. Even then the demand on communications and control is intense, hence the 'real time' intelligence processing aspect. It is not enough to know that at a certain hour there *was* a suitable target at a specific location; the information has to be instantly available to the commander before it can be worth launching a strike with these inevitably expensive weapons, which are certain to be in short supply.

Emerging technology is a perfect example of a **Force Multiplier** technique by which NATO hopes to offset the fact that its members do not have the political will either to buy enough tanks and aircraft, or to conscript enough soldiers, to offset the Warsaw Pact at existing levels of technology. Even if the techniques work, and few have yet been tested adequately, ET offers a solution only given two assumptions. The first is that the required force structure can be financed within the existing political constraints on defence budgets. The second is that the Warsaw Pact is unable to develop a counter-technology. Available experience, for example in **Electronic Warfare,** suggests that technological counters can be developed to even the most sophisticated target acquisition systems. (See also **Electronic Counter-Measures.**)

Sceptics argue that the technology will be so advanced that it will present very serious problems of maintenance (at a time when much existing weaponry is already beyond the competence of the average soldier), that far more of the weapons will be needed than optimists calculate, and that any technological edge NATO does gain will be eroded rapidly. If this is so, NATO will initiate an **Arms Race** in which it will end up with the same numerical disadvantage simply at a higher technological level. However, the potential capabilities of emerging technology weaponry happen to fit very neatly into the **Deep Strike** and **Follow on Forces Attack** doctrines which NATO has developed to deal with the Soviet **Echeloned Attack,** and have therefore gained a certain bureaucratic momentum in the USA.

Enhanced Radiation Weapons

The neutron bomb, as it has always been known in lay terms, is technically an enhanced radiation nuclear weapon. As the popular name implies, the neutron bomb emits an unusually large proportion of the energy released on detonation as radioactive particles, mainly neutrons. Most nuclear weapons engineering has striven to maximize immediate blast effect, and to cut down on energy release through radiation, both to increase the physical destruction through shock waves and heat, and to reduce **Fall-Out** and longer-term radiation damage. Enhanced radiation weapons operate on precisely the opposite design philosophy. They seek to minimize blast and heat causing physical destruction to the area surrounding the point of detonation, and try to boost the emission of high-energy radiation, particularly particles with short half-lives.

The reasoning behind the design of enhanced radiation weapons is quite straightforward, and reflects a military need which is itself the opposite of the purpose that underlies the orthodox nuclear weapons designs. The usual objective for a nuclear weapon is to destroy physical structures—buildings, concrete silos and bunkers, bridges and factories, and so on. The human death rate associated with an ordinary nuclear explosion is usually regarded as **Collateral Damage,** and steps may be taken to minimize it. Enhanced radiation weapons are intended not for strategic operations against fixed enemy physical plant, but for tactical strikes against its armies. In this latter case it is precisely human death, the destruction of the troops themselves, which is desired. This could be achieved just by using very high-**Yield** ordinary weapons, but in this context the consequent physical destruction is a disadvantage to the attacking army, especially in an environment such as a **Central Front** battlefield in a highly-urbanized

Europe. The ruins of cities, the destruction of roads and bridges, the roadblocks caused by felled trees and the fires and smoke resulting from even a small nuclear blast would delay troops advancing to take advantage of the shock caused by the use of **Battlefield Nuclear Weapons.** The ideal is a weapon which will emit a very powerful wave of very short-lived radiation. The physical environment will remain suitable for fighting over, and the ground will rapidly become safe from lingering radiation so that troops can move in and occupy it with little resistance from the enemy forces who, if not dead, will be suffering from the effects of radiation sickness.

The USA developed neutron bombs for this purpose and attempted to deploy them in Europe during the 1970s. The unpleasant human effects of the weapon offended the public sensibilities in much the same way that **Chemical Warfare** weapons do, and the outcry forced President Carter to abandon plans for their deployment, although the USA still holds stocks which could be introduced in the context of a war. The French government is also known to have authorized their development.

Equivalent Megatonnage (MTE)

In the media, and in most lay discussion, the power of nuclear weapons is expressed in terms of **Megatons,** or sometimes kilotons. A megaton is equivalent to the explosion of one million tons of TNT, a kiloton is equivalent to one thousand tons. This measure can be very misleading. In the early days of **Nuclear Fusion** weapons, as opposed to merely **Atomic Bombs** using **Nuclear Fission,** multi-megaton range **Yields** were commonly tested. In fact most of the weapons carried on bombers, and in early ballistic missiles, were rated at or above one megaton. Since the development of **MIRV** technology the USA, the UK and France have largely given up deploying warheads with individual yields as high as a megaton, although the USSR still has a large number of multi-megaton warheads on its **ICBMs.**

The typical yield of a single warhead on a **Multiple Re-entry Vehicle** missile is between 200 and 400 kilotons. The **Polaris** missile carried in the UK's submarines, for example, has three warheads of 200 kilotons each. This makes it appear as though NATO has a relatively weak armoury. The problem is that megatonnage is an inadequate measure in this area because the relationship between destructive effect and megatonnage is not linear. In fact there is a pattern of decreasing return in terms of damage for each extra kiloton of explosive power, largely because much of the energy released travels upwards into the atmosphere. In addition, the destructive effect of even small nuclear explosions is so great that few targets could withstand even as much as a megaton, unless they were hardened (see **Hard Target**). Nuclear strategists work, instead, with the concept of equivalent megatonnage (MTE) which, by a simple formula, re-expresses the power of warheads below one megaton to take account of this phenomenon. The formula is:

MTE = yield raised to the two-thirds power (or MTE = $Y^{2/3}$).

Thus if the 200 kiloton theoretical yield of the Polaris warhead is expressed as a proportion of one megaton, the yield in the equation is: $\frac{200,000}{1,000,000}$ = .2 megatons. Raising this to the two-thirds power gives .345 megatons, or 345 kilotons. Therefore the three 200 kiloton warheads in the Polaris missile do not, in reality, add up to 600 kt, or only 60% of a one megaton warhead. Instead, the total *equivalent* yield, .345 MTE times three, equals 1.035 MTE, so that the nominal 600-kiloton warhead total has a much larger effective yield. The formula, if applied to theoretical yields above one megaton, demonstrates why it is unneccessary, and wasteful, to build very large warheads. If one applies the formula to a five megaton warhead the MTE is only 2.9. The biggest nuclear weapon tested (by the Soviet Union) had a theoretical yield of 50 megatons, but the equivalent megatonnage, 50 raised to the two-thirds power, is 13.2 MTE. In fact some physicists suggest that for large yields the appropriate correction factor may be as high as .5, which would give an MTE for the theoretical 50-megaton yield of only seven.

Escalation

In modern strategy escalation has a vital role both as an analytic category and as an actual policy. Using the analogy of a ladder, strategists think of potential conflict arrayed as a series of steps from minimum to maximum violence. Before an actual war between NATO and the Warsaw Pact broke out one would perhaps expect an 'escalating' succession of increasingly hostile acts. As a result of a crisis, quite probably outside Europe, the pattern common from previous large-scale conflicts of this century might suggest the following developments: the exchange of diplomatic notes, followed by hostile and bitter public denunciations, the arrest of citizens of the opponent on charges of spying, economic pressure, attempts to intimidate by uttering threats, followed perhaps by partial mobilization, by placing nuclear forces on high alert status, and so on.

The strategic concept extends this expectation into warfare itself, and predicts that a **Central Front** war would follow almost pre-set stages. At first the war would be purely conventional. Then it might be escalated to the use of **Battlefield Nuclear Weapons,** and later of **Theatre Nuclear Forces.** From such geographically-confined nuclear war, events might escalate to **Central Strategic Warfare** in which limited and selective strikes were carried out, before reaching the top of the ladder and resulting in massive nuclear exchanges against cities and industry.

As a descriptive **Scenario** (a theoretical analysis of 'what might happen') such escalation is perfectly plausible, although it may rest too heavily on the assumed **Fire-Break** between conventional and nuclear warfare. As a policy, which it becomes when referred to, as NATO often does, in terms of **Escalation Control** or **Escalation Dominance,** it is much less satisfactory. The general advantage of using escalation as a strategy is to be gained when a power is able to dominate the enemy at each step in the ladder, forcing it to make the choice of whether to move up to the next stage of violence and destruction, and knowing that, if it does so, the enemy will again be at a disadvantage. This has led to the belief that NATO must at least match Warsaw Pact capacity at each level, so that the latter would not be tempted to try to seek victory by moving on to the use of, for example, theatre nuclear weapons. With matching capacities the Warsaw Pact, if it could not prevail at the conventional level, would be deterred from increasing the risks, and escalation would have been controlled. The theory would be somewhat more suitable were it not NATO, rather than the Warsaw Pact, which is liable to have to 'go nuclear'. Indeed, escalation control is sometimes used to make even this seem a strength: if the Warsaw Pact knows that NATO would have to escalate to nuclear attacks, it may not take advantage of any conventional superiority.

There are two major problems about basing strategy on this escalation ladder. Firstly, it means that rather arbitary distinctions based on the geographical location of targets, or the mechanisms used to destroy them, become constraints and even guides in war planning, and thus come to take priority over the attainment of actual political or military goals. There is a danger that an attack on a particular target, such as an air-base in the western Soviet Union, would not use the weapon best designed to destroy it, which might be a theatre nuclear missile, because the appropriate escalation level had not been reached. Therefore operations from that base might be allowed to continue until the conventional war had been lost; the base might then be destroyed at a later stage, by which time it may be of little strategic value.

The second problem with escalation strategy is that it may be entirely a figment of Western strategic thinking. What little information is available about Soviet military planners suggests that they do not think in these terms at all. Certainly **Soviet Strategic Thought** puts great emphasis on surprise, and on pre-emption. Their nuclear war doctrine appears to be that the only hope of prevailing is to be the first to strike. If this is true, it is quite improbable that they would be content to proceed conventionally, accepting high casualties in the process, and wait for NATO to shift the gearbox of war. If it seemed at all likely that an escalation to a nuclear level was going to happen, it would fit Soviet planning far better to use nuclear weapons themselves, at an early stage, on vital targets such as NATO air-bases, in order to further their otherwise conventional war effort.

Escalation has its attraction to Western strategic thinking because it acts as an extension of **Deterrence** into war itself. Western thought has tended to concentrate almost entirely on war deterrence doctrine, at the cost of having very little to say about what to do if deterrence fails. Another reason for escalation seeming an appropriate model for a future war is probably that the major NATO partner, the USA, is geographically separate from the likely main battle area and could avoid sustaining any direct damage from a central front war, as long as it does not escalate too far. Finally, the official doctrine of NATO, **Flexible Response,** itself assumes a ranking of possible military reactions rather than a predetermined strategic plan.

Escalation Control

Escalation control is the key idea in NATO strategy for a possible future war with the Warsaw Pact. The basic concept of **Escalation** is that levels and types of force can be arrayed along a continuum of violence and undesirability. All combatants would prefer, other things being equal, to fight at a lower rather than a higher level of escalation. For the sake of simplicity it is possible to think in terms of four successive levels of violence in a NATO – Warsaw Pact conflict. The first is a purely conventional war, the second involves the use of **Battlefield Nuclear Weapons,** and the third and fourth respectively see the use of longer-range theatre missiles (see **Intermediate Nuclear Forces**) and finally **Central Strategic Warfare.**

The essence of NATO's strategy of **Flexible Response** is to match the Warsaw Pact at each level of escalation. This would deprive the latter of being able to move up a level of violence if they were not winning at the existing one. For example, if NATO had no short-range or tactical nuclear weapons and was, nevertheless, managing to hold its own in a conventional war, the Warsaw Pact could escalate by using tactical nuclear forces. It would know that NATO could not respond in kind. This Warsaw Pact move would force upon the USA the burden of having to escalate to a still higher and more dangerous level, that of using longer-range theatre missiles to regain its position. This it might be very unwilling to do for fear that the war could slip completely out of control. In any case if the Warsaw Pact had **Parity** with NATO in theatre weapons (as it does) NATO could not rely on gaining anything by this move to the third level.

Thus a power 'controls' an escalation ladder at the point at which it can increase the violence of its actions without the enemy being able to counter at that level. NATO is generally thought not to have the parity of force at either the conventional or short-range nuclear levels to be able to claim to exert escalation control at those stages on the ladder. This presents serious problems in that NATO might be faced with the choice of either escalating to a very advanced level of nuclear confrontation or admitting defeat. A closely related term is that of **Escalation Dominance,** which would represent the ultimate in escalation control. In this version the dominating power would have superiority at every level of violence.

Escalation Dominance

Escalation dominance, similar, but not identical in meaning to the term **Escalation Control,** means that one side in a war has superior force capacity not only at the current level of **Escalation,** but at successive levels. Thus it is in 'control' in the sense that the enemy cannot escape from a weak position in, for example, the conventional phase by moving on to the use of **Battlefield Nuclear Weapons,** because it would still be at a disadvantage. Furthermore, the dominating side *does* have that choice. So if it is suffering more casualties than it thinks acceptable it can try to terminate conflict by changing the intensity level to, perhaps, the use of **Theatre Nuclear Forces,** where its relative advantage may be even greater. The key to escalation domination is that the dominating side must have superiority at every level.

Essential Equivalence

Essential equivalence is part of the family of concepts developed during the **'New Look'** of US strategic policy that began with **Countervailing Strategy** in 1974, under US Secretary of Defense, James Schlesinger (see **Schlesinger Doctrine**). It is one of the possible definitions of **Parity,** which is accepted as a major **Arms Control** goal by the USA, in the sense sometimes given to it of implying a situation with 'no unilateral advantage to either side'. Thus, in the terms of the Schlesinger doctrine, essential equivalence meant that there should be no area of nuclear capacity in which the Soviet Union has an advantage over the USA; the USA had to equal the USSR in all three legs of the nuclear **Triad,** rather than simply in total warheads or **Throw-Weight.** In particular, Schlesinger felt that the USA had to build the type of heavy and highly-accurate **ICBMs** needed to execute **Limited Nuclear Options.** This was so that there would be no possibility of a Soviet nuclear strike to which the USA could not respond in like manner.

Although essential equivalence embraces parity as a goal, and is therefore less dangerous than the search for **Strategic Superiority,** it is clearly a doctrine which increases **Arms Race** pressure because of the automatic need to develop equivalents to any **Weapons System** that the opponent has. Essential equivalence, which has been accepted by all subsequent Secretaries of Defense, has led to, among other things, the building of the **MX** missile to rival the Soviet heavy ICBM force. The problem of essential equivalence is that it is a reactive posture—whatever the others do, we must do (see **Action – Reaction**). It seems likely that the USSR has not adopted quite this policy, because it feels more free to develop weapons that fit its own strategic doctrines rather than to cover against the eventualities threatened by the opponent's preferred strategy.

EuroGroup (see NATO)

Euromissile

Euromissile is a media phrase invented to cover the sorts of missiles that figured in the **Arms Control** negotiations between the USSR and USA over their mutual modernization of **Theatre Nuclear Forces.** As such it covers the Soviet **SS-20,** and the American **Ground-Launched Cruise Missile** and the **Pershing II** missile. More generally it refers to nuclear weapons, above the tactical or **Battlefield Nuclear Weapon** level, intended for use on targets in Eastern or Western Europe.

European Defence Community (EDC)

In the late 1940s and early 1950s the problem of defending Western Europe against the perceived 'Soviet threat' did not so obviously seem something that could be solved only by NATO as it does, perhaps, now. European integration was seen as the post-war solution to most problems, and 'communities' were being set up to deal with the economy and with the production of coal, steel and atomic energy. A similar approach was attempted for dealing with military problems. Hence there was an important, and almost successful, move to set up a European Defence Community on analogous lines to, for example, the European Economic Community. The EDC would have replaced NATO as the primary source of Western Europe's defences, something which the Americans, who had not at this stage thought of their NATO role as meaning a permanent overseas commitment, would have welcomed.

However, a series of political conflicts over the design of the EDC, including the questions of who should control it and, above all, which nations could be members, eventually made setting it up impossible. Some wanted a genuine European army, fully integrated and with no specific national loyalty. However, this was too extreme for those attached to their separate military traditions, and the plans turned towards a system of national military contingents, much like the structure of NATO. The major obstacle was over the membership of West Germany. At this stage West Germany was still legally disarmed by the Allied powers of the Second World War. It became increasingly clear, not only to those planning the EDC, but even more so to NATO which, as it already existed, was facing real rather than theoretical problems, that West Germany would have to be rearmed. The personnel requirements, in the face of the apparently huge Soviet army, believed to consist of 96 divisions at that time, were simply impossible to meet while West Germany's population was excluded from defending its own country. French opposition to the inevitable inclusion of West Germany in the EDC finally led to its treaty, which was all but ready to enter into operation, being defeated in the French parliament in 1954. The EDC plan was subsequently abandoned and, as the alternative solution to Western Europe's defence problem, West Germany was integrated into NATO in 1955. (See also **Western European Union.**)

European NATO

European NATO refers, at face value, to all members of the North Atlantic Treaty Organization except for the USA and Canada. However, in practice it is usually taken to exclude all but the Scandinavian and Benelux countries along with Italy, West Germany and the UK and, despite its ambivalent membership, France. Although European NATO is not, as such, a formal sub-grouping within NATO, there *are* institutions, most notably the **Western European Union** and the **Independent European Programme Group** which serve to express and co-ordinate any specifically European interests which might exist.

From time to time there are calls for the strengthening of the 'European pillar' of NATO, often with the idea that this could somehow represent a third force in international politics that could provide some kind of a balance for the USA – USSR tensions. This assumes that there is more in common between the defence policies and priorities of European members of NATO than between a European and a North American member, which is by no means always true. While there may be a case for a European 'third force' outside the NATO framework, it is hard to see how the co-ordination of the European members *against* the leadership of the USA can do anything but decrease the cohesion of NATO.

European Nuclear Disarmament (END)

The European Nuclear Disarmament movement was a response to the public concern about the deployment of new **Intermediate Nuclear Forces (INF)** in Western Europe during the post-1977 **Euromissile** debate. Founded in 1980, and based in the UK, END has more wide-ranging aims than the British **CND** (Campaign for Nuclear Disarmament). It has called for a nuclear-free Europe in which no country would deploy or allow the deployment of nuclear weapons (see **Nuclear-Free Zone**). As such it made a real effort to identify common interests between the non-Soviet Warsaw Pact countries and the non-American NATO countries. Contacts were made with peace movements in Eastern Europe, but some suspicion fell on END because these activities coincided with a period when the USSR was trying to play up Western European anti-nuclear sentiment to bolster its position in the INF negotiations with the USA. No one suggested that END was in any sense contaminated by Soviet infiltration, but the coincidence of their activity with Soviet propaganda damaged END's credibility with some parts of the electorate.

Inevitably, END has a strong following in those countries, such as the Netherlands, which have never had any nuclear weapons of their own, but have been forced, on a point of NATO solidarity, to accept US deployment of the new **Cruise Missile** bases. END faces considerable difficulty, even more so than CND, in achieving its objectives. Whereas CND largely confines its activities to pursuing the unilateral removal of the UK's independent nuclear forces and of US missiles based in the UK, END needs to win the support of electorates in several countries for similar aims. Furthermore, because of its stress on the removal of US nuclear weapons from Europe and its advocacy of a central European demilitarized zone it is seen as implicitly anti-NATO.

Eurostrategic

Eurostrategic is a concept which has to come into circulation since the **Euromissile** debate began in the late 1970s. It demonstrates the highly relative, even subjective aspect of some strategic thinking. The problem is that, for a long time, **Strategic** weapons were defined as the **ICBMs** and bombers in the arsenals of the USA and USSR. Even then there were areas of uncertainty or disagreement. Both France and the UK understandably regarded their missile forces, especially the **SLBMs,** as 'strategic' weapons. They were, after all, more or less comparable to those carried by American or Soviet submarines. However, the USA has always listed the British **SSBN** squadron as a **Theatre Nuclear Force (TNF),** to emphasize its formal commitment to NATO usage.

The lack of any real substantive meaning to the idea of a strategic weapon became clear when the USA threatened to deploy (and later deployed) **Cruise Missiles** and **Pershing II** ballistic missiles in Europe to offset the Soviet **SS-20** batteries. The USSR had already possessed a similar capacity, with its SS-4 and SS-5 missiles, for some time, and introduced the SS-20 as a modernization; it was principally the accuracy of the SS-20, rather than its range, that was crucial. The new NATO weapons mainly replace fighter-bomber aircraft which had never been intended to attack targets outside the immediate battle area around the **Central Front.** The new NATO TNFs have the range to hit targets in the USSR as far east as Moscow, and possibly the accuracy to destroy **Hard Targets** such as Soviet **C³I** bunkers. Not surprisingly, the USSR has argued that these are therefore just as much *strategic* weapons as US ICBMs in the Midwestern missile silos. As the USSR's European-deployed SS-20 missiles cannot reach the USA, the imbalance does indeed raise problems of fairness in **Arms Control** negotiations.

What offsets this somewhat, but makes the strategic/non-strategic distinction even more complicated, is that the SS-20 missiles most certainly *can* destroy Paris or London. Thus from the viewpoint of a European, including in this definition a Russian, *all* nuclear ballistic and cruise missiles with ranges over a few hundred kilometres are 'strategic'. What may be a 'theatre weapon' to one or both superpowers is, in its effect, indistinguishable from a strategic missile to the inhabitants of France, Britain and Italy. (For those in West Germany and the Benelux countries the argument is entirely otiose: **Battlefield Nuclear Weapons** are 'strategic' when the whole of one's country is the battlefield.)

Hence the concept of Eurostrategic has developed, in an extremely arbitrary manner, to include some **Weapons Systems** that do not fit neatly into previous categories. Strategic weapons are now defined, for the purpose of superpower arms control negotiations, as those that can strike the homeland of one superpower from the homeland of the other. This increasing arbitrariness, among other factors, increasingly calls into doubt the typical **Escalation** theories ofen used in nuclear warfare **Scenarios.**

Extended Deterrence

Extended deterrence is a strategy where one nuclear-armed country tries to deter another not only from attack on itself, but also from attacking its allies. It is, therefore, the basis of NATO's **Flexible Response** doctrine. This ultimately rests on the American **Nuclear Umbrella,** the threat that a Warsaw Pact attack on Western Europe will, if necessary, lead to **Central Strategic Warfare** between the USA and the USSR. America, at least in theory, is not the only country capable of offering extended deterrence to its allies. Implicitly the USSR's nuclear capacity deters NATO from attacks on other members of the Warsaw Pact. Sometimes it is claimed that a justification for French and British independent nuclear forces is that if the USA, for some reason, removes the nuclear umbrella the Federal Republic of Germany, and other non-nuclear **European NATO** members, would benefit from the extension of Anglo-French deterrence to cover them (see **Independent Deterrent**).

This latter argument is somewhat fanciful. If the USA, with its huge nuclear capacity, would not put itself at risk to defend Europe, it can hardly be believed that the much smaller nuclear forces of France or the UK would have **Credibility** as a replacement. In fact one of the principal justifications always used by the French for building its **Force de Frappe** has been precisely that no country can credibly extend deterrence. They maintain that it just does not make sense to believe that a nuclear-armed nation, which might otherwise be safe from nuclear attack, will put itself at risk of nuclear devastation on behalf of another. Instead of extended deterrence, this theory, which is by no means unique to French analysts, stresses that a country owns nuclear weapons solely for the purpose of self-**Sanctuarization.**

As with most theories of **Deterrence,** the concept of extended deterrence lacks precision. For example, what is the USSR supposed to be deterred from doing by exactly what threat? Flexible response implies that the

Warsaw Pact can be deterred from any serious aggression against European NATO territory by US strategic weapons. However, it is this aspect of the nuclear umbrella that is increasingly doubted, not only in Europe, but among analysts and politicians in Washington. A restricted application of extended deterrence theory is visible in the posture that defending the vital interests of a country may require a nuclear strike, even though the enemy has not directly attacked its territory. The French government has, in recent years, hinted at such a strategy in an attempt to strengthen Franco-German political and military relations. Thus the French argue that their own security would be so threatened by the defeat and occupation of West Germany that, in their *own* interest they would have to use nuclear weapons against the Warsaw Pact. The trick of gaining credibility involves somehow defining one's own interests in such a way as to make the extension of deterrence appear not as an act of loyal risk taking, but as a purely self-interested strategy. This line of argument, though somewhat thin, is similar to the original **Tripwire Thesis** of the **Massive Retaliation** doctrine.

F

Fail-Safe

Fail-safe is a term used to describe the security procedures used by the USA, and presumably by other nuclear powers, to attempt to prevent the unintentional launching of a nuclear attack. There has always been a horror of a nuclear **Accidental War** being triggered either by mechanical malfunction, by an unauthorized attack made by officers below the proper level in the **National Command Authorities (NCA)** or, literally, by a deranged person. To protect against such an eventuality there are numerous mechanical and human checks built into the launch command system. These range from very complex code-words and authentication procedures to mechanical requirements for two or more officers independently and simultaneously to turn certain keys before a missile can be armed (see **Dual-Key**). The logic behind the systems, and hence the term, is that they are all set up so that any failure in the correct procedures makes launch impossible; in the event of an authentication procedure not working, all is 'safe'.

No mechanism can be guaranteed to be absolutely incapable of malfunction, but nearly all experts believe that the danger of accidental nuclear war has now receded almost to indetectability. What is less clear is whether specific acts of nuclear release, once a war has generally 'gone nuclear', can be controlled as effectively. In a **Scenario** where a **First Strike** by one superpower effectively decapitates its enemy, destroying all command and control procedures and killing most political leaders, would it be impossible for a much more junior officer, in charge of a silo complex or a missile submarine, to take the initiative to fire? The problem is complex, because a system that *did* prevent nuclear retaliation after a successful first strike would actually make such a strategy more attractive, while anything that facilitates last-resort initiative risks the strength of the fail-safe procedures.

Fall-Out

Fallout is the secondary consequence of a nuclear explosion, consisting of the dust particles and water droplets thrown up into the atmosphere which have been made radioactive by the initial explosion. The amount of fall-out, and the area which it covers, depends on a variety of factors, of which the principal one is whether the explosion was an **Air Burst** or a **Ground Burst.** An air burst produces considerably less fall-out because the nuclear fireball does not actually touch the ground, so there is much less irradiation of dust and other material particles. A ground burst, on the other hand, because it causes a huge crater, vaporizing the soil and scattering the irradiated material, creates a very large amount of fall-out. Other factors which mitigate against this simple picture of air bursts as 'clean' and ground bursts as 'dirty' include atmospheric conditions and wind patterns. An air burst produces less fall-out, but in dry conditions with strong wind, what it does produce can be spread throughout the upper atmosphere over very large areas, whereas the fall-out from a ground burst *may* be contained over a fairly small area around ground zero (see **Designated Ground Zero**).

The full effect of fall-out cannot be known with any precision both because of the limited available experience and because of the variable atmospheric conditions. Studies of the effect of a **Counter-Force** strike on the US **ICBM** fields produced fatality rates varying from only slightly over two million to nearer 20 million, depending on wind patterns and season. The danger of fall-out comes from the very long half-life, often in the order of several decades, associated with many of the poisonous elements created by a nuclear explosion. Apart from the direct deaths caused in the immediate aftermath of a nuclear explosion, radiation poisoning produces long-term health risks, especially from cancers and genetic defects, for decades after the event. As a crude rule of thumb the total death rate among those surviving the nuclear strike would probably be at least twice that of those killed by the initial blast damage and fire-storms. This excludes any estimate of the numbers who might die in the **Nuclear Winter** which some scientists believe would follow a major nuclear war, and of those who might die through starvation, crime and so on in the devastated cities.

Fighter Command (see Bomber Command and Strike Command)

Fire-Break

In the language of nuclear strategy a fire-break is a point in time, or a stage during an **Escalation** process, at which the whole nature of a war can change. It might happen, for example, that a fire-break occurred after the initial use of **Battlefield Nuclear Weapons**, providing an opportunity for **Intra-War Bargaining** and a chance to avoid moving on to the use of **Theatre Nuclear Forces** or **ICBMs.** At a higher level of escalation a fire-break might come after the exchange of two or three ICBMs; in such circumstances both sides might assess their damage sustained, consider their strategy and seek further rounds of intra-war bargaining.

Fire-Power

Fire-power, not surprisingly, is a measure of the amount of munitions (bullets, shells, mortar bombs and so on) that a military unit can deliver on a target. A prominent part of the recent history of warfare has been the increasing dominance of fire-power, at the expense of the importance of manoeuvre, tactics, or even courage and **Morale.** Although the growing trend first showed with technical developments such as the breach-loading rifle and rapid-firing field gun in the last third of the 19th century, it was only during the First World War that military experts became convinced that sheer fire-power decided battles. The impossibility of advancing in strength against the machine-guns of 1914 – 18 and the suppressive power of artillery bombardment were the conclusive factors.

Since then the fire-power of ordinary **Infantry** units has increased enormously. All major armies equip their ordinary infantry with personal weapons capable of firing automatically at a rate equivalent to hundreds of rounds a minute. Small squads carry portable anti-tank and anti-aircraft weapons as powerful as the specialist weapons that would have been available only to artillery units during the Second World War. One consequence is that recent experience, particularly in the Falkland Islands conflict, has been that munitions stores are used up very much faster than before. Stocks of weapons at all levels of sophistication are small throughout NATO, probably at a level that would only supply full-scale war on the **Central Front** for a matter of weeks. The second consequence is that the potential destructiveness of conventional war has increased to the point that a country, such as West Germany, which might become a battlefield would be devastated by even a short non-nuclear war.

First Strike

A first strike in nuclear strategic terminology is the launching of nuclear weapons by one side before the enemy has taken such a step. Despite the apparent obviousness of this definition, it refers to what is becoming an increasingly confused area. The original clarity of the first strike/**Second Strike** dichotomy depended on the simplicity of ideas about nuclear war in the 1950s and 1960s. At that time it was assumed that a nuclear war would take the form of a single exchange, in which each side would fire its entire arsenal in one salvo. It was also usually assumed that each side would aim for a comprehensive list of target types, trying simultaneously to destroy the military, political, and industrial power of the enemy, and possibly also deliberately targeting the civilian population. This was the sort of **Scenario** in which it was calculated that 80 – 100 million people on each side would die from immediate blast effects, and, later, tens of millions more from **Fall-Out.** Apart from indicating who was, in some sense, 'to blame' for Armageddon, the first strike/second strike distinction carried little theoretical importance.

The distinction began to be somewhat more important when the **Mutual Assured Destruction (MAD)** doctrine was developed in the 1960s. In order to ensure destruction of the enemy it was necessary to calculate what could be destroyed in one's own country by an enemy first strike. It was assumed that such an attack would have to be 'ridden out', as the 'receiving' country would not be in a position to retaliate until after it had suffered the first strike. This led to the policy of guaranteeing a secure second strike capacity, meaning that enough deliverable nuclear warheads would remain after the attack had been received to inflict an amount of damage which would be unacceptable to the enemy. The consequence of this theory was that weapons were identified and built for two specified purposes. One set of weapons, basically the **SLBMs,** were reserved for this guaranteed counter-strike. However, neither superpower wanted to be in a position where delivering that counter-strike was all they could do. Consequently a second (but not secondary) set of weapons were maintained. Logic tells us that they must be for the first strike role, although, not surprisingly, neither the USA nor the USSR likes to talk about them in this language.

At this stage in the development of nuclear doctrine it was somewhat unclear why the first strike forces were needed, because it was still assumed that a war would involve a massive spasm attack (see **Spasm War**), in which urban – industrial centres would certainly be included, even if they were not the only form of target. In fact neither NATO, whose **Flexible Response** doctrine has prevented them ever making a **No First Use** statement, nor the USSR, which never accepted MAD as a security policy in the first place, could eschew a first strike capacity.

From the early 1970s onwards nuclear strategy has come to be considerably more sophisticated, especially with the ideas of **Limited Nuclear Options**, as enshrined in policy statements such as PD-59 and, more recently, nuclear **War-Fighting** theory. The result is that it is perfectly plausible to picture one side launching a few missiles against carefully-chosen military or administrative targets, and the other retaliating in a similar manner, or possibly not bothering to make a nuclear retaliation at all. In such a context a 'first strike' might have nothing but chronological meaning because it would not be intended as a sudden massive knockout blow, and might not be deterred by the second strike capacity in the traditional meaning of the term. Indeed, the best **Deterrence** against embarking on a limited nuclear strike which would be the first launched by either side would be the enemy's own 'first strike' capacity. It is a fact that the USSR, for example, could destroy selected politico-military targets in the USA *without* having to deliver so powerful a blow as to force its President to release US **Assured Destruction** forces.

There has been an argument in the USA that the Soviet Union was seeking, or already has, a special sort of first strike capacity (see **Strategic Superiority),** but on the whole the terminology is now now more misleading than helpful. A good example of this is the argument in the UK as to whether or not the US **Cruise Missiles** are 'first strike' weapons. The opponents of cruise deployment say that they are, because they want to further their opposition by making British acceptance of cruise seem both morally untenable and dangerous. (Dangerous because, if the **Ground-Launch Cruise Missiles** *are* first strike, the argument runs, that makes them a more urgent target for the USSR to destroy.) The UK government insists that they are not, and that they cannot be used for a first strike. However, the government's claim is based on an entirely outdated concept of a first strike. They maintain that the missiles cannot be a first strike weapon because they are so very slow, taking hours to get to the USSR. Therefore the USSR would have plenty of time to launch a retaliatory raid before the cruise missiles arrived over their targets.

It is true that the problem with a traditional first strike was the need to let it happen before retaliating because **Launch On Warning** and **Launch Under Attack** were discredited policies. But there could be no suggestion that an all out attack of the traditional first strike scenario would, in any case, be carried out by cruise missiles. The truth is that NATO's policy most certainly *does* involve a preparedness to be the first to use nuclear weapons, that cruise missiles are NATO-dedicated under **SACEUR's** command, and that if they cannot be used for a first strike, they are probably of no practical military value whatsoever. Neither side is right, they have simply become entwined in a dated technical language.

What is needed is a recognition that the purposes to which weapons are to be put, the military and strategic aims which they are designed to pursue, are the relevant categorizing variables. In much the same way that the concept of theatre versus **Strategic** weapons has broken down with the advent of the **Eurostrategic** concept, the first and second strike distinctions make little sense in a multi-strike war-fighting context.

First Use (see No First Use)

Flexible Response

The NATO doctrine of flexible response, which rests heavily on **Escalation** theories, was developed as a way around the contradictions of maintaining the **Credibility** of the **Massive Retaliation** doctrine and of building up an adequate conventional defence against the Warsaw Pact. The idea of flexibility was never really spelled out by NATO, which has made a virtue of uncertainty by arguing that the Warsaw Pact will be unable to plan properly, because it will never know how the West will choose to respond. How flexible response developed historically is easy to relate. The Kennedy administration, when it took office in 1961, was determined to drop the massive retaliation strategy, under which NATO had only one response to any form of Soviet incursion in Western Europe—to launch a major nuclear strike against the USSR. The only real alternative to massive retaliation was a huge conventional rearmament drive by NATO, particularly its European members, to offset the Warsaw Pact numerical superiority. It was precisely the need to avoid having to do this that had made massive retaliation so popular in the first place.

It took until 1967 before the USA could persuade **European NATO** to adopt an alternative. In the famous NATO document MC 14/3, the then British Secretary of State for Defence, Denis Healey, was instrumental in selling what had originally been a British doctrine, **Graduated Deterrence,** under a new label. Graduated deterrence, now flexible response, merely means that NATO will meet a Warsaw Pact attack with a relevant level of force. It would not move directly to **Central Strategic Warfare,** unless, of course, that was how the USSR had initiated war, but nor would it be tied to operating only on the level at which the USSR had chosen to fight (and thus has much in common with **Escalation Control** and **Escalation Dominance).**

The trouble with MC 14/3 is that it remains a paper strategy, because the nuclear guarantee is still essential to it. No pretence has ever been made that a prolonged conventional war will be fought. It is not only that NATO does not, in fact, have war stocks to fight conventionally for more than a few weeks. This aspect at least could be remedied were Western European governments sufficiently motivated. It is more that Europeans do not view the prospect of a conventional war with modern weapons as an acceptable strategy at all. The original objection to flexible response was not just economic meanness. It was felt strongly that a reduction in the nuclear threat seriously increased the risk of war. Hence flexible response has never been intended to *remove* the threat of escalation: it is no more than a compromise between an American unwillingness to risk its cities to prevent a relatively minor invasion of Western Europe, and a European refusal to fight an updated version of the Second World War. All that flexible response means is that nuclear responses firstly may be less than central strategic, and secondly need not come straight away.

Follow on Forces Attack (FOFA)

Follow on Forces Attack has become the official NATO doctrine for fighting a **Central Front** war against the Warsaw Pact. It is largely a development of the **Deep Strike** tactic, and to a lesser extent the **Air-Land Battle Concept** devised by the US army and air force. FOFA is a response to the Soviet war plan of **Echeloned Attack,** and, like Deep Strike, is partly dependent on the development of **Emerging Technology** weapons. Because the Warsaw Pact has a considerable numerical superiority over NATO in troops, and even more so in tanks and other conventional weaponry, it might normally be expected to use its forces to hold the NATO line in most places and throw one or two highly-concentrated and overpowering armies at selected points, in order to smash its way through, according to the traditional principle of **Concentration of Force.** However, the danger of **Battlefield Nuclear Weapons** being used against such concentrations has led the USSR to develop an alternative approach, in which the full force of the Warsaw Pact will be arrayed in two or three lines; these will attack NATO lines in sequential waves all along the front. FOFA and the theories from which it derives is a plan to attack the second and subsequent echelons by air-power when they are still days away from the main land battle; should this succeed the NATO armies would only have to deal with the first echelon troops.

The doctrine was, somewhat unwillingly, accepted by the **European NATO** members in the mid-1980s. However, this acceptance may have little significance as the emerging technology needed to carry out FOFA will, if it ever appears, largely be owned by the USA. Whether in practice the West German or RAF ground attack squadrons, which may be desperately needed by the embattled armies, would ever be released for deep **Interdiction** raids is less than certain.

Force d'Action Rapide (FAR)

In the early 1980s French defence policy, under the socialist government of President Mitterrand, began to move away from the traditionally aloof **Gaullism** towards a potentially more pro-NATO orientation. One sign of this was the creation of an army force of about 40,000 troops organized for rapid intervention either in Europe or elsewhere. The Force d'Action Rapide, or rapid reaction force, consists in part of specialist troops, marines, paratroops, and mountain troops who were not otherwise organized into the main army corps structure. It also contains troops who already have an assignment in a table of organization but who could be quickly mobilized and transferred to the command of the FAR. Although, as yet, inadequately equipped, particularly in the helicopters needed for rapid troop transport, it is seen by France's allies as demonstrating a much greater willingness to come rapidly to the aid of NATO forces in an emergency. It also represents an increased interest by France in having a **Force Projection** capacity to intervene **Out of Theatre,** although French interventions in, for example, Chad, have not so far involved the FAR.

Force de Frappe

The Force de Frappe was the original name for the French **Strategic** nuclear force when it was first deployed in the shape of some 40 medium-range bombers, the Mirage IV, between 1964 and 1966. Since then the French have developed a full nuclear **Triad** of bombers, **Intermediate-Range Ballistic Missiles (IRBMs),** and ballistic missile submarines (see **SSBNs**). The overall strategic force is now more often described as the Force de Dissuasion, which can be translated very roughly as 'Deterrent Force'. The original Force de Frappe suffered from a general perception of lack of **Credibility,** because of the considerable problems which the bombers would face in penetrating Soviet airspace and the relatively small nuclear blow which the surviving aircraft would be able to deliver. Since the mid-1960s the strategic force has been developed considerably. The naval component now consists of five nuclear ballistic missile submarines (the French acronym is SNLE, for Sousmarin Nucléaire Lanceur d'Engins, equivalent to **SSBN,** and the missiles are known as MSBS, for Mer-Sol Balistique Stratégique, the equivalent of **SLBM**). The missiles are single warhead one-megaton models, but they are to be replaced with multiple warhead versions.

The land-based missile force consists of two squadrons, each of nine IRBMs, that is roughly equivalent in size to the MSBS, placed in silos on the Albion Plateau. The air force is now being re-equipped with **Stand-Off Bombs** like a modernized version of the old British **Blue Steel** missile, which will notably increase its penetration capacity.

As the French force has been expanded and updated, its credibility problems have declined, and when, by the end of the century, the full modernization programme has been carried out it will represent a very serious threat to the USSR. Indeed, the combination of the modernized French force and the British force, if the latter eventually acquires its four Trident II submarines, would easily reach the **Assured Destruction** parameters required by the traditional **McNamara** policy for the USA in the 1960s.

Force Multipliers

A force multiplier is any piece of technology which allows a smaller body of soldiers, or a smaller set of tanks, artillery, fighter aircraft or whatever to defeat a larger force of similar type. The concept is new, the phenomenon as old as warfare, particularly if innovations in tactical doctrine or military theory are also classed as force multipliers. The development of the phalanx formation by Greek spearmen, which allowed them to overcome much larger numbers of less well-coordinated enemies, is an early example. At the purely technological level a familiar historical example would be the ease with which small units of British redcoats, with bolt action rifles and Maxim machine-guns, could defeat masses of primitively-armed natives in imperial battles.

Force multipliers are currently much sought after by NATO as a way of offsetting the numerical superiority in troops, tanks and artillery enjoyed by the Warsaw Pact. Throughout NATO's history there has been such an imbalance but, at most times, Western forces have been able to rely on a technological edge which multiplied the effectiveness of its own forces and equipment. Since the mid-1970s this advantage from higher technological capacity has declined, and possibly vanished. Thus there is an urgent renewed search for force multiplying mechanisms, of which **Emerging Technology** is the best example. This search for technical advance leads to a particular form of **Arms Race,** because the development of a new **Weapons System** or more rapid communications network, by indicating the practical advantages of the relevant breakthrough encourages the opponent to make similar developments. An additional consequence is the considerable increases in industrial espionage and political attempts to prevent the export of technology which could have even a remote military utility to the USSR.

It is claimed by some, particularly members of the **Military Reform Movement** in the USA, that this concentration on higher and higher technology to offset numerical inferiority is self-defeating. The more advanced the weapons systems become, the less of them even the USA can afford to buy. Instead it is argued that the West should concentrate on building very much larger numbers of simpler and cheaper weapons, rather than relying on force multipliers.

Force Projection

Force projection is the military capacity to transport army and air force units to a distant spot and, if necessary, land them under fire in order to engage the local enemy. Such capacity is the classic requirement of a country which wishes to be able to exert control on a world-wide scale. Because of its specific nature it is not acquired automatically by a nation being powerful in general military terms. For example, the total military strength of Germany in the last quarter of the 19th century, with a huge well-trained army based on **Conscription,** was in some senses superior to that of Great Britain, with a smaller volunteer army. However, Britain had the capacity, because of its navy, to send troops anywhere, and was much more able to police a far-flung empire than Germany would have been. The whole concept of force projection is well summed up by the idea popular at the beginning of this century that the British army was a 'bullet to be fired by the Royal Navy'.

As the rivalry between the superpowers extends world-wide, and as both countries increasingly see dangers to their vital interests as being mainly outside the traditional European **Central Front,** there is growing concern with force projection capacity on both sides of the Iron Curtain. The USA which, like the UK has traditionally been a sea power rather than a land power, has, until recently, had a decided advantage over the Soviet Union in this area, despite the Warsaw Pact's overall numerical superiority, because the USSR has not had a force projection capacity.

Force projection capacity is not merely a matter of the size of naval or **Air-Lift** facilities, because special types of ships are required. The US advantage, which the Reagan administration planned to enhance, consists of having powerful aircraft carriers and a large specially-trained Marine Corps, equipped with amphibious warfare ships and specifically-designed armour and artillery for easy transportation and landing. Naval air power may be the key, above all else, to force projection, because of the need to fight at great distances from a home base against a local enemy with its entire military infrastructure close by. The military jargon here calls for an ability to 'insert' forces, to seize and hold a beachhead, possibly for some time, with light troops until reinforced by seaborne transport with the entire paraphernalia of modern war supplies. Clearly this is the sort of operation where even a very powerful country, such as the USA, can be at a serious disadvantage for some time against a generally much weaker, but locally well-integrated military structure. The advantages of local **Air Superiority,** and the difficulties of establishing an invasion force are so well known that the absence of supporting specialized naval assets can totally inhibit force projection. An excellent recent example is the great risk the UK took in reinvading the Falkland Islands when it lacked proper aircraft carrier forces. The British task-force came close to being defeated by the Argentinian air force operating from relatively near-by mainland bases.

The USA became sharply aware of its own weakness in force projection in the mid-1970s, and has made some effort to remedy this by the creation of a Rapid Deployment Force (known by its official title of Central Command, or **CENCOM**), by experimenting with **Light Divisions** and by planning much greater air-lift and **Sea-Lift** capacity. As US strategic thinking moves towards the idea of a **Global Strategy** and, indeed, as its strategy even for response to a **Central Front** war takes on aspects of **Horizontal Escalation,** these are clearly necessary. A major spur towards these developments has been the vast increase in the size of the Soviet fleet, turning it into a **Blue Water Navy** for the first time in over 70 years. There is, however, considerable debate as to the extent that this Soviet naval development involves force projection capacity, partly because it is unclear whether or not they are really building up conventional aircraft carrier forces. Nevertheless there is no doubt that the USSR is more concerned than it used to be about being a global power.

Force Ratios

Force ratios, simply the ratio of troops or equivalent pieces of equipment between two sides, occupy an important, if much debated, place in military theory. Since the publication of the **Lanchester Equations** by a British engineer during the First World War, a whole school of mathematical analysis of combat has developed. It has been greatly influenced by civilian operations research, in which force ratio calculations are also crucial. The subject generally is lacking in clarity because high-quality data is so scarce. Much military planning has, by default, had to be based on computer simulations of likely outcomes in conflict situations, and assumptions about the effect of force ratios in particular contexts are hotly debated in the professional literature.

Because data on the effectiveness of opposing **Weapons Systems** is even more scarce, force ratios are also the most commonly-used measure for the relative strength of the Warsaw Pact and NATO. Most of the evidence for the conventional superiority of the Warsaw Pact over NATO is in terms of such rather simple-minded ratios. For example, the ratio of tanks between the two superpowers is about 3:1 in favour of the Warsaw Pact, and the ratio of troop forces something like 1.3:1. The problem is that such numbers are virtually meaningless taken outside a specific context. The 3:1 tank ratio does, most certainly, mean that NATO would be defeated were it to try to invade Eastern Europe, but then, as NATO is expressly a defensive alliance, this should be irrelevant. A more appropriate example might be that traditional military wisdom calls for a 3:1 advantage as the minimum necessary for the success of an invading force, which might suggest that the Warsaw Pact is *not* conventionally superior to NATO to the point that it would be assured of victory. This argument is not meant to be a substantive judgment, but to indicate the danger of attaching too much importance to force ratios unless backed by the complex **Scenario**-dependent analyses of military operations research.

Formal Strategy

Formal strategy is a term used rather loosely to describe much of the nuclear strategic thought developed in the first 20 or so years after the development of the **Atomic Bomb.** To some extent it is synonymous with 'American strategic theory' and would embrace the work of thinkers as diverse as Thomas Schelling, with his semi-mathematical approach in books like *The Strategy of Conflict,* Herman Kahn with his detailed **Scenarios** for **Escalation** ladders in *On Escalation* and *On Thermo-Nuclear War,* and other more intuitive but still rigorous writers. The formal strategists were originally much influenced by the operations research and **Game Theory** approaches to solving strategic and tactical problems, and were the great creators of complicated scenarios. **Think-Tanks,** especially **RAND** with its systems analysis approach, were often the breeding ground of formal strategy. Their real achievement was in the perception of the automaticity of nuclear war, with its neat tables of megatonnage effects and seductively rational strategies that could be logically derived from concepts such as the **First Strike/Second** Strike distinction.

There was, and is, nothing wrong with formal strategy despite the sneers that it has sometimes attracted from proponents of other analyses. It is clearly better that very cold and logical thought be applied to something as terrible as nuclear war. The problems arise because, by its very nature, formal strategy has to assume utterly rational behaviour on the part of political and military decision-makers in pursuit of their tactical goals. Yet nothing is more likely to characterize the state of mind of US, British, French and Soviet leaders once nuclear weapons explode over their countries than sheer panic and terror. Furthermore, many of the core concepts of formal theory, for example what constitutes 'unacceptable damage' in **Deterrence** theory, are not in fact matters of clear scientific evidence but purely subjective speculations. What is needed is not the abandonment of formal strategy, which in many cases is, indeed, simply the formalization of age old strategic instincts, but an effort to build into the theories both uncertainty and, if possible, empirical evidence about the core psychological assumptions.

Fortress America

Fortress America is a phrase used to characterize the essence of the **Isolationism** present in US foreign policy debate. The idea is that the USA does not, and cannot, have anything to fear from invasion of its own territory, that it is militarily immune, and can therefore withdraw from any alliance or entanglement in external affairs. Nowadays the phrase is more likely to be used, pejoratively, by opponents of isolationism than by its proponents. Manifestly the advent of **Strategic** nuclear weapons has undermined the previous confidence that Americans could have about their immunity. However, in reality there has never been a period in this century when the isolationist position has been valid, because of the USA's dependence, as an exporting nation, on the preservation of its trading links and routes.

Forty Nations Conference (on Disarmament)

This conference, which is technically called the Conference on Disarmament (CD) must not be confused with the CDE. The full title of the CDE, the **Conference on Security and Confidence-Building Measures and Disarmament in Europe,** is often shortened to Conference on Disarmament in Europe. The CD meets in Geneva under the auspices of the United Nations, and consists of representatives from 40 nations, with a strong Third World contingent. It covers a wide range of **Disarmament** and **Arms Control** issues, and reports annually to the UN General Assembly. The overwhelmingly most important issue with which it deals is the attempt to get a world-wide ban on all **Chemical Weapons.**

Forward Based

Military units are 'forward based' (or forward deployed) when they are kept permanently in or near the area they are expected to have to fight in should a confrontation occur. Thus the **British Army of the Rhine** and the US Seventh Army are forward based, as units are kept permanently in West Germany, rather than in their home countries. Forward basing is sometimes politically controversial in that it is often believed that it costs the country providing the troops more to keep them forward based than if they were home based. From time to time, therefore, those wanting to cut defence budgets, yet unwilling to accept actual reduction in military preparedness, call for home basing rather than forward basing. This has, intermittently, been a particularly strong demand in the USA, and has also occurred in British politics.

The argument is less clear-cut than it may seem. Admittedly there can be extra costs associated with forward basing, although these tend to fluctuate with exchange rates. There are also enormous costs, in terms of battle readiness as well as finance, attached to home based units that have primarily an overseas combat role. The most obvious costs are those associated with the **Sea-Lift** and **Air-Lift** facilities needed to transfer units rapidly into a **Theatre of War.** The USA, in fact, has systematically failed to make adequate provision for this task. It keeps at least 10 of the army divisions which are committed to NATO in the continental USA, and does not have adequate lift capacity to get them to the **Central Front** in less than 10 days.

Failing to have forward based units also incurs additional training costs. Given the rapidity with which a major war in Germany is expected to develop, it is vital that troops be familiar in advance with the terrain and conditions of the potential battle area. If most units were home based, extensive training manoeuvres, with all the costs of transportation, would be required. Finally, in the short term, neither the USA nor the UK would save money by bringing their forces home, unless they were also disbanded, because both lack adequate barracks and training facilities in their own countries for the units currently kept abroad, which in both cases represent nearly one-third of their armies.

Forward Defence

Although forward defence has a general meaning in strategic discussion, its normal use nowadays is to refer to the specific defence plan advocated by NATO for a **Central Front** war. In this context forward defence means that NATO armies will attempt to stop any Warsaw Pact invasion of Germany at the border, and try to win the war there, without surrendering any territory. This runs contrary to many doctrines of war-fighting, which hold that fixed linear forward defences, where there is a commitment to stay in place along the whole front, are disastrous and yield far too much initiative to the enemy. The analogy most often drawn is to the static trench warfare of the First World War, although it must be said that forward defence, in this case, was enforced on the opposing armies by stalemate, and that generals on both sides tried desperately to create a war of manoeuvre.

Because NATO forces would almost certainly be outnumbered by troops from the Warsaw Pact, most military professionals would prefer a defence based on giving up ground to the enemy, letting them commit themselves to particular axes of attack, and defeating them by manoeuvring NATO troops onto the flanks of the attacking forces (see **Manoeuvre Warfare**). In addition to this tactical preference, NATO has a particular need not to be tied to a fixed forward defence. This is because much of the full strength of NATO depends on reinforcements from North America which could take about 10 days to arrive at the battlefield. NATO commanders would prefer to be able to fight a battle of slow tactical withdrawal, holding up the Warsaw Pact forces as much as possible, but conserving most of their strength until the reinforcements arrived, and then launch a major counter-attack against tired Warsaw Pact armies at the limit of their supply lines. This is described in military language as 'trading space for time', and is the standard response by unready armies caught by surprise. It was, for example, exactly how the Soviet army defeated Germany in the Second World War.

Unfortunately for NATO this sort of flexible defence is politically unacceptable to the West German government. Ever since the Federal Republic of Germany was admitted to NATO, in 1955, its governments have operated with the premise that there can be no voluntary withdrawal from, and no trading of, German territory to 'buy' time for the arrival of reinforcements. The entire effort must be aimed at preventing the Warsaw Pact taking *any* West German territory, and thus all troops must be committed from the beginning to holding the line as far east as is at all possible. While the doctrine is much attacked by non-German military analysts in NATO, it is quite understandable, and far less absurd than some of its critics, especially Americans in the **Military Reform Group,** present it as.

It is not just a matter of West German pride or patriotism. Even a conventional battle, with modern **Fire-Power,** is going to devastate the land it is fought over. Trading space for time means, in effect, turning the whole of West Germany into such a **Battlefield,** so that the economic and human costs of winning the war could end up, for West Germany, rather worse than the costs of losing it. Furthermore, advocates of retreating to await reinforcements tend to ignore the fact that there is very little ground to trade—at its widest West Germany is only about

480 kilometres from its eastern to western borders. Whatever the merits may be of manoeuvre warfare versus forward defence, for the West Germans there is no room to negotiate on the issue. As one German general has put it to an audience of American generals, 'forward defence is the price of the alliance'.

Forward Deployed (see Forward Based)

Forward Edge of the Battle Area (FEBA)

The traditional concept of the 'front line' in warfare, familiar from the First World War, has become outdated. **Interdiction** tactics mean that aircraft and very long-range artillery, multiple rocket launchers and even chemical and nuclear weapons will be targeted deep beyond an imaginary front line between ground forces. Indeed, because of the complex demands of **C³I** and the danger from the enemy's own interdiction forces, targets such as command bunkers, supply depots and air-bases behind the lines will be possibly more important than the traditional target of front-line troops. As a consequence the image of a linear front between two armies, with the war zone limited to a narrow strip on each side, is anachronistic. Hence the concept of a battle *area,* with the line where one's ground troops directly face the forces of the enemy being seen as the forward edge of a deep **Battlefield**. If the entire battle zone was looked at, FEBA would be in the middle. To the extent that such a 'front line' exists at all, and modern doctrine such as the US army's **Air-Land Battle Concept** plays down its likelihood, it will certainly no longer be the only area to see intense and decisive combat. (See also **Defence in Depth.**)

Fratricide

Fratricide is an untested assumption about what might happen to nuclear warheads fired at the same target. The theory is that the first nuclear explosions of a salvo aimed at the same or near-by targets will destroy or render useless any warheads in the immediate area that have not yet been detonated. Such a situation could occur, for example, if one superpower tried to destroy the **Silo**-based missiles of the other in a **First Strike**. Because silos are **Hard Targets,** very accurate hits by powerful warheads would be required. Given the likely **Circular Error Probables** involved, to ensure that at least one warhead detonated sufficiently close to the target it would be necessary to aim two or more at each silo. The fratricide thesis is that the explosion of the first warhead to arrive would render the subsequent ones impotent. Extending this thesis, if all the targets were close together it might be impossible to destroy more than one or two because of the fratricide effect on other missiles. Ideas like these lay behind the **Dense Pack** theory of how to base the new American **MX** missile (see **Basing Mode**). Much of this argument depends on totally unknown variables, such as the range of the fratricide effect and the consequence for nuclear weapons of the **Electromagnetic Pulse** from a nuclear detonation.

The concept can be used, slightly more generally, to refer to the danger that one's own nuclear explosions might damage other attacks one is making. An instance sometimes discussed is the effect on bombers making airborne nuclear attacks of having to fly through an atmosphere contaminated by earlier nuclear explosions.

Freeze Movement

The Freeze movement is the closest there is to a US counterpart of British **CND** (Campaign for Nuclear Disarmament) in terms of size, public support and political success. Even so, the aims of the two organizations are very different indeed. Freeze arose at the end of the 1970s as the American reflection of the anti-nuclear sentiment sweeping Western Europe in the wake of the **Euromissile** issue. It is much more an **Arms Control** movement than a nuclear abolition campaign. The title itself refers to the central plank of its objectives—that the USA and the USSR should immediately stop any further deployment of nuclear weapons, freezing their stocks at current levels. From that position arms control negotiations should progressively reduce the total of all nuclear weapons.

At one stage Freeze had a considerable degree of public legitimacy in the USA, with Democratic candidates for the presidency in 1980 and 1984 at least flirting with acceptance of its position. Indeed, a nuclear freeze motion was passed by the House of Representatives in the 1982 – 84 Congress, although it was not successful in the Senate. This would have been a unilateral move, freezing US deployment in the hope that the USSR would follow suit. Freeze does not, however, take the position that the USA should unilaterally *abandon* nuclear weapons, a position that, in the foreseeable future, could never command the support of even a substantial minority of the American electorate. An attempt was made to import Freeze into the UK as a moderate alternative to CND, but with little or no success.

Frigate

A frigate is the basic workhorse ship of most modern navies. Roughly the size of a Second World War **Destroyer,** it is used primarily for **Anti-Submarine Warfare (ASW),** but can carry out most general naval activities. In large fleets, such as the US navy, frigates have a more restricted role because of the number of ships in the much heavier cruiser and even battleship categories. However, for a navy like that of the UK, which is still the third largest in the world, the frigate is becoming virtually the only operating element in the surface fleet, except for a very small number of light aircraft carriers. The consequence is that the British navy is designed almost entirely as an ASW force, and would not be able to engage substantial elements of the Soviet navy in a sea battle.

Front End

The front end is the section of a **Ballistic** missile which remains after the propellant stages have been disengaged. It is this section which leaves the earth's atmosphere and continues the ballistic flight through the **Mid-Course Phase** and the **Terminal Phase.** The front end of a more advanced missile would consist of components such as **Warheads,** guidance computers and **Decoys.** The unit which houses all of these constituents is known as the **Bus**. Because the front end is a separate entity, otherwise identical missiles can have very different capacities and functions. The **Trident** missile that the UK is buying from the USA will have an entirely British-designed and -built front end, and so will be different from the American version. In the same way the **Polaris** missiles in the current generation of the British **SSBN** force are vastly different from those of the mid-1970s, although the missile parts themselves are the same; this is due to the fitting of the new **Chevaline** front end.

G

Gaither Report

The Gaither Report, the official title of which was 'Deterrence and Survival in the Nuclear Age', was an internal US government report written for President Eisenhower in 1957. It represented the first serious acceptance by the US government that the USSR had become a really serious nuclear rival to the USA. In particular it stressed that the development of Soviet **ICBMs** was such that by 1959 they would probably be able to destroy almost the whole of the **Strategic Air Command** on the ground. It strongly recommended building a much better early warning system, and accelerating America's own ballistic missile programme. Although an earlier report, the Killian Report of 1955, had warned of the possibilities of a Soviet surprise attack, it was the Gaither Report that presented the danger of what came to be known as the **Missile Gap** and spurred on the development of American nuclear weaponry. Some members of the group which wrote the report took the view that this Soviet capacity made the Eisenhower doctrine of **Massive Retaliation** impossible, but they were in a minority and the President stuck firmly to his policy.

Galosh

Galosh is the NATO code-name for the Soviet **Anti-Ballistic Missile (ABM)** defensive system. Deployed in the late 1960s, the Galosh system has never been thought to be very effective. It has never reached the full 100 interceptor missiles that would have been allowed under the **Anti-Ballistic Missile Treaty,** which formed part of the **SALT I** negotiations, and the 64 missiles the USSR originally deployed are believed to have been reduced to only 32. Galosh was designed primarily to defend Moscow against a limited nuclear attack. While it could never have been effective against an American strike, which could afford to 'waste' warheads to exhaust the system, it might have served against the level of force that France or the UK deployed at the beginning of the 1970s. It was always a very cumbersome system, involving nuclear warhead-carrying missiles with only a 300-km range, and there is some suggestion that the detonation of such a missile, even if it destroyed an incoming warhead might actually damage the rest of the Galosh system itself (see **Fratricide**).

It was because of the inefficiency of systems like these that the two superpowers were prepared to sign the ABM Treaty; the USA has dismantled the whole of its initial system, the Nike-Hercules batteries around Washington, DC, and the USSR has never developed Galosh fully. However, new technologies may now make an effective ABM shield, such as the US **Strategic Defense Initiative,** possible, and it is widely believed that the USSR is developing a much more sophisticated, non-nuclear defensive missile system with which to replace Galosh. Again, they may well find it worthwhile deploying a partially-effective system if it would seriously reduce the threat posed by the UK or France, even if it was still believed to be impossible to defend against a major US strike.

Game Theory

Game theory is the application of mathematical reasoning to problems of conflict and collaboration between rational self-interested parties. Developed in the 1940s by the Austrian mathematicians von Neumann and Morgenstern, it has been applied to dozens of problems in political science, strategic theory and even moral philosophy. In the 1950s it attracted much attention from strategic theorists and defence analysts, and still has some influence as a mode of analysing the interactions between powers in a potentially nuclear confrontation. The main point of game theory is that, given assumptions about parties' preferences (usually called 'utility schedules'), possession of information and psychological tendency to risk, it is possible to deduce how each will react to actions or possible actions of the other. The really crucial assumption is that each actor is entirely self-interested and completely rational.

Game theory is most often demonstrated at an elementary level through a paradigm known as the 'Prisoners' Dilemma', which is valued by teachers because it arrives at a solution which, intuitively, would not occur. It supposes that two men have been arrested on suspicion of having jointly robbed a bank. They are held in separate cells by the police, who do not have enough evidence to prosecute unless one of the suspects confesses and implicates the other. Both men know this, but cannot communicate with each other. If one asks an observer what will happen there is an obvious tendency to say that both men will be silent, and avoid punishment. A game theorist, rather, will ask what the 'pay-off matrix' looks like. The pay-off matrix is a table that shows the consequence to each suspect of the interaction of his own and his partner's independent decisions. Suppose that the sentence for burglary is 20 years in prison, but that the police will intercede with the court on behalf of one who confesses, and have the sentence cut to five years. The dilemma facing the prisoners is that the best outcome, not being convicted, is only available if he can trust his partner. So if burglar X decides to trust Y, but Y fears X may not be trustable, Y may confess, accepting five years as better than 20, while X gets the worst of all possible worlds, the full 20 years given to a criminal who does not co-operate with the police.

The text-book answer to the Prisoners' Dilemma is that both co-operate, to *minimize* the worst that can happen, rather than trying for the outcome which is maximum in their presumed utility schedule. This yields what is known as the **Minimax** strategy, taken by game theorists to be the most probable outcome in such a game interaction. Obviously much depends on the pay-off matrix—so that if the difference between the sentences for those who confess and those who do not is trivial, the probability of opting for a minimax strategy is very much less.

Despite early hopes, the use of game theory to strategic analysis is not great. However, it does serve to focus attention on the way that the determinants of decisions are the interaction pay-offs, and that any strategic decision must be taken with an eye to the opponent's most likely action. One reason that game theory is held to apply to nuclear strategy is that nuclear **Scenarios** often involve a quality present in the game theory example; the superpowers cannot wait to find out what the enemy actually has done before they make their own decision. Nuclear confrontation, therefore, may lack an **Action – Reaction** dynamic, and depend, as with the prisoners, on looking at a matrix of 'what will happen to me if . . .'. Nuclear scenarios can be set up in game theory terms quite easily, and they can be made to yield quite unexpected consequences. (See also **Formal Strategy.**)

Gaullism

Gaullism has come to mean much more than a description of a particular style of foreign and defence policy practised by General de Gaulle as President of France from 1959 – 69. It is now applied to any similar policies, whether practised by a French government or by the government of any other country. The essence of Gaullism can be demonstrated by two crucial elements of de Gaulle's own policies as President. One was the building of the **Force de Frappe,** and the other was the French departure from NATO's integrated military command structure, which led to the removal of all NATO bases from French soil. Other parts of de Gaulle's policy, such as the independent foreign policy towards the USSR and his blocking of British membership of the EEC, are also relevant to a study of Gaullism.

In all these policies his motive was to maximize the independent status of France, and to ensure that France's vital interests, especially in the security arena, were given priority and were not compromised by alliance obligations. His nuclear defence policy rested on the **Tous Azimuts** doctrine, and on his conviction that **Extended Deterrence** was meaningless. The removal of French troops from direct NATO control involved a rejection, technically illegal under the NATO treaties, of the obligation automatically to come to the aid of a NATO member under attack. Instead it was made clear that France would fight in Europe only to the extent, and only where and how, its perceptions of its own interests dictated. In strategic terms Gaullism is best summed up as the use of nuclear force for purposes of **Sanctuarization,** and the unashamed use of national power, whether military or economic, solely for the national self-interest.

Since the late 1970s, and especially since the Parti Socialiste's presidential and legislative election victories in 1981, France has shown some signs of moving away from this doctrine. Although French nuclear **Declaratory Policy** is never made very explicit, there have been serious hints that the French would be prepared to use their nuclear forces as **Extended Deterrence** over West Germany, as an alternative to the US **Nuclear Umbrella.** Other military moves, such as the creation of the **Force d'Action Rapide,** suggest a much closer integration of French troops with NATO command. In addition to this there has been a general *rapprochement* with the USA, which had always been the chief *bête noir* of de Gaulle's Western diplomacy.

The extension of the concept of Gaullism to other countries is sometimes seen, for example, in justifications for the UK enhancing its strategic nuclear force but reducing its ground troop commitment to NATO. Such a policy is described as Gaullist because of the retreat from alliance obligations to further the national interest.

General Staff

The idea of a general staff principally originates from the famous Prussian military reforms which transformed an inefficient army, which Napoleon had easily defeated, into the foremost military machine in Europe by the middle of the 19th century. The Prussian, later German, General Staff has been admired, though seldom fully imitated, by military thinkers in most countries. What it essentially consisted of was a relatively small group of very highly-trained and intellectually-able officers, drawn from the main officer corps relatively early in their careers, who then spent the rest of their professional lives in a central unit. This unit, the General Staff, had the full-time job of planning for all the possible wars that Germany might face, in the most minute detail. They were not combat commanders, and saw little active service with the troops; their professional loyalties, rather than being to their own regiments or branches of the services, were to the General Staff, and therefore to the armed forces, or even the nation, as a whole. Because they worked and planned together on a permanent basis they were able, once war broke out, to communicate easily and with trust and understanding, thus avoiding the confusion, rivalries and disagreements endemic to higher command in the armed forces of most nations. (See **Inter-Service Rivalry.**)

In the First and Second World Wars the ability of the German military machine to co-ordinate its efforts and to put its plans into effect was in marked contrast to that of the allied armies opposing them. Although the Soviet Union has adopted something like the general staff conception, the German model has had few other imitators. In most of NATO, for example, staff posts are held for a couple of years at a time by officers who rotate through a variety of military occupations. Therefore real expertise and, even more crucially, the communal loyalty of a general staff, never develops. For this reason many members of the US **Military Reform Movement** concerned, in the mid-1980s, at the inability of the **Chiefs of Staff** properly to control the US military, advocate the creation of a full-time general staff. For a variety of reasons, some having to do with the individual services wanting to preserve their autonomy, and some relating to fears of loss of civilian control to so powerful a body, the idea of a full general staff has never been acceptable to either the USA or the UK.

Geneva Conventions

The Geneva Conventions are a series of treaties signed in Geneva at various times between 1864 and 1949. The agreements were attempts to mitigate the horrors of war both to members of the military and to civilians. There are two main elements to the conventions. The first is to set standards for treating wounded enemies more or less as well as one treats one's own. The other major element, and probably the better known, sets standards for the treatment of prisoners of war. The various early treaties were largely flouted, especially in the two World Wars, leading to a new round of talks and the signing, in 1949, of four conventions developing the principles of the earlier treaties. These deal with ameliorating the condition of wounded forces in the field and at sea, with the treatment of prisoners of war, and with the protection of civilians in time of war. A majority of states have now ratified these conventions. Two protocols were added in 1977, one of which dealt with the rights of civilians and combatants embroiled in **Guerrilla Warfare.**

Global Strategy

Global strategy, which is similar to the concept of **Maritime Strategy,** is often proposed by some schools of strategic thinkers in the USA. Essentially it is the claim that the USA has world-wide vital interests, and should ensure that it has the capacity to protect them all. While this may seem both obvious and uncontroversial, it carries two major implications.

The more general policy implication is that Globalists are, in fact, opposed to the prevailing **Atlanticist** strategy in which the US commitment to NATO is paramount. A global strategy would rank Western Europe as no more important to the USA than several other theatres, in particular that known as the Pacific Basin, and consequently would involve a considerable reduction in resources **Forward Based** in Europe. Globalism is almost a form of armed **Isolationism,** in as much as it treats the USA as a party with no permanent ties or unconditional obligations, but unlike traditional isolationism it accepts that military intervention can be in America's own interest.

A second implication is on the technical and procurement aspects of defence policy. The types of military capacity required for a global intervention strategy are very different from those needed for the forward deployment of heavy divisions in Europe. As with other movements in US defence thinking, globalism calls for extensive **Sea-Lift** and **Air-Lift** facilities, for **Light Divisions,** and for the sort of rapid reaction forces envisaged under the **CENCOM** plan. Although globalism concurs with the general positions of the **Military Reform Movement,** the two are not necessarily synonymous.

Graduated Deterrence

Graduated deterrence was an early version of **Flexible Response** developed by British strategic thinkers during the 1950s. It arose through unhappiness with the possible dangers of the Eisenhower doctrine of **Massive Retaliation.** It admitted openly that the threat to react to *any* Soviet incursion into NATO territory by an all out **Strategic** nuclear attack on the USSR lacked **Credibility.** The theory had various expositions, and perhaps never really added up to much more than a sobering caution about the utility of nuclear weapons. At its simplest it advocated the use of **Battlefield Nuclear Weapons** for limited military purposes where the USSR had not itself made a nuclear attack. Stronger versions insisted that a nuclear response could only ever credibly be threatened against a nuclear attack, although a **No First Use** statement was never in itself part of graduated deterrence.

Some of the political leaders prominent in the development of graduated deterrence theory, notably the British Labour government's Secretary of State for Defence, Denis Healey, were later influential in helping the US administration to convince European members of NATO of the virtues of the flexible response doctrine. Although the Kennedy administration had come to power in 1961 determined to move away from the strategy of massive retaliation, it was not until 1967 that these ideas gained acceptance in Western Europe. No advocate of graduated deterrence was successful, however, in persuading a British government to make the conventional arms improvements that might have made massive retaliation less necessary as NATO's response. They were also rejected in Western Europe, on the grounds that they would make war more likely by weakening the threat of **Assured Destruction**—exactly the same reason why there was such initial opposition to flexible response.

Gravity Bomb

A gravity bomb is not, although the jargon may make it sound otherwise, a special new high-technology weapon. On the contrary it is a term used simply to mean a good old-fashioned sort of bomb that is just dropped from an aircraft. In America one might also come across the term 'iron bomb', with a similar meaning. The point is that, because nuclear weapons are now predominantly carried as **Warheads** on missiles, **Stand-Off Bombs** or other advanced **Delivery Vehicles,** there is a tendency to forget the large number of nuclear 'warheads' which are not really warheads at all, but straightforward 'bombs'. The entire British independent tactical nuclear armoury, for example, consists of gravity bombs to be dropped by fighter-bombers, such as the Tornado. Much of the French independent tactical force is held in a similar form. Reliance on gravity bombs is becoming increasingly dangerous as the sophistication of anti-aircraft defences, even at the **Infantry** company level, makes close approach to a target very much harder than in the past. It was largely for this reason that NATO initially planned to modernize its **Theatre Nuclear Forces,** which once depended almost exclusively on gravity bombs delivered by American F-111 fighter-bombers.

Greenland – Iceland – UK Gap (GIUK)

The Greenland – Iceland – UK gap refers to the only access the Soviet navy has from the western USSR into the Atlantic Ocean. Unless ships and submarines coming out of Soviet northern ports go through the Denmark Strait between Greenland and Iceland, or through the Norwegian Sea between Iceland and Scotland, they would have to pass through the English Channel. Clearly the last of these options, narrow and shallow, would be an enormously dangerous route, within very short range of NATO air and sea bases. Consequently much effort is put into surveillance and patrolling of these two maritime 'gaps' in what is otherwise continuous NATO-controlled land. Effectively, if NATO navies can 'stop the gap', Soviet naval power will be contained. **SSBNs** (nuclear missile submarines), **Hunter-Killer Submarines** for attacking NATO convoys, and surface units could not be deployed, and NATO's **Sea Lines of Communication** would be safe. Probably the most vital part of NATO's **Anti-Submarine Warfare** effort is in this area, and the maritime and underwater surveillance systems (see **SOSUS**), even in peacetime, operate at full capacity. The need for

Soviet shipping to pass through these relatively narrow gaps makes it very easy for NATO to track the deployment of Soviet sea power, thus making any surprise attack by the Soviet navy extremely difficult to achieve. The vulnerability imposed on the USSR by the GIUK gap is a very good example of the way in which the USSR, compared with the USA, suffers from a geopolitical disadvantage.

Ground Burst

A ground burst, in contrast to an **Air Burst,** is where a nuclear warhead detonates only at, or near, ground level. (More technically it is where the height above ground of detonation is less than the radius of the ensuing fireball.) Ground burst detonations are needed when the purpose is to create an exceptionally heavy shock wave over a relatively small area. Typically this would be where the user was trying to destroy a **Hard Target,** such as a missile **Silo** or deeply-buried command and control bunker (see **C³I**). Targets like this would require a blast overpressure (that is a pressure over and above the normal pressure of the atmosphere) of perhaps 2,000 pounds per square inch (psi). This could only be obtained by a combination of a high degree of accuracy, a ground burst or 'contact detonation' and a very large explosive **Yield.** In contrast, ordinary unreinforced concrete or brick buildings would be vulnerable to a blast wave of between 5 and 10 psi, and humans in the open would probably be killed at 5 psi. As the area covered by any overpressure from a single bomb is larger the greater the height of the explosion, air bursts would be more suitable when the aim was to create extensive damage to non-hardened targets. An accompanying hazard to ground bursts is that they create an enormous amount of **Fall-Out** as the dust blown into the atmosphere by the detonation is made radioactive (see **Radiation**) and carried into the upper atmosphere to be distributed over a very large area by wind and rainfall patterns.

Ground Zero (see Designated Ground Zero)

Ground-Launched Cruise Missile (GLCM)

Ground-launched cruise missiles (the acronym, GLCM, is pronounced 'glickum') are politically the most controversial, but militarily the least important of the general family of **Cruise Missiles.** Although they could be equipped with conventional warheads, as **Sea-Launched Cruise Missiles** mainly are, the GLCMs in operation with NATO forces are used entirely in a **Theatre Nuclear Force** role. They are intermediate-range weapons capable of travelling several thousands of miles, but are very slow compared with **Ballistic** missiles and could be vulnerable to **Surface-to-Air Missiles** and other defences. The advantage of GLCMs is that they are highly mobile, being fired from self-powered launch vehicles, and can thus be driven around the country and hidden relatively easily. The US-owned GLCMs deployed in Europe are at the centre of the **Euromissile** debate and, in mid-1987, it seemed likely that they would be removed as a result of progress in **Arms Control** negotiations. (See also **Double Zero Option, Intermediate Nuclear Forces** and **Zero Option.**)

Group of Soviet Forces, Germany (GSFG)

The Group of Soviet Forces, Germany is the main **Forward Based** and battle-ready force which the USSR ranges against NATO. It is as big as the whole of the NATO forward deployed armies put together. As it is expected that only a few East German divisions would be of much importance to the Warsaw Pact alongside the Soviet army in a **Central Front** war, this should not be taken to imply quite the conventional force disparity which some Western analysts claim exists. Nevertheless, the GSFG is very powerful. It consists of five separate armies, totalling 20 divisions, very heavily equipped with tanks and artillery. In addition the GSFG has attached to it an entire **Tactical Air Force** as big as the combined NATO air forces when fully deployed. It is the GSFG which would form the bulk of any surprise attack that the USSR might decide to make against NATO, although in the event of a (more likely) fully-mobilized attack it would be reinforced by divisions from the western USSR as well as from elsewhere in Eastern Europe.

Guerrilla Warfare

Guerrilla warfare, under one title or another, has been practised for as long as its distant relative, warfare conducted by formally organized and disciplined armies. Relatively weak or poor countries, rebel groups within a national society, or those formally defeated on the main battlefield cannot continue an armed struggle against an organized and rich country by fielding orthodox army units and fighting pitched battles. However, the organization and **Morale** of formal armies can be damaged, perhaps destroyed, by using tactics such as a series of hit-and-run raids or by infiltration of their lines. By never forming into large units and allowing themselves to be trapped into fighting pitched battles the guerrillas are able to avoid the damage that massed **Fire-Power** and superior numbers of a regular army would inflict. Very large armies can have their effective personnel substantially reduced by having to garrison hundreds of villages and towns, by having to send guard detachments with every supply convoy, and generally having to operate in an inconvenient environment. Classic examples of guerrilla warfare include the Spanish partisans fighting the Napoleonic armies in the Peninsular War, and the Franc-Tireurs who kept up the fight against the Prussian armies after the latter's victory in the 1870 Franco – Prussian War. Major later examples include the French resistance movement during the Second World War and, above all, the Vietcong during America's war in Vietnam.

It is widely believed that a drawn out guerrilla war will defeat an orthodox army, but history produces little evidence for this. The Napoleonic armies were only defeated by Wellington's British army, and the French resistance was really of use only as a support to the 1944 Allied invasion. The US army was almost invariably successful in actual operations against the Vietcong, which was almost completely destroyed in the 1968 Tet Offensive; thereafter it was the regular army of North Vietnam which the Americans were fighting, an army that had already shown its worth in the entirely orthodox campaign against the French in 1954. The reason for the apparent success of guerrilla warfare is that it is usually practised in countries where a foreign 'army of occupation' has no support among the indigenous population. Although there is generally no significant military success, the costs of continuing to garrison a foreign country and maintain a presence by force is often too much for the *political* will of the occupying power, which may decide to withdraw, or accommodate the guerrilla leaders, although militarily undefeated.

Gun-Type Bomb

A gun-type bomb is one similar to the first **Atomic Bomb** dropped by the USA over Hiroshima. One of two basic alternative designs for such a weapon, it is generally less efficient than the alternative, which is a spherical bomb like the one, nicknamed 'Fat Man', which was used on Nagasaki. The gun-type gets its name because it involves two separate sections of uranium at either end of a long metal tube. The larger piece has a hollow core, and the nuclear reaction is caused by shooting the smaller piece, engineered to fit rapidly into this core, propelled by an explosion from the far end of the bomb casing. The action is like firing a bullet into a target. The design is the simplest way of keeping the two pieces, both of sub-critical mass, apart until detonation, and then bringing them together so rapidly that an explosive chain reaction starts, rather than the heat of a slower reaction just melting them.

The alternative design, known as an implosion bomb, solves the problem of keeping the radioactive matter sufficiently dispersed to avoid the critical mass point by shaping it into a thin hollow sphere. The sphere is surrounded by shaped explosive charges which, when detonated, force the sphere (made of plutonium in the Nagasaki instance) to implode rapidly, concentrating the radioactive metal into a dense and critical mass. (See also **Nuclear Fisson** and **Nuclear Fusion.**)

H

Hard Target

A hard target, more properly a 'hardened' target, is one that is heavily protected against a nuclear blast. Typical hard targets are missile **Silos** and command and control bunkers (see **C^3I**). Targets like these are buried deep underground, and have walls built of steel-reinforced concrete several feet thick. Such objects would be destroyed only by enormous blast power, requiring a combination of extreme accuracy and very high **Yield** from nuclear warheads in **Ground Burst** detonations. As a rough guide, an ordinary brick or concrete building above ground would be seriously damaged by a blast with an 'overpressure' of between five and 10 pounds per square inch—that is, a five – 10 psi pressure over and above the normal atmospheric pressure. The command and control bunkers built by NATO in West Germany are reported to be safe up to blast overpressures of 2,000 psi.

The destruction of hard targets would be vital in a nuclear war, but would require very much more effort, in warheads committed, than ordinary industrial or even most military targets. The terminology is sometimes also applied in conventional warfare, so that a 'hard target' would be, for example, an armoured vehicle immune to an ordinary high explosive shell, in contrast to a soft or 'thin skin' target such as a truck or a human body.

Heavy Division (see Light Division)

Helsinki Conference (see Conference on Security and Co-operation in Europe)

High Level Group (HLG)

The High Level Group is a body of military and civilian officials in NATO which is subordinate to the **Nuclear Planning Group (NPG).** The NPG, consisting as it does of national defence ministers, cannot itself carry out the detailed policy planning, and depends on the HLG to formulate proposals which it usually ratifies. The HLG was, for example, highly influential in designing the policy which led to the deployment of the **Euromissiles,** and also to the **Montebello Decision** of 1983.

Horizontal Escalation

Horizontal escalation is a strategic option that has come to be talked of increasingly in the USA since the military build-up under the Reagan administration, and is particularly advocated by the US navy and other supporters of a **Global Strategy.** In a sense it is a logical development from **Flexible Response** and other thinking on how to respond to a conventional Warsaw Pact attack without having to 'go nuclear'.

Horizontal escalation is actually a very simple idea. If the Warsaw Pact attacks in one area, usually assumed to be the **Central Front,** flexible response, the current NATO doctrine, requires a matching response, or possibly an escalated response, in that area. But to limit retaliation to this would be to break an age old military axiom, that the enemy should not be allowed to take the initiative. Consequently the advocates of flexible response argue that NATO, or at least the USA, should be prepared to attack the USSR elsewhere in the world, as well as matching the original aggression. Thus, for example, US troops might advance suddenly into North Korea, bombing raids might be launched against the Asian Republics of the USSR, or Cuba might be invaded. The point is to make it clear to the USSR that they will not be allowed to choose a battleground safely away from their homeland or other vital interests. Such a strategy necessarily involves **Force Projection** capacity and also justifies the huge 600-ship navy planned by the Reagan administration. Rightly or wrongly, the implication of horizontal escalation is that any NATO – Warsaw Pact conflict would rapidly develop into global war.

Hot Line

The so-called 'hot line', which provides an avenue of direct communication between the US President and the Chairman of the Politburo of the USSR, was installed in the aftermath of the **Cuban Missile Crisis.** It is not, as is generally perceived by the public, a telephone, but a telex machine—in part because of the problem of potential ambiguity in verbal communications. The practice, which is in fact common in all high-level negotiation, is for each country to send its messages in its own language. The theory is that while a crucial translation mistake may be made in reading a message, at least it can be guaranteed that the original message was actually sent as intended. The hot line has never, as far as it is known, been used in a real crisis situation. Where these have occurred, the most well documented being the 1973 Arab – Israeli war, strategic communication has been handled obliquely. So in 1973, for example, President Nixon chose to indicate his determination to prevent Soviet intervention not by sending a direct message, but by putting all US forces, including nuclear forces, to a high state of alert.

Hunter-Killer Submarine

The modern generation of submarines comes in two broad categories: those intended to carry just ballistic missiles (the **SSBNs**), and those intended to destroy ships and other submarines. These latter submarines are often described as 'hunter-killers', although the official naval description is likely to be more innocuous, as, for example, in the British Royal Navy where they are described simply as 'fleet submarines'.

There are two principal roles for fleet submarines: to find the enemy's submarines and destroy them, hence the hunter-killer title, and to prey on the enemy's surface shipping, naval or mercantile. During the whole of the Second World War there was only one occasion when a submarine was attacked successfully by another submarine, and even then both were on the surface. The 'hunter-killer' submarine has been made possible by two major technical advances. Firstly, improved **Sonar** devices have enabled submarines to detect other submarines while both are submerged. Secondly, torpedoes have acquired many of the characteristics of **Emerging Technology** weaponry, thus making them truly 'smart' weapons (see **Smart Bombs**).

Fleet submarines are produced for navies in both nuclear and traditional diesel-powered models, while SSBNs are invariably nuclear-powered. The advantage of silence, crucial for remaining undetected while searching for an enemy submarine, is to be had with diesel power, but the range and endurance needed for SSBN patrols require nuclear power. In most navies there is a continuing debate between the relative benefits of diesel and nuclear power for the hunter-killer fleet, although the US navy has committed itself to nuclear power for all submarines.

Hydrogen Bomb

The hydrogen bomb, now more often referred to as a **Thermonuclear** bomb, was the second, and vastly more powerful, major development in the field of nuclear weapons. It relies on **Nuclear Fusion** whereas the **Atomic Bomb** depends on the weaker, but technically much simpler, **Nuclear Fission** process. There is a relatively low limit to the power that can be generated by a fission bomb because the technique involves the rapid bringing together of two pieces of radioactive material (usually uranium, but alternatively plutonium). Separately the two radioactive constituents are just sub-critical, but together they constitute a critical (that is, unstable) mass. Clearly the upper limits are set both by the size a piece of material can be and remain sub-critical, and the distance apart they can feasibly be kept in a portable device such as a bomb (see **Gun-Type Bomb**). A hydrogen bomb suffers from no such inherent limits, and theoretically could be made to yield as much energy as required. Certainly hydrogen bombs of above 50 **Megatons** have been tested, although almost certainly never produced for use, by the USSR.

There was considerable debate in the American nuclear weapons fraternity about whether the 'superbomb' (as it used to be called journalistically) should be built at all. Apart from moral concerns, it was thought by many that there was no military use for it, and that effort and scarce resources should go to the production of low-**Yield** atomic weapons for tactical use. The USSR's detonation of its first atomic bomb in 1949, years before it was expected to have this capacity, tilted the balance inside the US defence establishment, and President Truman authorized the development of thermonuclear weapons. It was possible to move rapidly to the development of the hydrogen bomb because much of the theoretical work had already been done during the **Manhattan Project** on atomic weapons. By 1953, when the USSR announced that it had tested an H-bomb, the USA already had a stockpile of such weapons at various yields. By 1957, at the latest, Britain had also successfully tested a thermonuclear device with a yield of at least one megaton.

I

ICBM

Intercontinental ballistic missiles are the very long-range missiles capable of hitting targets over ranges of more than 16,000 kilometres; indeed, Soviet and American ICBMs can reach any target in the opposing country from bases in their own heartlands. They were first developed after the Second World War, largely based on the technological breakthroughs of German scientists who had built the V2 rocket with which British cities were bombarded in 1944. (Both the USSR and USA competed to find and recruit these scientists and their plans.)

The first **Ballistic** missiles developed were only **Intermediate-Range Ballistic Missiles,** such as the US **Thor** missile that had to be based in Britain and Turkey to reach targets in the USSR. However, by the end of the 1950s both superpowers had started to deploy true ICBMs. The first generation, the American Titan being an example, had liquid fuel motors like the huge rockets used to launch satellites which were, indeed, developed from them. Despite being based in concrete **Silos,** they were potentially vulnerable to **First Strike** attack because of the time it took to fuel them. (Many of the Soviet missiles, even recent models such as the **SS-18,** and the SS-17 and -19, are still liquid fuelled, although the general design is less vulnerable.)

The first generation of ICBMs had single warheads, often with huge megatonnage (the Titan II was reputed to have a nine-**Megaton** warhead), and were highly inaccurate, with **Circular Error Probables (CEPs)** as high as 1 kilometre, and could be used only against large urban targets. Despite this relative primitiveness, Titan II was only due to be taken out of service during 1987. The next generation, represented by the US **Minuteman** class first deployed during the mid-1960s, are more accurate, carry **Multiple Independently-Targeted Re-entry Vehicles (MIRVs)** and, at least US models, are now entirely solid fuelled. Thus the US forces are typically capable of delivering three warheads in the 150 – 350-kiloton range, with a CEP of about 250 metres. Because they are solid fuelled they can be launched with very little warning.

America has produced relatively few versions of ICBM, and the third generation, the **MX** missile, was only being deployed during the late 1980s. The USSR, which keeps a far higher proportion of its warheads on ICBMs, rather than relying as the USA does on **SLBMs,** has tended to experiment more. In 1985, for example, there were at least six different types of Soviet ICBMs deployed, with two new models, the SS-24 and -25, being tested. On the whole Soviet missiles are heavier, with more and much larger warheads than the US models. This has led to US fears of a Soviet drive for **Strategic Superiority** and the so-called **Window of Vulnerability.** The USA, meanwhile, has led the way in producing MIRV capacity, and has built larger and more-accurate SLBMs.

It is largely a consequence of geography that no other country has fully-developed intercontinental missiles, as none faces quite the same range problems. The French SSBS-3, with a range of 3,500 km, can effectively be treated as an ICBM for their purposes. However, the French land-based missiles (they have only 18) are regarded as extremely vulnerable and the main dependence is on the missile submarine fleet (see **Force de Frappe**). China almost certainly has missiles with ranges of about 6,500 km. The only member of the **Nuclear Club** never to have deployed long-range land-based missiles of its own is the UK, which believes them to be too vulnerable for its purposes. Britain's **Independent Deterrent** force depends solely on one leg of the **Triad,** the SLBM.

The days of relying on huge static intercontinental-range missiles may be limited, because France, as well as the USSR and USA, is seriously planning for smaller, more mobile missiles, possibly with only one warhead each, which would be less vulnerable to an enemy first strike (see **Midgetman**). Nevertheless, short of major **Arms Control** progress, the large inventories of ICBMs, nearly 2,500 between the USA and USSR, will continue to be deployed for the foreseeable future.

Implosion Bomb (see Gun-Type Bomb)

Independent Deterrent

'Independent deterrent' is a phrase used to describe the British, and sometimes the French, strategic nuclear forces. All five publicly-acknowledged members of the **Nuclear Club** actually have independent nuclear forces, but this point does not need stressing with the USA, the USSR or China, as their forces are self-designed and built entirely from their own resources. Using the same criteria it is also unusual to specify that the French force is independent.

The reason why the idea of an independent deterrent is important in British politics is that, at several times in post-war history, the UK has had to depend on the USA in order to remain a nuclear power. Although Britain was involved in the design of the first **Atomic Bombs,** it was not a foregone conclusion that it would ever construct its own nuclear force, and the decision so to do was taken, by the 1945 – 50 Labour government, without fully realizing the technological and financial challenge involved. For most of its history the British nuclear deterrent has been politically controversial, with the Labour Party both at the beginning of the 1960s and in the mid-1980s being formally opposed to its retention. Some believe that one of the reasons why the British nuclear force has never been completely acceptable across the political spectrum, unlike the French **Force de Frappe,** is precisely the doubts concerning its real independence.

The first generation of the British deterrent force, the **V-Bomber** squadrons armed with **Gravity Bombs,** was more or less entirely a British development, although there was some engineering advice from the USA. The second generation, the **Polaris** submarines, however, were armed with missiles bought from the USA, because Britain had cancelled its own **Blue Streak** land-based missile project and lacked the technological base for a submarine missile programme. This dependence on the USA to provide, test, and supply spares for the British **SSBN** force led to claims that it was not truly independent. These claims were further strengthened by the fact that part of the USA's terms for selling Polaris was that the UK formally commit the force to NATO. Nevertheless, the UK government retains the right to withdraw them and use them for its own purposes in extreme cases of national interest.

The third generation system, the **Trident** submarines, if they are deployed by the Royal Navy, will also suffer from this dependence on the USA for missiles and their upkeep. Therefore the charge has again been raised that the British independent deterrent is 'neither British, nor independent'. The debate is entirely a political one, with little real substance. The USA could not in practice restrict British use, and the NATO commitment may well be essentially nominal. It has come to be standard practice for British politicians to stress the NATO link in any case, as a way of justifying retaining a nuclear capacity which might otherwise be attacked even more widely.

Independent European Programme Group (IEPG)

The IEPG is an offshoot of NATO which exists for the purpose of defining areas of weapons procurement where there is a common Western European interest in co-operative production. It is argued that competition between the several national armaments industries makes each of them inefficient and too vulnerable to competition from the huge US arms producers. If the procurement plans of several Western European countries can be coordinated, the resulting purchase orders should be big enough to ensure more economical production through the increased volume of turnover. In addition, the pooling of research and development capacities is also more likely to be able to rival the Americans.

There have been real successes for the IEPG. An indication of its increased political weight has been the holding of meetings at the secretary of defence level, whereas the usual representation is of the procurement directors from the member states. However, the IEPG faces two particular problems. Each European country's military has its own definitions of the ideal weapon, and they naturally dislike the thought of getting an aircraft, tank or **Frigate** which represents a committee compromise rather than a product designed with a specific task or tasks in mind. Secondly, several members, notably France, West Germany and the UK, have their own economic interests in keeping the whole range of national military technologies viable. They do not wish to become specialists in only some areas both because they hope for international arms sales and because, once lost, a technological capacity cannot easily be built up again. This could render the country in question vulnerable if it were ever put in a position where it needed to build a **Weapons System** entirely by itself.

Infantry

The infantry, the lightly-armed foot soldiers, are one of the three traditional sections of an army, along with the cavalry and the artillery. Increasingly the distinction is less clear than it used to be, with the tendency to put infantry soldiers into **Armoured Fighting Vehicles** so that not only do they no longer have to walk everywhere, but also many do not even dismount from their transports to fight. Most military thinkers would still believe, however, that infantry retained a vital and distinct role, which is often described as 'holding the ground' once it has been taken, even if armour and air-power is the predominant way of driving enemy forces away from the battlefield in the first place. Traditional infantry forces, with little or no mechanized transport and no armoured protection, have largely vanished from modern armies, except in special roles such as **Airborne** troops.

Infrastructure

Infrastructure refers to the physical support facilities needed to protect, supply and operate military forces, especially high-technology units such as **Tactical Air Forces.** A military organization, for example NATO, with large **Forward Based** strength deployed in West Germany from the forces of other countries, and even larger reinforcement needs during mobilization, is necessarily deeply concerned with infrastructure provisions. Air-bases have to be provided with hardened shelters for aircraft, and **C³I** bunkers must be prepared, regardless of any absence of immediate likelihood of their being used. Supply depots and, in the case of the USA, the special storage requirements for their forward-positioned **POMCUS** equipment, all add to the need.

The problem is that while infrastructure is vital, and could even mean the difference between defeat and victory, it is an 'unglamorous' defence expenditure. Neither the active duty military nor the civilian voters are likely to show much enthusiasm for its acquisition. An officer has less chance of being promoted for supervising the building of an empty air-base than for commanding a fighter squadron, and both politicians and electorate find it easier to vote in favour of increased money for tanks and parades than for pouring concrete. One of the most important recent developments in NATO has been the success of the **Long-Term Defence Programme** since 1977 in improving infrastructure provisions in Europe.

Institute for Defense Analyses (IDA)

The Institute for Defense Analyses is one of the less well known of the American defence **Think-Tanks,** the most famous of which is **RAND.** Although it is seldom heard of, even in the USA, the IDA may well be among the most influential think-tanks. It draws its importance from being run, unlike the others which belong to individual services (RAND is, for example, air force property), largely for the **Office of the Secretary of Defense.** As such it helps the central political administration of the **Department of Defense** in its permanent struggle to control the planning and procurement of the individual services.

Intercontinental Ballistic Missiles (see ICBM)

Interdiction

Interdiction means the use of military force of some kind to prevent transportation of supplies, equipment and troops past a particular point or along some route. Typically, in modern warfare, it will refer to the use of air-power to destroy bridges, major railway junctions or other **Choke Points** well inside enemy territory, thus preventing not only supplies, but also reinforcements from reaching the battle area. During the Vietnam War, for example, a major role of US air-power was to bomb the routes, especially the famous Ho Chi Minh trail, hoping to starve the Vietcong guerrillas inside South Vietnam of all their supplies. Interdiction is a vital element of US and NATO plans for a **Central Front** war, and is enshrined in official doctrines such as **Deep Strike** and **Follow On Forces Attack.** Because of the **Soviet Doctrine** of attacking with several lines of armies, several days apart, it is hoped to use interdiction to prevent the scheduled arrival of these **Echeloned Attacks.** As **Emerging Technology** weaponry seems to offer the hope of highly-accurate conventional attacks hundreds of kilometres inside enemy territory, interdiction could be a decisive tactic in any future war between the superpowers.

Intermediate Nuclear Forces (INF)

Intermediate Nuclear Forces, also called **Theatre Nuclear Forces (TNF),** have been defined for the purposes of **Arms Control** negotiations taking place in the mid-1980s as those missiles with ranges between 1,000 and 5,500 kilometres. A further refinement to the definition can see them referred to as Long-Range Intermediate Nuclear Forces (LRINF), to distinguish them from a category of Short-Range Intermediate Nuclear Forces (SRINF) with ranges of 500 – 1,000 km, which would previously have been thought of as straightforward INFs. This is a first class example of the flexibility of weapons categories which allows them to follow the development of arms control negotiations. INFs (and, indeed, LRINFs and SRINFs) are intended for use on primarily military targets during the early stages of nuclear **Escalation** in a NATO versus Warsaw Pact war.

It was the modernization of INF capacity by both the USSR and the USA from the late 1970s onwards, with the deployment first of the Soviet **SS-20,** and then the US **Pershing II** and **Ground-Launched Cruise Missile** forces, which led to the **Euromissile** crisis. The **Zero Option** proposal of President Reagan, which failed to prevent the deployment of the new generation of INFs from 1981 onwards, was revived after the Reykjavik Summit of 1986, when renewed energy was put into the achievement of an INF agreement. Soviet offers to remove their missiles in the 500 – 1,000 km range (see **SS-22 and SS-23),** in what became known as the **Double Zero Option,** serve as further illustration of the vagueness of the term 'intermediate nuclear forces'. All that can be asserted with confidence about the term is that it comes between **Strategic** weapons, or **ICBMs** and **SLBMs** owned by the superpowers, and very short-range **Battlefield Nuclear Weapons.** It may be that terms such as **Eurostrategic** and **Long-Range Theatre Nuclear Forces** are more useful than INF, but even these cannot escape completely from the confusion caused when the range and possible use of missiles overlap the boundaries between arms control categories.

Intermediate-Range Ballistic Missile (IRBM)

An IRBM is a fully-fledged **Ballistic** missile in exactly the same respects as an **ICBM** except that it does not have the range to hit the USSR from a base in mainland America, and vice versa. The term is not applied to weapons in the tactical battlefield and short-range theatre categories, so an IRBM would have a range of at least 1,000 kilometres. The earliest ballistic missiles were IRBMs. For example, the US **Thor** missile had to be based in countries such as the UK or Turkey for it to be able to attack targets in the USSR. The **Cuban Missile Crisis** of 1962 arose when Khrushchev sought to deploy IRBMs in Cuba, which would have brought them within range of the US heartland. The term IRBM is rarely used nowadays, such is the wealth of available language for strategic weaponry. However, many missiles, described variously as **Theatre Nuclear Forces** and **Intermediate Nuclear Forces,** and including the US **Pershing II** and the Soviet **SS-20,** are effectively IRBMs, as are the French land-based strategic nuclear missiles.

International Institute for Strategic Studies (IISS)

The IISS is the UK's principal civilian military **Think-Tank** and has an impressive world-wide reputation. When it was founded in 1958 it was simply the Institute for Strategic Studies, but it was felt that such a single-country perspective reduced the value of its analyses. Consequently it was transformed into a genuinely international institute and changed its name in 1964. Although not large by American standards, it has a professional staff as well as visiting academics and others, drawn from many countries, and there is no doubt that this international aspect gives some of its publications an authority which they might not otherwise have. The single most important undertaking of the IISS is probably the annual publication of *The Military Balance,* which is regarded world-wide as being as close as is possible to an authoritative statement of the armed forces and defence expenditure of every nation.

Interoperability

Interoperability has become a major concern for NATO forces as they strive to transcend the conventional superiority of the Warsaw Pact. NATO's problem is that it is not one unified military force at all, but, on the **Central Front** alone, 10 separate forces, all with their own procurement policies and many with their own armaments industry to support. This means that incompatibilities between equipment, and in operating procedures, can very seriously reduce the efficiency level of the total NATO force below what its sheer numerical capacity would seem to provide.

An often-quoted example is that of refuelling ships of NATO member navies. There may well be a **Frigate** squadron of ships from the British Royal Navy, and from the navies of the Netherlands and of West Germany, operating together. The bore of the pipes used for refuelling are all of different diameters, making it impossible for the ships to aid each other, and indeed calling for several, rather than only one fleet auxiliary to support the squadron. Similar problems are legion—radio sets that cannot be inter-tuned, the need for many different calibres of ammunition for weapons of the same generic type, incompatible radar identification systems so that no one can be sure whether an aircraft is friendly or not, and so on. The cumulative effect is to increase the effective Warsaw Pact **Force Ratio** superiority considerably.

Efforts are continually being made to remedy this problem, but it can only properly be dealt with at the political level. Real interoperability would require a substantial integration of the Western European and US armaments industries, and would involve the military procurement officials in NATO countries being allowed to standardize on one type for each **Weapons System,** regardless of where it is made. Despite the increasing tendency for major projects, like the Tornado **Multiple-Role Combat Aircraft,** to be given to multinational consortia, the serious problems of lack of interoperability show no general signs of being overcome because of the divergent national economic interests at stake.

Inter-Service Rivalry

It is likely that all military structures in history have suffered to some extent from rivalry between the separate services. Certainly we know that naval and army strategies in the Greek city states were at odds because of the rival class interests represented by the two services. When defence analysts talk of inter-service rivalry they are not referring to some competition for glory that might even have beneficial consequences. In the modern era inter-service rivalry is overwhelmingly about defence procurement funds. Even under governments committed to high levels of defence spending, the costs of modern **Weapons Systems** are such that success by one service in winning the funds to buy a much-needed new weapon will inevitably cut down what the other services can gain. A current example in the UK is the way that the RAF's planned purchase of the Tornado **Multiple-Role Combat Aircraft,** followed by the Royal Navy's procurement order for the Trident submarines, together have systematically deprived the army of funds required for tank modernization.

This rivalry tends to lead to a system where each service fights for its own preferred projects, and the central authorities are seldom in a position to create a rational overall policy based on independently-assessed national needs. Partly as a consequence, the development of strategic doctrine itself becomes distorted; doctrine tends to follow the procurement of various weapons systems, rather than vice versa. It is quite probable, for example, that the size and multiplicity of the US nuclear stockpile has more to do with the fact that both its navy and air force were determined to be carriers of the national deterrent, than because there is real strategic justification for the nuclear **Triad.** As a more general example, the development of a **Global Strategy** for the USA is stressed, to a large extent, because it gives the US navy a justification for its large fleet, which the European land war emphasis of the **Atlanticist** strategy does not.

Inter-service rivalry, breeding both economic and strategic inefficiency, is the problem which most concerns those wishing to convince both Britain and the USA of the advantages of creating a **General Staff.** In fact, rivalries can even exist inside services; the US navy is notorious for being split between the naval air-power lobby, pushing for major investment in **Carrier Battle Groups,** and the submarine service, to the extent that an officer's career prospects can be dependent on coming from the branch currently best represented at senior levels.

Intra-War Bargaining

Many of the more sophisticated theories of nuclear strategy attach a great deal of importance to planning for gaps or **Fire-Breaks** between rounds of nuclear exchange. It is hoped that carefully developed strategic options and targeting policies will allow for bargaining and negotiation in such periods to terminate the war before it reaches **Mutual Assured Destruction** levels. Intra-war bargaining, and the allied concept of intra-war deterrence, hinge on the idea that each side initially would refrain from destroying targets of especial value to their enemy.

The argument is that if, for example, the USA ensures that it does not hit any major civilian targets in the first round of **Central Strategic Warfare,** it can hope to deter the USSR from retaliating, and persuade it to settle the war on relatively favourable terms because of the threat to launch **Counter-Value** strikes in a further **Escalation** if the bargaining breaks down (see **Nuclear Hostages**). This is all part of the more general move towards an idea of nuclear **War-Fighting,** which is beginning to replace the older assumption that a nuclear exchange would involve a once and for all launch of a **Spasm War** nature against every target the enemy holds dear. It requires very carefully designed **Limited Nuclear Options,** and considerable effort to avoid **Collateral Damage.** Even US advocates of intra-war bargaining or deterrence admit that there is a serious danger of destroying **C^{3}I** facilities. They specify that the USA must actually refrain from targeting the Soviet leadership, to ensure that there remains someone with sufficient **National Command Authority** to negotiate and bargain.

The idea of intra-war deterrence is simply that even if a nuclear war breaks out, it need not follow that one must suffer the worst attack the enemy could possibly inflict. As long as the attacked country does not retaliate by, for example, destroying the attacker's capital city, an element of restraint for the next round would be retained, as there would still be prized centres liable to attack. As such, intra-war deterrence is part of the broader concept of **Damage Limitation.**

The critics of both theories of intra-war behaviour object that expectations of rationality on the part of leaders, in the face of national fear, confusion and rage, are unreal. They doubt that nuclear war could be controlled in such a sophisticated way. It is noticeable that such conceptions play no part in public **Soviet Strategic Thought** about nuclear war. The possibility of controlling nuclear war is denied, and the **Declaratory Policy** appears to be that, in such an eventuality, the USSR will not spare any targets.

Intra-War Deterrence (see Intra-War Bargaining)

IRBM (see Intermediate-Range Ballistic Missile)

Iron Bomb (see Gravity Bomb)

Isolationism

Isolationism is the name usually given to the foreign policy that the USA operated between the two World Wars. It means what the name implies – that the USA would isolate itself from the rest of the world and accept no obligation to involve itself in either peace-keeping or military alliances. The policy has a long tradition in American history and, indeed, the very first foreign policy of the new republic after the War of Independence was isolationist. The return to isolationism after the USA's involvement in the First World War, against the desires of the then President, Woodrow Wilson, stemmed largely from a sense that European powers were corrupt and decadent, and that there was no reason at all why young Americans should die to prop up such societies. A major consequence of this policy was that the fledgling League of Nations lacked the one thing that might have helped it to maintain peace, a powerful but largely disinterested member, as the Senate refused to ratify the USA's membership.

It can be seen just how strongly isolationist beliefs were held when one remembers that a President as powerful as Franklin D. Roosevelt, although personally convinced that the USA should intervene in the early years of the Second World War, was not able to secure the support of Congress until the USA was itself attacked by Japan. Since 1945 the USA officially has eschewed isolationism, replacing it with a clear international commitment, enshrined in policies such as the **Truman Doctrine** and formal alliances such as NATO. From time to time commentators believe they can detect a resurgence of isolationist sentiment, particularly in relation to problems within NATO. The sentiment has a similar motivation to that of the 1930s—a belief that Western European powers are selfish in not spending enough on their own defence, and that they are too weak to stand up to the USSR or properly to support the USA in its international role (see **Burden Sharing** and **Global Strategy**). Again the implication is that the USA should not strain its own economy, or risk its own citizens, on behalf of 'undeserving' foreign powers.

J

Joint Chiefs of Staff (JCS)

The Joint Chiefs of Staff, usually simply called the 'Joint Chiefs', represent the highest level of military command in the US **Department of Defense** and act as the President's top military advisers. The JCS committee consists of five people: the senior naval officer, known as the Chief of Naval Operations, the Commandant of the US Marine Corps, and the Chiefs of Staff of the army and the air force, with the committee being headed by the fifth officer, the Chairman of the Joint Chiefs, a Presidential appointee who can be drawn from any of the services. Along with the Secretary of Defense, and the senior civil servants in the **Office of the Secretary of Defense,** they are unifying elements in the huge and divided American military machine.

Over the last few years the Joint Chiefs have come to be seen as increasingly ineffective, particularly as they tend to represent their separate service interests rather than trying to forge an optimum universal service policy for the country's overall defence (see **Inter-Service Rivalry**). Part of the problem is that each member has a joint role. Each service naturally looks towards its head for leadership and protection, but at the same time that officer is meant to be part of a collective military 'cabinet' transcending service parochialism. In addition they suffer from two major restrictions imposed by the Congressional legislation that set up the system. Firstly, they have only a very small staff of officers, which are drawn from the separate services. There is little chance for this staff to develop any collective inter-service spirit, as each officer serves only a short time before returning to the appropriate service, upon the approval of which any career advancement depends. Secondly, the members of the Joint Chiefs as a group, and the Chairman individually, have no direct command authority. Presidents often bypass the JCS on operational questions, giving instructions directly to the senior generals or admirals commanding the regionally-distributed forces, through whom the JCS have to work even if they *are* consulted.

These drawbacks have led members of the US **Military Reform Movement** to press for the creation of a **General Staff** system in the USA. Although this is unlikely to happen as such, reforms intended to strengthen the hand of the Chairman of the JCS have recently been passed. Comparable reforms have also been made in the British **Ministry of Defence.**

Just War Theory

At least as long ago as the early days of Christianity there have been attempts to produce philosophical and theological theories limiting the right to war. The just war theory, which has its origins in medieval Catholic political theology, but has been broadened and developed by thinkers from many other traditions, is probably the dominant example. Although the theory is rich and complex, its essence can be reduced to fairly simple terms.

To start with, traditional approaches have tended to distinguish two broad areas of debate: when it is morally acceptable to go to war at all, and what forms of warlike activities are permissible if the general war effort is itself 'just'. The former question, on the acceptability of war at all, Jus ad Bellum in the traditional terminology, has latterly developed a more restricted meaning. At one stage, for example, there was some sense that one nation could make war on another to punish it for actions that did not directly affect the first nation. Alternatively, it has been considered in the past as acceptable for one country to attack another in retaliation for some hurt it has suffered which was, in itself, much less serious than full-scale war. Most modern thinkers would limit the right to go to war to self-defence alone, while it is probable that an alliance war, in which war is made on a third party which has attacked an ally, would come under the general category of self-defence. Most of the other restrictions that the original just war theory imposed, such as having 'right intentions', and that the war must be declared by a legitimate authority, have little meaning in the modern context.

Probably it is with regard to the second question, Jus in Bello or the limits to be set within legitimate acts of war, that modern theory has most problems. The governing dogma in this area is covered by two restrictions: an act is only legitimate if it abides by the criteria of **Proportionality** and discrimination. Proportionality essentially means that the evil of the action, the number killed, for example, must be proportionate to the military end in sight, and to the value of winning the war. If a campaign can only be carried out in a particularly brutal manner, and if winning that campaign is not truly vital to overall victory in the war, it would be regarded as 'unjust' on the grounds of being unproportional to the aim. Alternatively, the tactics required to win a war might be so appalling that the war itself was not worth winning. For example, the attacker might have had the rather limited aim of annexing a province. If the attacked country's claim to that province were not particularly strong, the virtue of self-defence might not justify risking the lives of the troops of the aggressor and the defender in a prolonged siege.

Even more restrictive, given the nature of modern weapons, is the criterion of discrimination, which effectively rules that only enemy soldiers, or those who can in some way be seen as voluntarily guilty of causing the war, may be killed, and that no act of war that will expose enemy civilians to risk can be legitimate. Both the proportionality and discrimination doctrines effectively rule out weapons of mass destruction, meaning that nuclear war cannot be 'just'. In fact, had they been followed, they would probably have forbidden any strategic warfare of the forms used in this century.

K

Killian Report (see Gaither Report)

Kiloton (see Megaton)

Kinetic Energy Weapons

The destructive effect of projectiles can be of two sorts. Through most of history the damage has been caused by the sheer impact of one solid object, a spearhead, arrow, musket- or cannon-ball, against another. From the mid-19th century onwards there was a shift towards warheads which had little impact value but caused their damage by exploding. These weapons, essentially the artillery shell or **Gravity Bomb,** caused damage in two ways. Firstly, the heat and shock waves of the explosion caused damage and injury. Secondly, impact damage to objects in the vicinity of the explosion was caused by shrapnel, either from the shell or bullet casings, or from metal shards specifically incorporated into the projectiles for that purpose.

The development of exploding, shrapnel-producing projectiles was of great importance. The classic example is probably the development of 'canister shot' in the late 18th century. With this a cannon fired not a solid 'ball', which would kill only those in the direct line of fire, but something rather like a huge shotgun cartridge, which spread musket balls in a wide arc around the line of fire. Exploding weapons such as these continued to be vital so long as humans were the primary battlefield target. With the increasing importance of anti-tank warfare, however, the principal need changed. The main objective became to hit a single large heavily-armoured object so hard that it could be stopped, whether or not its human passengers were hurt. Consequently there has been a return to designing weapons where a very heavy, but not necessarily large, projectile is fired with maximum possible speed against the target. These are known as kinetic energy weapons. The combination of weight and velocity produces far more damage than the highest of explosives could. Most anti-armour weaponry now follows this principle.

Kinetic energy weapons have become still more fashionable with the advent of serious research on the **Strategic Defense Initiative (SDI).** The problem of destroying missile warheads in space is peculiarly difficult because the near vacuum environment experienced by **Re-entry Vehicles** make all but the nearest of misses by even a nuclear warhead useless. However, the warheads, which are physically frail, are very vulnerable to direct impact with any hard object, however small, if it is travelling fast enough. One part of the SDI is based on this principle; large numbers of solid objects would be fired into the paths of the incoming missiles. Sometimes called 'smart rocks', the projectiles do not depend on complex and expensive explosive warheads and can be manufactured and fired in large numbers. These are kinetic energy weapons just as much as the heavy solid anti-tank rounds, made of depleted uranium, with which modern armies are equipped.

Korean War

The Korean War, which lasted from June 1950 to July 1953, was the first major occasion in which the USA, with contingents from 14 other United Nations members, fought against an established communist power bloc. It resulted from an invasion by the communist North Korean regime south of the 38th parallel, which had been fixed as an arbitrary dividing line at the end of the Second World War. UN attempts to reunite communist North and democratic South Korea had failed. Initially the landing at Inchon by a large American force, followed by an invasion of North Korea, drove them back. However, the pursuing UN clashed with communist Chinese forces on the Chinese border, and a very large Chinese army pushed the UN forces back into South Korea, with the southern capital, Seoul, being captured for the second time. Eventually the UN forces, predominantly American, were able to drive the communists back across the 38th parallel, where the war fell into a stalemate. The cease-fire line agreed in 1953 roughly follows the 38th parallel, and a demilitarized zone, supervised by UN forces, separates the two countries.

The Korean War had two main effects on US thinking. Firstly, Americans regarded their war effort as a failure, and it represented a considerable shock to military **Morale** which, perhaps, sowed the seeds of defeatism a decade later in Vietnam. Secondly it was a deeply unpopular war, fought by reservists who had thought that their Second World War experience was the last time they would have to fight. Casualties were very high, with over 140,000 Americans (and 17,000 from other UN contingents) killed, compared with 55,000 in the Vietnam War.

These factors combined to encourage President Eisenhower, who came to office in the last months of the war, to enunciate his **Massive Retaliation** doctrine. One of the dilemmas of the Korean War was that while the USA enjoyed an almost complete nuclear monopoly at the time, they found no way of using the weapons, and were forced to fight a conventional war for which they were ill-prepared. It was also partly for this reason that massive retaliation, threatening nuclear war against civilian targets rather than against conventional forces, became the main line of US strategy.

Krasnoyarsk (see Anti-Ballistic Missile Treaty and Phased Array Radar)

L

Lanchester Equations

The Lanchester equations are the pioneering, and best known, results of what is now called operations research, as applied to military activities. Lanchester was an engineer who, during the First World War, produced a series of mathematical analyses of **Weapons Systems** and their relationship to manpower. His actual interest, largely forgotten now, was in the use of aircraft in warfare, but he studied a set of problems about the relationship between personnel and equipment in armies. Although the technical details of his work have long been surpassed, his basic insight is still taken very seriously by military planners. Shorn of the mathematical formulae, his main equation essentially says that numbers are more directly important. To make up for a lack of sheer numbers by technology (that is, by equipping troops with highly-sophisticated and lethal weapons), requires a very large investment because **Lethality** increases much more rapidly with every troop unit than it does with every unit of extra *efficiency* per troop unit. The exact scale of the statistical relationship is largely irrelevant; Lanchester called it a 'square law' meaning that the lethality increased by the square of the number of troops, but by a much lower proportion as the capacity of each troop was increased. The modern implication of the Lanchester laws, and similar more modern analyses, is that the idea of compensating for numerical inferiority with technological **Force Multipliers** is likely to prove more expensive than is often thought.

Laser Weapons

Laser weapons are devices that produce tightly-focused beams of very high energy electromagnetic radiation. 'Laser' is, in fact, an acronym for Light Amplification by Stimulated Emission of Radiation. The radiation can, in principle, be of any wavelength, although those with which we are more familiar from civilian research emit radiation in the infra-red, visible and ultra-violet sections of the electromagnetic spectrum, and are usually referred to as optical lasers. These depend on a chemical reaction to stimulate the emitted radiation. One technique on which much research is being done in the USA is the idea of an X-ray laser which could pump out an incredibly powerful beam. These, however, would require a nuclear explosion to trigger them, and all the power of that explosion would effectively be focused into a tight beam, rather than being diffused in all directions, as in a standard nuclear detonation. The main interest in laser weapons comes from the research and planning for the **Strategic Defense Initiative:** any laser beam-producing technology with enough power to be of military use would be too bulky to have much battlefield use.

Launch on Warning

Launch on warning is a strategic policy sometimes suggested as a way of avoiding the **Vulnerabilty** of land-based missiles to a **First Strike** attack. There is a theoretical problem associated with the fear of an enemy gaining **Strategic Superiority,** in which the bulk of a country's land-based missiles and many of its bombers might be destroyed on the ground by a surprise nuclear attack. The time required to give a launch order and have it confirmed and executed would take sufficiently long that a well-coordinated simultaneous attack would not allow for the firing of any missile between the first explosion over a silo and the destruction of most of the defender's weapons. The reason the problem occurs is that it is assumed that no country would ever fire a retaliatory strike before it had actually received at least one nuclear detonation on its soil (see **Launch under Attack**). The dilemma with piloted bombers is not so extreme. Those on alert would take off as soon as any appreciable risk of their destruction was perceived, and could be recalled if necessary. Missiles, however, are unlike bombers in that they cannot be recalled on reaching a **Fail-Safe** point, and would have to wait for confirmation of the attack before being launched.

The alternative is to launch at least a partial strike at some stage after the detection of an enemy attack, and before the warheads detonate over the silos. Hence the idea of launching 'on warning'. Clearly the major difficulty is one of certainty of detection. At most, the USA would have only 25 minutes warning of a Soviet strike, assuming that the **Boost Phase** exhaust emissions of the attacking missiles were detected in the first few minutes of flight. In practice the size and trajectory of the attack would also need to be known, and this would reduce the response time available. Adequate certainty would probably require confirmations from the **Ballistic Missile Early Warning System** radars, some 10 minutes into the flight time. Even less time would be available if the attacking missiles had been launched from submarines, when the total flight duration would be about 12 minutes.

Even if the USA or the USSR were prepared to take the terrible risk of firing a major strike before absolute confirmation that they were about to receive a major nuclear attack, it is not certain that launch on warning could ever be a safe policy. If the complex confirmation procedures, such as checks and codes necessary to ensure that accidental or unofficial launches cannot occur, are to be observed, then both valuable time and a detailed network of highly-vulnerable communication channels must be used to obtain proper clearance from the appropriate **National Command Authorities.** Once again, launch would inevitably be delayed too long to escape the incoming missiles. If the authorization system were relaxed the risk of accident or deliberate warmongering would be raised. Communications are a particular problem, because if launch were delayed until at least some warheads had exploded, the **Electromagnetic Pulse** would be liable to hamper seriously rapid communication, putting the remaining missiles further at risk. Even if a sizeable number of the defender's missiles were launched during an attack, it has to be remembered that, in the first few minutes of their trajectory, missiles are actually *more* vulnerable to nuclear explosions going off around them than they would be sitting in their silos. There seems to be very little possibility of a safe 'launch on warning' policy operating successfully, and it is believed that both the US and Soviet military leaderships have accepted that it cannot be relied upon, even if it is still frequently suggested as a policy.

Launch under Attack

The dangers of reacting prematurely to what seems to be a nuclear attack, the **Launch on Warning** strategy, are such that no country plans, or at least admits to planning, for such an eventuality. Instead the US doctrine (the USSR has never made it clear what its policy is) involves waiting until Soviet missiles actually detonate over America before retaliating. This is usually referred to as the strategy of launch under attack, implying a rapid response, but one delayed until there is no danger of mistaking the nature and scale of an attack. Alternatively the USA is sometimes referred to as being prepared to 'ride out an attack' before retaliating.

The consequence of accepting launch under attack is that a country's **Strategic** forces must either be very large, or very well protected, in order to be confident of having a secure **Second Strike** capacity. Because of this there is a real fear in certain US circles of the USSR gaining **Strategic Superiority**—being in a position to destroy the bulk of US land-based missiles in their silos. It is also because the US rejects launch on warning that they keep the bulk of their warheads in submarines **(SSBNs),** while the major part of the USSR's strategic force consists of multi-warhead **ICBMs** in heavily-defended silos.

The distinction between launch on warning and launch under attack refers only to missiles which, not being capable of recall, cannot be committed without certainty. In contrast the **Strategic Air Command** bombers would be ordered to take off as soon as there was warning of an attack. This presents a major strategic problem to the USSR. Even if the Soviet Union managed to destroy most of the air-bases (by firing submarine-launched missiles from near the US coast) and the US ICBM silos (with its own long-range ballistic missiles) simultaneously, the USA would still retain its **Assured Destruction** capacity in the form of the SSBNs.

Layered Defence

Modern defences against air attack have to cope with a large number of missiles with very sophisticated target-seeking warheads which can be launched from ranges of well over a hundred miles. As a result it is inevitable that at least some attacking weapons will 'leak through' any one defensive shield. Increasingly defence systems depend on multiple layers of detection and destruction. In the naval context, for example, a **Carrier Battle Group** will have at least three layers of defence. The first will be fighter aircraft, from the carriers, which hope to intercept attacking bombers before they can launch their air-to-surface (ASM) missiles. However, the detection range of the fighters will probably not greatly exceed the range at which the bombers can launch their ASM salvoes. As the attacking missiles approach the fleet, specialized **Area Defence** ships, such as the new US Aegis cruisers, will attempt to destroy the incoming missiles with **Surface-to-Air Missiles (SAMs)** of their own. At still closer ranges the **Point Defence** systems on individual ships will attempt to destroy remaining enemy weapons that are targeted on them. Even this last stage may have two layers, with point defence SAM systems first and then, at very close range, with rapid-firing machine guns with automatic radar control. The same concept of layered defence is crucial to the **Ballistic Missile Defence** schemes currently being designed as part of the **Strategic Defense Initiative.**

Lethality

The lethality of a weapon is a measure of the amount of damage which it can be expected to inflict on a specific type of target. The concept can be applied to all weapons—there is no reason why the lethality of an infantry rifle should not be measured—but it is usually restricted either to nuclear weapons, or to the more technologically-sophisticated among conventional armaments. The lethality coefficient is mainly a combination of two factors, accuracy and **Yield,** or, as with **Kinetic Energy Weapons,** some other measure of the force applied.

Accuracy is normally measured in terms of **Circular Error Probable (CEP),** while yield is calculated according to the **Equivalent Megatonnage (MTE)** of a warhead. The formula for the latter is MTE = nominal yield$^{2/3}$, which effectively increases the equivalent yield of a small weapon of, perhaps, a few hundred kilotons, in relation to the impact of a **Megaton**-level warhead. As lethality is measured for nuclear missiles, it is inversely proportional to the square of the accuracy of a weapon, but directly proportional only to the$^{2/3}$ power of the yield: this explains the vital role of low CEP weapons in modern nuclear arsenals. For example, a relatively small warhead, such as a **Trident** missile of perhaps 200 kt and CEP of 100 metres, may well be more effective in destroying a **Hard Target** than a megaton-level weapon with a much higher CEP. The advent of **Precision-Guided Munitions** has led to very much higher lethality in modern nuclear weapons, even though the individual warheads are, in general, much smaller than in the past.

Light Division

One distinction that has become increasingly important, particularly in the USA, between types of **Divisions** is that between a light and a heavy division. From time to time fierce debates occur inside American military, and even political, circles on the appropriate balance between these two forms of fighting unit. In recent years the **Military Reform Movement** has been particularly vocal in demanding that the US army switches away from a predominance of heavy divisions, while some professional officers have counter-argued that there is already too much of a bias towards light divisions. This apparently highly-technical debate has serious underlying strategic and political origins. The basic difference is that a light division is primarily a traditional **Infantry** division, with only a small amount of light supporting weapons in terms of artillery, and no armoured component other than light reconnaissance vehicles. It is also mainly 'foot-mobile', with very little in the way of motorized troop carriers. The essence is that it should be easily and rapidly deployable, entirely by air, to fight anywhere in the world. In contrast, a heavy division, as its name suggests, has a high armoured element; its infantry is equipped with **Armoured Fighting Vehicles,** its artillery and anti-aircraft elements are very strong, and it has sizeable engineering and other support elements, all well equipped. Such a division cannot be made **Airmobile,** and consequently takes a very long time to deploy and is dependent on major **Logistics** support. The only way to use such divisions in a hurry, unless they are actually stationed in the **Theatre of War** during a crisis, is to have all their equipment pre-positioned, and to fly in only the soldiers themselves. In order to have an adequate number of divisions in Europe should a war break out without months of warning, the USA has developed such a system, known as **POMCUS.**

What can be seen in the heavy versus light division debate is, in fact, the much more general clash between the **Atlanticist** and the **Global Strategy** positions in US foreign policy. The heavy divisions are designed to fight a war in Europe against heavily-armoured high-technology Warsaw Pact armies. Not only can they not be easily transported elsewhere but, the military reformers argue, they would not be particularly useful anyway. The development of the argument by advocates of the global strategy is that the high-technology war in Europe is extremely improbable, and it is very much more likely that the USA will face a series of limited conflicts, of a **Guerrilla Warfare** nature, in the Third World. The establishment of the Rapid Deployment Force (see **CENCOM**) under President Carter, is a response to this argument.

The current US ground forces breakdown is 13 heavy to 15 light divisions, which those mainly concerned with an Atlanticist strategy think already inadequate for a European war. They also challenge many claims made by proponents of the light divisions, including the assumption that such divisions are much cheaper, and that they are, in fact, all that much more mobile. Finally the basic premise, which is that intervention in the Third World, or, as NATO puts it, **Out of Theatre** operations, do not require heavy divisions, is itself challenged.

Whatever the correct technical answer, there is no doubt that a further shift towards light divisions would both seriously undermine the US NATO capacity, and give considerable impetus to those who want the USA to be able to intervene militarily throughout the world. The assumption that supposedly very expensive heavy divisions are only kept by the USA to fight in Europe fuels the argument that **European NATO** is failing to pay its share, and is essentially 'exploiting' the USA (see **Burden Sharing**). Some sort of compromise may be reached via American experiments into what are called 'high-technology light divisions'. At the moment, however, the debate is a very good example of how apparently purely technical arguments frequently reflect underlying political positions.

Limited Nuclear Options (LNOs)

Limited nuclear options are US targeting plans for missile, or possibly bomber, attacks on specific targets. They range from single **Shot Across the Bows** demonstration strikes, through smaller or larger **Counter-Force** measures and attacks on **Other Military Targets** to, presumably, restricted **Counter-Value** strikes that could be used in response to a Soviet intimidatory attack on one or two US cities. The idea is to have a whole range of carefully prepared responses to any form of Soviet nuclear use (see **Flexible Response**).

LNOs first came to public knowledge as part of US **Declaratory Policy** in 1974, when President Nixon's Secretary of Defense, James Schlesinger, announced a new strategy that moved America away from simple **Assured Destruction** targeting (see **Schlesinger Doctrine**). This was the culmination of a process that had begun earlier in that administration. The President had, several times, rhetorically raised the question of whether it was really adequate to leave the USA with no alternative in the face of a Soviet strike but to launch a nuclear **Massive Retaliation.** The doctrine has continued to be popular with successive administrations, notably President Carter's, whose policy document **PD-59** was the clearest indication that the USA was preparing for counter-force strikes.

In practice, the American **Single Integrated Operations Plan (SIOP)** has, from its very inception in the late 1950s, always contained a variety of options falling short of massive retaliation. In many ways the Schlesinger **New Look** and Carter's counter-force policy were shifts in emphasis rather than actual policy changes. However, it is not clear whether the USA had either missile/warhead combinations or the **C³I** capacity in the mid-1970s to carry out such a limited option, however well-selected the targets were. As a result of Carter's initial planning and Reagan's military funding increases the USA now probably does, or shortly will have, the weapons, specifically the **MX** ICBM and the **Trident II** SLBM, capable of destroying **Hard Targets** with minimum **Collateral Damage,** which is the principal requirement for a counter-force option.

Since the development of some support for the idea of nuclear **War-Fighting** in the last decade the concept of LNOs has become even more valuable. The current generation of the SIOP—SIOP-5—is thought to be particularly well endowed with such plans. Although, to many, the idea of providing the President with a set of responses short of Armageddon seems only sane, the new strategies have come under attack from those who believe that anything which makes nuclear war more controlled, makes it more thinkable and therefore more likely. As a result the Nixon administration had considerable difficulty in getting the policy accepted by Congress, and the MX missile has continued to be a source of recurrent conflict between the administration and Congress.

Limited War

Limited war can obviously be only a relative term. At times it is used in strategic discussions to mean virtually anything short of **Central Strategic Warfare.** War can be seen as limited in two dimensions: the area and number of participants, or the means used to wage it. A third variable that once would have been important refers to the goal of the opposing sides. Thus the Second World War was *not* limited not only because of its geographical scope, but also because the object of the Allied powers was the unconditional surrender of the Axis powers, while Germany's war aim appears to have been the total and permanent control of the whole of Europe. In previous eras wars were fought more typically for limited aims: Louis XIV's wars, for example, had the specific aim of making the River Rhine the eastern border of France.

In the modern period it is more likely that 'limited war' will refer to a confrontation that does not involve the superpowers or, if it does, one that is fought away from the **Central Front** and which does not involve nuclear weapons. The problem with the concept is that war is only ever 'limited' from an external perspective. So, for example, many would regard the Arab – Israeli wars as 'limited'. They only involved second- and third-rank powers, did not use nuclear or chemical weapons, were geographically restricted and short in duration. However, the Israeli High Command would not have viewed them as limited, as the very continued existence of the state of Israel was in question. Perhaps the emptiness of the term 'limited war' is best demonstrated by recent discussion of 'limited *nuclear* war', by which is meant a central front war where nuclear exchange is restricted to the European continent west of the USSR border. Such a war would, of course, be Armageddon for Europeans. Similarly even a purely conventional war, given the effect of modern **Fire-Power**, would be so devastating inside East and West Germany that it could not be, in any meaningful way, regarded as limited. (See also **Limited Nuclear Options**.)

The only useful application of the concept of limited war is for military planning by the superpowers, who have to be ready, at least in the US perspective, to fight what is sometimes described as 'one-and-a-half wars'—a major central front war, and a smaller war, limited geographically and in weapons usage, elsewhere in the world.

Lisbon Force Goals

When NATO was founded, in 1949, it was not immediately apparent that it would be an alliance reliant upon predominantly nuclear weapons. The USA had a virtual monopoly of nuclear weapons, but still only a very small inventory, and military planners had not yet come to see them as more than a particularly powerful form of conventional bomb. Thus, for the first few years, NATO planning was focused on matching the conventional strength of the USSR, and on preparing for an intense but conventional war. Throughout the 1950s estimates of Soviet strength stood somewhere between 140 and 170 **Divisions,** and in the intense **Cold War** atmosphere engendered by the **Korean War** deliberate expansionism by invading Western Europe seemed a plausible Soviet goal.

The NATO members therefore committed themselves, in 1952, at a conference in Lisbon to build up their forces to a position which would be adequate to hold such an invasion. They set their target at what was thought to be the minimum satisfactory level of 96 divisions. This would have involved nearly doubling the NATO troop strength in only two years. It soon became apparent that the goal was unattainable and, as tensions eased with the truce in Korea, the political will to enforce the economic sacrifices that would have been necessary in a Europe still recovering from the Second World War evaporated. In the USA the Eisenhower administration wanted to concentrate on economic growth and avoid the inflationary pressure that such rearmament would have placed even on the strong American economy.

It was the abandonment of the Lisbon force goals (they were reduced to a target of 30 divisions) that effectively set NATO on its nuclear dominated path. The US administration tried to redress the force imbalance by transferring huge numbers of **Battlefield Nuclear Weapons** to Europe, the USSR not having deployed such systems at this stage. Generally the strategy shifted to seeing NATO, and particularly the US troops in Europe, in the light of the **Tripwire Thesis.** The troops were there to ensure that the USSR could not just march across the border. The Soviet Union would have to fight, and this would signal their real intent and determination. Such a clear sign of Soviet aggression would immediately trigger the US **Massive Retaliation** strategy, which would destroy the Soviet Union with nuclear weapons.

Despite the years of experience, despite the official policy change to **Flexible Response**, and despite Americans continually urging **European NATO** to enhance their conventional forces so as to exercise **Containment** of the Warsaw Pact without recourse to nuclear weapons, the strategic planning of NATO was determined by the early abandonment of the Lisbon force goals. One irony that has become clearer since the mid-1960s is that the estimate of Soviet strength was almost certainly excessive, and the Lisbon goals consequently were set too high. Had a plausible goal been set, and approximately achieved, NATO might never have started on the nuclear course that it has taken.

Logistics

Logistics is the military science of supply, storage and delivery of all the necessities of an army, from ammunition, fuel, reinforcements and weapons to clothing, food and medicine. While unglamorous, logistics is often the key to victory in battle, and even more frequently the essential skill of long-term strategic success in a campaign or war. Many of the great military leaders in history have been noted as much for the care they took over supply and transport as for their tactical skills, Napoleon and Wellington being famous as a pair of opponents who had this in common. As military technology developed, and as campaigning came to be a matter of huge armies fighting protracted battles, the capacity of soldiers to live off the land and carry most of their munitions with them disappeared. Many examples could be employed to illustrate this. Both in Rommel's desert campaigns in North Africa and in Patton's advance through France during the Second World War the supply of petrol for tank armies was as decisive as any more traditionally 'military' factors; the current model of **Main Battle Tank** in the US army has a fuel consumption of five miles per gallon. Furthermore, an **Infantry** soldier is able to carry perhaps 60 rounds of rifle ammunition, while modern rifles are capable of being fired at a rate of several hundred rounds a minute.

There is no doubt that logistical competence and preparation will be of vast importance in any NATO – Warsaw Pact conflict in Europe, where NATO's ability, to take just one example, to store and then move ammunition, is as much of a worry to military planners as is the general force imbalance between the two sides (see **POMCUS**). With the pace, mobility and **Fire-Power** of modern warfare no army unit can hope to carry fuel, ammunition, medical supplies or even food for more than two or three days' consumption. Indeed, it is estimated that the ratio of logistics personnel to front-line combat soldiers may be as high as 10 to one.

Long-Range Intermediate Nuclear Forces (see Intermediate Nuclear Forces)

Long-Range Theatre Nuclear Forces (LRTNF)

Long-range theatre nuclear forces are now more usually referred to as **Intermediate Nuclear Forces.** The name change has come about as a result of the complex and lengthy **Arms Control** negotiations which have occupied the USA and the USSR ever since the **Euromissile** issue developed, from 1977 onwards. A variety of names have come into circulation because of an essential ambiguity about exactly what it was that was sought to be controlled. Everyone is clear that the superpowers' intercontinental ballistic missiles **(ICBMs)** and submarine-launched ballistic missiles **(SLBMs)** belong in one class of weapons. Clearly opposed to this is the category of tactical or **Battlefield Nuclear Weapons.** There is a fundamental difference between these two groups: an ICBM or SLBM owned by one superpower can hit the homeland of the other, whereas tactical missiles, with ranges up to perhaps 500 kilometres, could not reach far beyond the **Battlefield** area. **Theatre Nuclear Forces** cover weapons intermediate to these, with ranges varying from 500 to 5,500 km. While an American theatre missile (or fighter-bomber) would present a danger to Warsaw Pact forces as far back as, for example Poland or western parts of the USSR, and a Soviet equivalent could reach NATO's rear lines, it is still the case that the superpower heartlands are safe from them.

The Euromissile crisis came about initially because the USSR deployed the **SS-20** missile, ostensibly for potential use against military targets in the NATO rear area. However, as it has a range of 5,000 km it could, in fact, destroy any target in Western Europe (and as one model carries a 650-kt warhead, the SS-20 could execute **Counter-Value** attacks on French or British cities just as effectively as an ICBM). In response the USA, at the request of Western European members of NATO, began to plan the deployment of **Ground-Launched Cruise Missiles** and **Pershing II** ballistic missiles. These have rather shorter ranges, but still of up to

2,500 km, and can thus be seen as potential threats to Soviet territory. It was this uncertainty about the potential use of missiles which led to the introduction of the term **Eurostrategic.**

'Long-range theatre' is probably an entirely accurate description of the purpose for which both sides designed their weapons. The Pershing II missile, for example, carries a rather small warhead, of not more than 50 kt, but with very high accuracy enabling it to be used against super-**Hard Targets**, such as command bunkers. The SS-20 is very much less accurate, but is almost certainly intended for use against NATO air-bases and similar military targets. Neither side has the slightest need to use its theatre forces if it wishes to destroy European cities, or targets inside the USSR. Nevertheless, the propaganda war from 1977 onwards obfuscated what was, in essence, a normal process of modernization which both sides had decided on independently. The phrase 'intermediate nuclear forces' has no substantive meaning, and the multiplication of categories merely confuses discussion about the purposes of the weapons systems.

Long-Term Defence Programme (LTDP)

The Long-Term Defence Programme was adopted by NATO in 1977 as an attempt to offset the conventional superiority which the Warsaw Pact was believed to have. This is usually set at a ratio of about 2 or 2.5:1 in the Warsaw Pact's favour. The member nations each agreed that they would increase their defence expenditure by 3% a year, in real terms, until 1985. The achievement of this target, it was thought, would remedy the imbalance. The programme failed for three reasons. The simplest was that most member states just did not maintain the promised increases. Only the USA and the UK unambiguously carried out their commitments, and the latter relaxed the increase at the earliest legitimate date. Even if all member states had complied with the requirement, it is unlikely that the LTDP would have attained its primary objective. Simply increasing expenditure in such a blanket fashion in no way guaranteed that particular needs of NATO, as opposed to particular interests of each individual member state, would be satisfied. In practice most of the extra money that was spent seems to have been taken up in paying for projects which the relevant governments were already committed to, rather than in buying any extra capacity above existing plans. Finally, no monetary target can deal with the problem of 'shooting at a moving target', as NATO planners are prone to describe it: the USSR was not likely to freeze its own expenditure to allow the West to catch up, and it did not do so. (See also **Interoperability.**)

M

McNamara

Robert McNamara is possibly the single most influential person in the short history of nuclear strategy. He was the Secretary of Defense for two Presidents, John F. Kennedy and Lyndon B. Johnson, and thus ran the USA's defence policy from 1961 – 68. During this period two major changes were made in US policy, reversing the earlier stance taken by the Truman and Eisenhower administrations, which had presided over the beginning of the nuclear age. McNamara's style was quite unlike most political heads of defence departments anywhere. He had been a very successful President of Ford Motor Company before Kennedy brought him into government. In office he applied the lessons of cost-effectiveness and rational planning learned in a major industrial corporation to an area where they had been notably lacking. It is often said that he, alone of all civilian Secretaries of Defense, had full control over the military as well as the civilian staff of the **Department of Defense.**

When Kennedy and McNamara came into office the USA was still committed to the doctrine of **Massive Retaliation.** It was the official strategic policy both as a response to a Soviet strike on the USA, and for dealing with a Soviet invasion of Western Europe. The nuclear targeting programme was a mass of independent and often conflicting selections by different commanders, although at the end of Eisenhower's administration the first attempt to create a **Single Integrated Operations Plan** had begun. The whole US posture was based essentially on the idea that the USA had, and would always retain, **Strategic Superiority** over the USSR.

McNamara pushed the **Pentagon** into developing the strategy of **Mutual Assured Destruction,** in which the aim of always having more nuclear power than the USSR was abandoned in favour of building and retaining a secure **Second Strike** capacity at a level that would guarantee the destruction of the enemy, whatever action had been taken. The actual megatonnage levels set by this doctrine were used as much to control the demands of the forces, especially the **Strategic Air Command,** for more weapons as to deter the USSR. McNamara's other major change, although it took much longer to enforce, was to shift the NATO doctrine from reliance on massive retaliation, under which the conventional forces were regarded according to the **Tripwire Thesis.** Although his new doctrine of **Flexible Response** was first suggested in 1962, it was not until 1967 that he managed to persuade NATO formally to accept it, and he was never really successful in getting **European NATO** to increase its conventional force strength to a point where reliance on the early use of nuclear weapons could be given up.

He was equally famous in American defence circles for importing a new management style into the Pentagon, often referred to as systems analysis. Weapons programmes and even military tactics were subjected to a combination of cost analysis and operations research more familiar in manufacturing industry (see **PPBS**). This has not been an entirely admired process, because it led to what most of the military thought was far too much interference by bright young civilians in areas that were none of their business. One school of thought, for example, holds that much that was wrong with the way the USA executed the war in Vietnam stemmed from this development.

In recent years McNamara has again come to the fore in nuclear debate by joining with other leading strategists of the 1960s in calling for NATO to abandon its doctrine of first use of nuclear weapons (see **No First Use**). He has gone so far as to say that although he built the main US nuclear weapons systems, such as **Minuteman** and the **Poseidon** ballistic missile submarine fleet, he always advised Presidents in private that there were no circumstances where he felt the use of nuclear weapons could ever be rational.

Maginot Mentality

The Maginot Line was the complex and costly fortified defensive line built by the French between the two World Wars to defend against German invasion across their eastern border. When built it was thought to be the very height of modern military technology, and was believed to be impregnable. It was, at the same time, a symbol of French military, and perhaps social, attitudes—that they would in future adopt a static defensive posture rather than manoeuvring and grasping the initiative by taking the war directly to the enemy. The shamefully rapid defeat of the French army and the British expeditionary force, in May 1940, by the numerically inferior invading German army has often been blamed on this strategy. Not only is the Maginot Line itself deemed to have failed but, more generally, the entire attitude of static defence is seen as having made the Allied forces quite unable to cope with a rapidly manoeuvring army which exploited the advantages of surprise, initiative and shock.

Many critics of modern NATO strategy have accused it of suffering from a similar doctrinal malaise, called the 'Maginot mentality'. Particularly disliked by these critics is the doctrine, insisted on by the West Germans, that invading Warsaw Pact forces must be held at the frontier, that the entire line of the German frontier must be defended equally strongly, and that no territory whatsoever may be surrendered. This, usually called a fixed **Forward Defence,** is seen as repeating the French mistake by ceding the initiative to the Warsaw Pact. They would be able to assemble their forces where they liked, break through, and manoeuvre behind the forward defensive lines, cutting off communications and penetrating deeply into the NATO rear area (see **Operational Manoeuvre Group**). The strongly fortified points would be by passed rather than fought for. Because these tactics worked so well for the Germans in 1940, bringing about the collapse of French military resistance without the need to fight any major pitched battles, it is thought that NATO is in danger of re-enacting a basic mistake in the art of war. The more extreme critics, in fact, maintain that fixed linear defence can *never* succeed against armies trained in **Blitzkrieg** tactics.

While there may be something in the criticism, it is vastly oversimplified and ignores several major points. Firstly, the Maginot Line itself did not actually fail—no German units even attempted to penetrate it. What happened was that French defence funds ran out before the Line could be completed, and that the Germans simply went round the ends of it. Had the Line been completed it is less clear that the strategy would have failed. Secondly, fixed defence *has* worked many times in history, and is practised with great success today by the Israeli army on large parts of their borders. It is also worth noting that while the greatest exponent of blitzkrieg warfare was the German army in the Second World War, its West German successor is committed to NATO's forward defence doctrine. Finally, the demand for forward linear defence is a condition for guaranteeing West German faith in its membership of NATO. The political impossibility of the West German army accepting a **Manoeuvre Warfare** doctrine which would turn large parts of the Federal Republic of Germany into a **Battlefield** is not difficult to understand.

Main Battle Tank (MBT)

Main battle tanks are the modern tanks that form such a major part of both the Warsaw Pact and NATO armies in Europe. Modern versions of the tank are highly-technological developments from the tanks of the Second World War. The latest models, such as the US Abrams tank, the M-1, are very much faster and more mobile. They use highly-advanced laser aiming devices, and stabilizing engineering, which allows them to fire with very great accuracy and at long range, even while on the move. The tanks have armour coating proof against anything but extremely-powerful anti-tank missiles and cannons.

The main disadvantages are expense (the M-1 costs around US $2 million) and the inevitable fragility of the high-technology engines and armaments. There is a school of thought that suggests that even the most advanced main battle tanks will always be vulnerable to relatively cheap anti-tank weapons (an infantry anti-armour missile costs only about £7,000), and that their ability to deal with well-equipped **Infantry** and, even more so, with specialist anti-tank helicopters, is low. As a result even more money has to be invested in providing **Armoured Fighting Vehicles** so that they can accompany the tank units to protect them from infantry and airborne attack. Despite these doubts the tank units are central to both Warsaw Pact and NATO tactics, and it is here that the Warsaw Pact numerical superiority is at its highest. The Warsaw Pact is estimated to have over 25,000 MBTs available for a **Central Front** war, against a NATO total of about 13,000. The USSR has not attempted to match the very high technology of the American MBT preferring, instead, to rely on numbers, but it must be remembered that most of the NATO tanks are of an earlier generation than the US Abrams. Even then some countries, like the UK, cannot afford properly to bring their own tank forces up to the standard of the generation which preceded the M-1.

Manhattan Project

The Manhattan Project was the code-name given to the programme to build the first **Atomic Bombs.** Initiated in 1942, the overall project took its name from the Manhattan District of Tennessee, in which the first development centre at Oak Ridge was located. While much of the development work happened elsewhere, especially at the Los Alamos laboratories and the Alamogordo test site, both in New Mexico, Oak Ridge was important because it was there that the huge chemical plants for enriching uranium as the explosive were set up. The project, although huge and incredibly expensive was handled with complete secrecy. Even Truman, who was the President who took the decision to drop the bomb, was never informed of it while he was Vice-President to Franklin D. Roosevelt. (See also **Nuclear Fission.**)

Manoeuvrable Re-entry Vehicle (MARV)

MARVs are the next stage in sophisticated warhead design after multiple independently targeted re-entry vehicles **(MIRVs).** While a MIRVed missile can deliver several warheads against separate targets, once ejected from the missile's **Bus** (or **Front End**) they follow predetermined and predictable courses. A MARVed missile, on the other hand, would eject as many warheads, each also destined for a separate pre-selected target, but which would follow a variable and changing course, making them much harder for **Ballistic Missile Defences (BMDs)** to destroy. No country admits to having developed MARVed missiles, although there have been claims that the **Chevaline** front end for the UK **Polaris** missile has a similar capacity. Unless BMD technology increases very considerably it is improbable that it would be worth investing in MARV capacity, although this is precisely the sort of response which the USSR might make were America's **Strategic Defense Initiative** to go ahead in full.

Manoeuvre Warfare

Manoeuvre warfare is a shorthand description of a basic battlefield tactic, currently somewhat fashionable among defence analysts, particularly those associated with the American **Military Reform Movement.** The somewhat over-simplified thesis contrasts **Attrition Warfare,** said by some to be the typical American, and British, tactical style with manoeuvre warfare. Attrition warfare relies on numerical and material superiority to batter through the enemy lines by killing troops and destroying equipment in a more or less head-on attack. The paradigm case of attrition warfare is the tactics of both the Allied and German armies in the trench warfare battles of the First World War. Here lengthy artillery bombardment preceded a frontal attack by heavily massed troops hoping to gain a decisive local numerical superiority. It is further argued that Allied successes in the Second World War, whether by British generals, such as Montgomery in North Africa, or by US generals in Italy and France in 1944, again depended on using massive technical and troop superiority to win by outright destruction of the overwhelmed German armies.

The contrast is that manoeuvre warfare depends on the skill of commanders and their deputies, rather than the sheer force of the units under them. As the name suggests, it depends on outmanoeuvring the enemy: moving rapidly to exploit temporary weak points, pouring troops through gaps in a line, attacking from the flank or from behind, and continually seeking to surprise the enemy. Rather than physically destroying the enemy, the idea is to throw it off balance, attacking where least expected, and cutting lines of communication, information and supply. Thus the enemy is prevented from using its forces in a decisive way. The model here is the German **Blitzkrieg** tactics which proved so successful in the battle for France in 1940. The armoured divisions advanced far more rapidly than the Allies expected, avoiding the French and British strong points and penetrating deep behind the 'official lines'. Allied armies were cut off from each other and isolated, with the High Command being unable even to know quite where the Germans were, still less to launch effective counter-attacks. In contrast to the German blitzkrieg, however, NATO would be likely to use manoeuvre warfare primarily as a defensive measure.

The principal motive behind switching from attrition warfare to manoeuvre warfare is that which drives most military thinking in NATO countries—the numerical superiority enjoyed by Warsaw Pact armies. It is argued that NATO cannot hope to win a war of attrition because they will never have the necessary troop and material superiority. However, supporters of manoeuvre warfare maintain that the qualities of initiative and skill, among junior officers as much as senior, are ones which the Western powers can hope to foster, and which the much more rigid command structure of Soviet armies hinder. Highly-mobile and adventurous tactics of infiltration, aimed at destroying the Soviet High Command's ability to control and direct the larger Warsaw Pact armies, offer the best chance of compensating for this numerical inferiority.

Doubters are prone to point to **Clausewitz**'s teaching that, in the end, a decisive battle can never be avoided, and, in addition, that the success of blitzkrieg did depend on spectacularly incompetent generalship and appalling **Morale** in the French army. It is far from clear that modern Warsaw Pact armies suffer from either of these. Nor is it obvious that the Soviet officer corps is in fact so stolid and lacking in initiative, and Western officers so eminently capable of being trained in the relevant skills. Although manoeuvre warfare is an over-simplified theory, it has had a considerable influence in the USA where the official operations doctrine, **Air-Land Battle Concept,** enshrines many of its tenets. More generally it may come to have consequences for the procurement policies of the USA, and probably of other NATO members, as it is likely to be incompatible with the armour-heavy structure of modern **Central Front** armies. It is certainly a total contradiction of the **Forward Defence** doctrine, which is a political necessity for NATO. (See also **Operational Manoeuvre Group.**)

Maritime Strategy

Maritime strategy is an alternative name for the **Global Strategy** espoused by many experts in the American defence debate, particularly those from the **Military Reform Movement.** It is opposed to the NATO-oriented strategy of **Forward Based** troops in Europe, and the construction of an army tailored for the putative **Central Front** war. Instead it calls for an acceptance that the USA has vital interests everywhere in the world, and that most of these call for a major **Force Projection** capacity and an army based on **Light Divisions.** These requirements call for a heavy reliance on naval power and on **Sea-Lift** capacity, hence the term maritime strategy. Following this strategic preference would involve a considerable reduction in the US commitment to NATO.

More generally, maritime strategy refers to any country, Britain being a particular example in the past, whose interests are widely scattered and which necessarily requires control of the oceans by its navy to protect trade, and relies less on continental land armies. In some ways the sort of armed **Isolationism,** backed by major naval power and sea-lifted light troops, called for by many in the USA mirrors the dominant British strategic thinking of the 18th and 19th centuries.

Massive Retaliation

Massive retaliation was the strategic doctrine adopted and publicly announced by the Eisenhower administration in 1954 as the basis for the USA's support not only for NATO, but for all the allies which the 1947 **Truman Doctrine** had committed the USA to defend. Thus the traditional conception of defence was effectively replaced with **Deterrence** and retribution. Instead of committing troops to fight a possibly lengthy war to defend an American ally invaded by the USSR or one of its client states, the new doctrine relied entirely on strategic nuclear force. The theory was that any real incursion over an ally's border would be met immediately with a major and devastating nuclear strike not on the enemy's troops, but on the Soviet homeland.

There were several interrelated reasons for this new approach, although the underlying cause was the comparative cheapness of strategic nuclear forces. NATO members had clearly been unable, or unwilling, to meet the demands to raise the large standing armies called for in the **Lisbon Force Goals.** If NATO was to present a credible deterrent to the Soviet Union some other way had to be found to dissuade the latter from the adventurism in Europe that seemed probable in the **Cold War** atmosphere of the time. It was for this reason that the **Tripwire Thesis,** by which conventional forces would be in Europe only to test the determination of a Soviet attack, but not seriously expected to prevent one, was born. The defeat of these troops, in particular the US contingents, would 'trip' the all out nuclear attack on the Soviet homeland.

Furthermore, the USA had just, inconclusively, withdrawn from its first war against a communist invader, in Korea. The **Korean War** had been deeply unpopular in America, and costly in both lives and money. The new policy was intended to avoid the need for any such engagement in the future, by setting the publicly-announced cost of similar 'proxy' conflict by the USSR so high that US anti-war sentiment could not be relied upon to weaken any response. Finally, Eisenhower was a fiscal conservative, determined to reduce government expenditure and, if anything, less prepared to tolerate high defence spending than most civilian presidents.

There were two main problems with the massive retaliation doctrine. Firstly, it lacked **Credibility.** Quite apart from any Soviet capacity to retaliate, it seemed to many highly improbable that the USA would really destroy millions of Soviet civilians to retaliate against a possibly minor border incursion by a minor ally of the USSR. It was not even considered much more likely that it would do so if, for example, the Soviet army expelled the West from Berlin. Thus NATO and other US allies were liable to be subject to pressure from **Salami Tactics,** against which massive retaliation would be impotent. Secondly, the doctrine could only work as long as the US had a near monopoly in intercontinental nuclear weapons (see **ICBM**). As soon as the USSR could develop the capacity to destroy a few American cities the threat of massive retaliation would weaken, and if the Soviet Union ever gained anything like nuclear **Parity** it would be totally empty. Yet even when announced it was known that this state of affairs was inevitable, and the USSR was already well on the way to developing a **Thermonuclear** capacity, having exploded its first **Atomic Bomb** in 1949, only four years after the Americans. Massive retaliation was abandoned overtly in 1967 when the USA eventually persuaded NATO to accept the **Flexible Response** doctrine; whether it was ever more than a **Declaratory Policy,** as opposed to an employment policy, is unclear.

MC 14/3 (see Flexible Response)

Megaton

The **Yield** of a nuclear weapon is measured in terms of the amount of a conventional explosive, such as TNT, which would have to be exploded to produce the same energy release. A megaton-level detonation is equivalent to exploding one million tons of TNT, a kiloton being equivalent to one thousand tons. When it is remembered that the typical terrorist car bomb only contains, at most, a few hundred pounds of conventional explosive, some idea of the grotesqueness of these figures might be appreciated. Extensive calculations have been made concerning the likely impact of megaton-level weapons, although there is obviously little direct experience. The **Atomic Bombs** dropped on Hiroshima and Nagasaki in 1945 were in the range of 10 – 20 kilotons, whereas the typical **Re-entry Vehicle** in a **MIRVed** missile carries at least 150 kilotons. One estimate suggests that a one-megaton **Air Burst** at a height of 2,500 metres would cover an area of 75 square kilometres with an overpressure of 10 psi (that is 10 lbs per square inch above normal atmospheric pressure). An overpressure of between 5 psi, which is equivalent to applying a weight force of 180 tons, and 10 psi would destroy most brick-built houses. The simplest methods of calculation suggest that the death rate from such an explosion would be roughly equivalent to the total population within the 5 psi area of ground zero (see **Designated Ground Zero**). (This does not mean that everyone in the area would die; effectively, assuming a constant population density, the number of people who survive inside such an area is thought to be equal to the number who die outside it.)

However, these calculations refer only to the immediate blast damage caused by the one-megaton detonation, and ignore both **Radiation** and heat effects—which are even harder to quantify. Some examples of the likely effects here are that second degree burns can be caused at a distance of 11 km, and third degree burns 8 km, from ground zero. (Second degree burns over 30% of the body will almost certainly be fatal.) To put these examples and calculations of damage and casualties into perspective, research suggests that only seven one-megaton warheads would be required to destroy the entire city of Moscow, which has a population of slightly over 8.5 million.

Mid-Course Phase

The mid-course phase is one of the three recognized phases in the flight of a **Ballistic** missile, and is usually referred to in an **ICBM** context. In the first or **Boost Phase** the rocket engines are burning and the missile is propelled by powered flight. As the engines cut out the missile enters the longest period of its flight, the mid-course phase, during which it is coasting on a predetermined ballistic trajectory outside the earth's atmosphere. If the points of launch and impact were in the USA and the USSR, or vice versa, this phase would last about 20 minutes. It is effectively 'in orbit' at this stage or, using alternative technical language, 'sub-orbital'. The mid-course phase, before a **MIRVed** missile ejects the **Warheads** for their descent during the **Terminal Phase,** is the prime opportunity for **Ballistic Missile Defences (BMDs)** to operate. A single hit will destroy the entire warhead load, and the missile is in a predictable path for long enough to allow tracking and aiming of interceptor devices. At the same time, because it is at its maximum height, hitting the missile requires the development of defensive technology which at least some analysts believe is not yet feasible. It is for operations in the mid-course phase that the more fanciful plans for space-based weapons have been developed, and earning the US **Strategic Defense Initiative** its nickname of Star Wars.

Midgetman

Midgetman is the popular name given to a US plan to build what is more formally referred to as a Small Intercontinental Ballistic Missile (SICBM). This programme, which had been discussed in US strategic circles from the early 1980s, was given considerable emphasis when the **Scowcroft Commission** recommended it as the next generation of US missiles, after the deployment of the **MX.** It is somewhat unusual in being favoured both by the **Pentagon** and by the **Arms Control** fraternity.

The Midgetman is designed both to be invulnerable to a **First Strike,** but also not to give the USSR any reason to fear it is intended for launching one. Thus it provides a contrast to the MX, which looks very much like a first strike weapon but fails, given the problem of **Basing Modes,** to be invulnerable. Midgetman would achieve the first criterion, invulnerability, by being a mobile missile, capable of being driven all around North America. It would be so widely dispersed that the SICBM batteries would not constitute a good target for a Soviet first strike. The quality that achieves this mobility, the smallness of the missile, gives it its nickname. But the smallness is also part of the second design criterion. The SICBM is intended to be a single warhead missile, because the development of **MIRV** technology, with its huge proliferation of **Re-entry Vehicles,** is seen as a major source of arms control problems. With only 100 missiles, each restricted to one warhead, and a **Throw-Weight** of only 450 kg, which will restrict the **Yield** even of that single warhead, it would be impossible for the USA to deliver a crippling blow to the Soviet **ICBM** forces.

The main question about Midgetman is a fundamental one: what is it for! It would probably not constitute, and is not needed as, a **Second Strike** or **Assured Destruction** weapon for **Counter-Value** targeting; it is deliberately *not* intended to be a first strike **Counter-Force** weapon. Therefore it has no very clear role. It is true that if the MX missiles were removed, and if one discounts the **Strategic Air Command** weapons, the SICBM, which is designed to have a relatively small **Circular Error Probable,** would be available for carrying out **Limited Nuclear Options.** However, it seems more likely that its real aim is to ensure the continuation of all three legs of the nuclear **Triad,** which has become an end in itself for many American nuclear planners. The future of the Midgetman programme is not clear in the present climate of American defence budgetary problems, and it is quite likely that its fate will be as a bargaining counter in arms control negotiations, particularly as the USSR has tested and may be deploying its own equivalent, the SS-X-25.

Military Reform Caucus

The Military Reform Caucus can best be described as the Congressional wing of the **Military Reform Movement.** It is a bipartisan group of members of the Senate and House of Representatives alarmed at the escalation of defence expenditure coupled with what they see as a continued weakening in US strategic capacity. The members are drawn from the more conservative elements of the Democrats and from the whole spectrum of the Republicans. They are not typical among defence critics because they all stand for a strong US military posture, and do not object to high defence expenditure as such. Their concern is, rather, for military efficiency. In particular they tend to focus on the overreliance on nuclear weapons resulting from American (and **European NATO**) conventional weakness, as well as on escalation in defence procurement costs.

Several of the most influential US Congressional politicians are either members of the caucus, or closely associated with it. These include Nancy Kassebaum, a powerful Republican Senator, and Sam Nunn, a Democrat, and the Senator perhaps most widely respected on defence matters in the whole of Congress. The caucus is well represented on both the Senate and House of Representatives **Armed Services Committees.** The implications to foreign policy of the reform movement make this caucus potentially crucial for US – European relations within NATO.

Military Reform Movement

The Military Reform Movement, which is a rather loosely-organized body of defence analysts in the USA, has, since the mid-1970s, become a very powerful voice in the American defence community. There is no single agreed set of doctrines that all members of the movement would subscribe to, and the motives and primary concerns of the various military reformers vary enormously. The following basic propositions would all be assented to by some members, though perhaps no single thinker would accept all of them:

1. The US military establishment has been woefully inadequate and unsuccessful since 1945.
2. A major reason for such inadequacy is the American preference for **Attrition Warfare** rather than **Manoeuvre Warfare.**
3. A second reason is the over-bureaucratic nature of the military establishment, and the careerist and managerial ambitions of the officer corps, which has replaced a 'service and leadership' attitude to soldiering.
4. At the same time, and partly for reasons connected with the above, defence procurement has gone on the wrong track, focusing on buying more and more technologically-advanced **Weapons Systems,** in smaller and smaller quantities but at greater and greater prices, so that the defence budget, although ever-increasing, actually buys less and less real security.
5. Consequently, the whole US strategy has outgrown existing military capacity and, in particular, the NATO commitment is both wrongly handled in tactical terms, and distorting the overall defence posture so that other vital US interests cannot be protected.

It is easier to list the complaints of the military reformers than their suggested solutions, and it is with the latter that consensus tends to break down. Some basic policies can be isolated however, most of them covered in further detail by other entries in this book. To start with, a major reorganization of the **Department of Defense** is called for, and many would also support the attempts to reform the **Joint Chiefs of Staff.** Most also urge a transformation in career development patterns and military education to improve the nature of the officer corps, developing skills and initiative, and building unit loyalty; military reformers often stress that officers should be 'leaders, rather than managers'.

On the procurement aspect of defence many urge the **Pentagon** to adjust its policy to one of buying far larger quantities of simpler and more robust tanks, aircraft and other equipment to offset the Soviet numerical conventional superiority and to avoid the cost overruns and maintenance problems associated with high-technology military equipment. Some also link this to a preference for 'light' and easily-transportable equipment, useful in the many non-European theatres which they see as vital to the USA, at the expense, if necessary, of reducing the European NATO commitment. A particular example of this is the pressure to develop **Light Divisions** rather than the armoured heavy divisions designed for a **Central Front** war.

This last point reflects the neo-**Isolationism,** or the preference for maritime or **Global Strategy,** common among members of the reform movement, as well as their anti-European attitudes based on the conviction that **European NATO** does not pay its 'fair share' of the common defence burden (see **Burden Sharing**). As can be seen, the concerns of the military reform movement are nothing if not broad, and are far from restricted to military technicalities. The movement represents a major strategic and foreign policy alternative to the whole development of US post-war history.

The movement is decidedly influential, with Congressional links in the **Military Reform Caucus** and important supporters within the US military itself, particularly in the US Marine Corps. Indeed, the latest version of the offical army operations manual, FM 100-5, sometimes referred to as the **Air-Land Battle Concept,** is heavily influenced by the reformers. They have very considerable support in the media, especially in influential papers such as the *Washington Post* and the *New York Times*, and may fairly be said to dominate policy discussion in many areas. Even if their ideas are never totally accepted, they can claim real successes, as with the recent Congressional Act strengthening the authority of the Joint Chiefs of Staff, and the US army's creation of extra light divisions.

Minimax

Minimax is a technical term within **Game Theory,** the mathematical analysis of conflict situations that has been found to have a certain utility in strategic reasoning. In a minimax strategy attempts are made to minimize the worst eventuality, rather than to maximize the best possible outcome. For example, a nuclear **Deterrence** posture that maximized **Damage Limitation** and rested on a secure **Second Strike** capacity could be said to minimize the damage to a country if its enemy launched an unprovoked attack. Such a strategy would not be the same as one intended to give a chance of total victory with a **First Strike.** The latter might leave the aggressor open to much greater damage if a miscalculation had been made, and is thus not a minimax strategy. Most simple game theory analyses end up predicting minimax as the strategy likely to be adopted, independently, by both sides, as long as they are unable to trust each other.

Minimum Deterrence

A minimum deterrent is a **Strategic** nuclear force, most probably of a secure **Second Strike** nature, such as a submarine-launched ballistic missile **(SLBM),** that is thought to be just enough to deter an enemy from mounting a **First Strike.** The British, and sometimes the French, strategic nuclear forces are regarded as minimum deterrents. Although very weak compared with those of the USA and USSR, it is thought that the British nuclear force could inflict enough damage on the USSR to make it very improbable that the Soviet government would risk making a nuclear attack on the UK.

Obviously there is no technical answer to the question of how much damage must be threatened to deter an aggressor (see **Assured Destruction**). Deterrence is inherently a psychological phenomenon in the mind of the would-be aggressor. Some argue that the existing UK nuclear force, the **Polaris** submarine squadron, is too small actually to deter, partly because its threat is too weak to have **Credibility.** In contrast, even senior US strategists have sometimes admitted that the explosion of just a few nuclear devices over a small number of cities would be so unparalleled a disaster that no government could ever risk embarking upon a strategy which invited such retaliation. The UK government has always used the argument that its strategic forces are at the minimum deterrence level to justify their contention that they have no obligation to offer to reduce them as part of any **Arms Control** agreement.

Ministry of Defence (MOD)

The British Ministry of Defence was the first genuinely integrated defence administration of a NATO member. In a series of reforms in the late 1950s and early 1960s the originally separate service ministries (the War Office, the Air Ministry and the Admiralty) were merged into one super-ministry to avoid duplication of defence preparations, and to facilitate rational economic planning. (There had been a Ministry of Defence before, but it had existed parallel to and separately from the service ministries, with no more than a co-ordinating function.)

Later developments, particularly since 1979, have increased the extent of integration. Since 1982 there have been no appointments of junior ministers with specific service roles, and the authority of the **Chiefs of Staff,** and specifically of the Chief of the Defence Staff, has been increased. Although **Inter-Service Rivalry** still exists, it is nothing like as intense as in other countries, notably the USA where separate service departments with their own political heads cohabit with a co-ordinating **Department of Defense.** The integrated MOD is able to consider overall national strategic interests, especially in budgeting and in the procurement of weapons, more successfully than where strategic planning has to be a compromise made only after separate plans and budgets have emerged from the various services.

Minuteman

The Minuteman family of missiles, named after the early guerrilla fighters against the British in the American War of Independence have been, and continued to be so into the late 1980s, the backbone of the US **ICBM** Force. Minuteman I, designed in the mid-1950s and deployed from 1960 onwards, was the first solid-fuel missile with intercontinental range and, like its successors, was housed in concrete **Silos** in the American Midwest, under US Air Force control. The advantage of solid-fuel missiles is that they can be fired very rapidly. Liquid-fuelled missiles, which continue to feature prominently in the Soviet ICBM force, have a lengthy preparation time and cannot be kept at permanent readiness.

Two subsequent generations of Minuteman have been deployed (all the Minuteman I stock has now been withdrawn). Minuteman II, which was a development of the original model, entered service in 1966, and still accounts for 450 of the US force of approximately 1,000 ICBMs. The third generation missile, Minuteman III, is really a completely new design, built during the 1970s, and makes up the remainder of the ICBM force. The Minuteman III missiles are equipped with **MIRV** technology **Front Ends** and carry three independently-targetable warheads, variously rated at **Yields** between 170 and 335 kilotons. They are not particularly accurate, with **Circular Error Probables (CEPs)** varying, depending on type, between 180 and 275 metres. This makes them only marginally useful against **Hard Targets.** The older Minuteman II missiles have not been MIRVed, and are reported to have single warheads of slightly over one **Megaton,** with a CEP in excess of 300 metres.

SALT II treaty conditions place an upper limit on the US ICBM force, so the deployment of **MX** and **Midgetman,** if the latter is ever built, will probably be accommodated by decommissioning the Minuteman II force. As these older missiles are only capable of **Counter-Value** strikes, adequately catered for by other legs of the nuclear **Triad,** their replacement by more accurate and, in the case of MX, MIRVed, missiles would be an overall enhancement of the US **War-Fighting** capacity.

MIRV

The development of multiple independently-targeted re-entry vehicles in the late 1960s was probably the single biggest escalation in the nuclear **Arms Race** there has been. It was led by the USA, and many American strategic analysts, now faced with **Arms Control** problems and Soviet nuclear **Parity,** wish their country had never developed the technology. A MIRV is one of a set of **Warheads** (otherwise known as **Re-entry Vehicles**) carried by a single missile which can be directed with great accuracy at any target in a broad area, separately from the other warheads in the missile's **Front End.** Thus a single missile is effectively multiplied, being capable of attacking several quite separate targets within what is called its 'footprint'. Obviously weight limitations on the **Payload** capacity of the missile mean that each of the MIRVs it carries has to be less powerful than the single warhead that earlier generations of missiles carried. Most MIRVs have a destructive power of under 300 kilotons, compared with earlier single warheads rated at or above one **Megaton.** However, the phenomenon of **Equivalent Megatonnage,** and the fact that most targets, except for **Hard Targets,** would be destroyed by **Yields** considerably less than a megaton usually make this restriction unimportant.

The number of MIRVs in a missile's front end varies considerably. Until recently the usual payload for a US missile, for example an **ICBM** such as the **Minuteman III** or an **SLBM** such as the **Poseidon,** was only three warheads, each of between 150 and 350 kilotons. The USSR's large ICBMs, such as the **SS-18,** can carry up to 10 MIRV warheads of around 900 kilotons. The newest generation of US missiles, the **MX** ICBM and the **Trident II** SLBM have the capacity to carry as many as 14 – 17 warheads, each likely to be of about 350 kilotons. The UK **Polaris** missile is not independently targetable, being a **Multiple Re-entry Vehicle,** although the Chevaline updating of its front end is rumoured to have given it MIRV-like capabilities. However, the new Trident system, due for deployment by the Royal Navy in the 1990s, will probably have eight MIRV warheads, and the French are also developing MIRV capacity for the generation of SLBMs which they will deploy at about the same time.

MIRV capacity has had several serious consequences for arms control, the most obvious being a problem of comparability. When missiles had single warheads it was relatively easy to know what the respective size of the two superpowers' nuclear arsenals was, and the **SALT I** treaty, which concentrated on missile launchers, was based on such arithmetic. Now, however, the number of missiles may be a very poor indicator of actual strength. In addition, the problems of designing **Ballistic Missile Defences** have been vastly increased because a small number of missiles carried, for example, by a few submarines, can launch hundreds of warheads. As they will be accompanied by **Decoys** this presents the enemy with a virtually insoluble problem. The British Trident fleet, for example, will be able to fire at least 256 warheads from just the two submarines (see **SSBNs**) expected to be at sea at any one time, while the full planned American SSBN fleet, even assuming only 60% of the craft were available, could strike at over 2,000 targets simultaneously.

Until recently the most serious threat to the USA has been from the USSR's very large ICBMs, because only they could manage the combination of explosive power and accuracy (see **Circular Error Probable**) that was required to destroy a hard target, such as a missile silo. However, the Trident II MIRV system is very much more accurate than its predecessors, and these missiles will present to the USSR the same threat of a **First Strike** that the USA has believed itself to be living under during the period known as the **Window of Vulnerability.**

Missile Experimental (see MX)

Missile Gap

At the end of the 1950s American confidence in their technological lead in nuclear weapons had been seriously eroded by a sequence of surprises: the speed with which the USSR tested first an **Atomic Bomb,** then a **Hydrogen Bomb,** and finally the shock of their **Sputnik** programme. This, combined with the low quality of US strategic intelligence at a time when satellite surveillance was not available, resulted in the USA massively over-estimating the size of the Soviet **ICBM** force. It came to be believed that there was a 'missile gap' of dangerous proportions, which made the US forces, particularly their aircraft-deployed **Strategic Air Command,** which still held the bulk of their warheads, terribly vulnerable to a Soviet **First Strike.**

The missile gap featured prominently in the presidential election campaign of 1960, when the Democratic candidate, John F. Kennedy, used it to establish his credentials in defence and foreign policy. The belief in the gap was sincere, but by the time of the biggest crisis in US – USSR relations, the **Cuban Missile Crisis** of 1962, the USA had learned from Soviet defectors that the gap was illusory. In fact some estimates suggest that by 1963, when the USA had several hundred **Minuteman I** ICBMs, the USSR had deployed only eight missiles capable of reaching North America.

Montebello Decision

The 'Montebello Decision' refers to the policy adopted in October 1983 by NATO defence ministers, meeting in Montebello, Canada, under the auspices of the **Nuclear Planning Group,** to reduce the stockpile of **Battlefield Nuclear Weapons** held in Europe. The decision was to withdraw 1,400 such warheads by 1988, and in addition to withdraw one extra warhead for each **Ground-Launched Cruise Missile** or **Pershing II** deployed. While the move was clearly intended as an **Arms Control** gesture, it is also true that the warheads, many of which were atomic mines known as **Atomic Demolition Munitions** were obsolete. What is less often realized is that the Montebello Decision also involved a commitment to modernize those stocks which were retained. Even the new levels which, combined with reductions made between 1979 and 1983, reduce the overall arsenal by 2,400 warheads, retain far more battlefield nuclear shells than could ever plausibly be used.

Morale

Morale is the virtually indefinable psychological element in military units that has a crucial effect on their efficiency, courage and general combat effectiveness. According to Napoleon the result of a battle depended three-quarters on morale and only one-quarter on the balance of forces. One aspect of morale, the soldiers' confidence in their leadership, is illustrated by a frequently-made comment on Napoleon himself—that his presence on a battlefield was worth several divisions of troops. More scientific evidence of the importance of this psychological element has been available since the Second World War. American studies of conflict in that war shocked military experts when they revealed that as many as 50% of soldiers in combat never fired their weapons at all. Later studies of the US troops in the **Korean War** tended to show similar results, with the additional finding that it was confident, educated and willing soldiers who were much more likely to fire. Clearly anything that encourages more soldiers to take an active part in battle would act as a **Force Multiplier.** A further example is seen in one of the primary failings of the Allied **Strategic Bombardment** offensive in the Second World War. Attacks on important industrial targets usually had little effect because most bombs were dropped far from the actual target; as each wave of aircraft came in bombs were dropped on the first fires spotted on the ground, rather than as near to the centre of the target as possible, with a resulting 'creep back' effect. Only élite squadrons, with high self-confidence and sense of purpose, could be relied on to make sure that they really were attacking the target.

There is not only no clear definition of morale, but also no obvious answer as to how it is built up. High morale generally exists where a combination of factors apply. There must be a real sense of competence; the military personnel must each believe in their individual ability to achieve the task set them. There must be great confidence in the leadership, so that orders are obeyed without question. There must be confidence and trust within each unit: individuals must believe that their colleagues will support them reliably. Finally, there must be a sense of validity in the purpose of the war; troops must believe, essentially, that they are fighting for 'right', and that the enemy is in the 'wrong'.

There are countless techniques for bolstering morale. All military structures rely to some extent on political indoctrination, with the Soviet military practice of attaching a senior political officer to each unit to carry out extensive ideological training being the standard example. Even where such formal mechanisms are missing the use of stereotypes of the enemy as callous and brutal is common, and anything tending to reduce this image is discouraged. The story of how British and German soldiers fraternized in no man's land during the first Christmas of the First World War is well known; less well known is the fury of the High Command on both sides, who forbade any repetition because it would tend to undermine the necessary hatred of the enemy fostered by rumours of 'atrocities' committed by them.

More positive morale building is well exemplified by the long-standing British regimental tradition by which soldiers are recruited from particular counties. Great effort, through insignia and ceremonial, is put into teaching the history of the regiment. This encourages communal identity and pride not just in the present, but also going back in the history of the regiment. (It is interesting that the US army, which has never had such a system, is now trying to create a general equivalent by ensuring that intakes of soldiers serve continuously together for as long as possible.) At a more practical level military services often invest much more effort in recovering their wounded and getting them rapidly to rear base hospitals than cold-blooded planning might warrant. This ensures that soldiers do not feel that they will be abandoned if injury removes their tactical value.

A second aspect, **Civilian Morale,** has a distinct impact on military morale. The sense that the troops are being backed with enthusiasm by the home population, who can then see themselves as valued and honoured defenders of the innocent civilians of their nation, is vital. Many senior US officers attribute the collapse of morale among units in the Vietnam War to the hostility of civilian public opinion in the USA, which treated returning soldiers, even though they were conscripts, as though they were personally responsible for fighting an evil war by evil means.

One unanswerable and controversial question about military morale is whether professional **All Volunteer Forces** are more or less likely to have high morale than conscript armies (see **Conscription**). What does seem to be clear is that the conditions under which military forces are engaged has a direct impact upon morale. A conscript army certainly requires a sense of justifiable political purpose unless its morale is going to suffer, whereas aspects such as regimental tradition and pride in one's own competence are often enough to motivate a professional unit.

Moscow Option

The Moscow Option, sometimes also called the 'Moscow Criterion', is the unofficial but widely-accepted plan for the use of the British **Strategic** nuclear force. As the **Polaris** squadron of ballistic missile-carrying submarines **(SSBNs)** is relatively so weak as to be of only **Minimum Deterrence** value, it is crucial that the damage it can inflict is seen to carry the maximum threat to the Soviet Union. Since the mid-1960s, at the latest, this threat has been defined as the ability to destroy Moscow in retaliation for a nuclear attack on the UK (see **Weapon of Last Resort**). Britain has never had a professed nuclear strategic policy (see **Declaratory Policy**), unlike the French and the Americans, but leaks, as well as reading between the lines of what official statements there are, makes it clear that this capacity is seen as absolutely vital. One strong piece of evidence for this is the great fear UK analysts have always had that the Soviet **Ballistic Missile Defences (BMD)** would render the Polaris squadron ineffective. For this reason the costly **Chevaline** programme to redesign the **Front End** of the Polaris missiles was carried out, in great secrecy, during the 1970s. The sole purpose of Chevaline is to ensure that the **Galosh** BMD screen around Moscow can be penetrated.

The Moscow Option seems to have been chosen for two related reasons. The first is the assumption that the most valued target, the one a country will least wish to risk, is its historic capital. The second is that for some time, perhaps even as early as the 1950s, there has been a strand in British strategic thought which holds that the real aim of a nuclear attack must be the destruction not so much of the enemy's industry or population, but its political structure, usually referred to as 'the organs of State Power'. Obviously destruction of the capital would go a long way towards achieving this.

Many critics of the Moscow Option suggest that it may be the *least* sensible target to pick. To start with, there is something odd about using the **Nuclear Club's** least powerful strategic force against the only city in the world that has an **Anti-Ballistic Missile** screen. Secondly, the real centre of State Power in a nuclear emergency will not be in Moscow at all, but in the heavily protected and scattered command and control bunkers (see **C^3I**). Even the USA doubts that it has the capacity to destroy these comprehensively. Finally, it is asked whether trying to 'decapitate' the enemy command, with the result that no **Intra-War Bargaining** could be carried out, is anything but a suicidal policy. This latter objection is currently somewhat irrelevant, because the Polaris force could only be used as a single-strike last resort retaliation, but it will be a considerably more powerful criticism if, once **Trident** is deployed, the Moscow Option is adhered to. Trident will give the UK the capacity to approach American **War-Fighting** strategies, with the use of **Limited Nuclear Options,** and much more subtlety will be required than is in the existing UK strategic plan.

Multi-Bipolarity (see Bi-Multipolarity)

Multilateral Force (MLF)

The idea of a multilateral nuclear force, which was popular under a variety of names from the mid-1950s to the mid-1960s, was an attempt to solve the problem of NATO's collective reliance on nuclear weapons when these weapons were almost entirely under American control. The Western Europeans had always been worried about the US **Nuclear Umbrella.** This was either because they feared, like the French and West Germans, that the USA would not, in the end, take the risk of using nuclear weapons to protect Europe, or, on the contrary, that the Americans would be too 'trigger happy' and that European authorities would not be able to control **Escalation.** The problem was exacerbated by the position of West Germany. Under the Paris Agreements of 1954 the Federal Republic was not allowed to develop its own nuclear weapons. However, it was also both the territory in which **Battlefield Nuclear Weapons** would be used, and the country most anxious for really effective nuclear **Deterrence** to protect it against the danger of becoming the **Battlefield** for a new and destructive conventional war.

It was felt that some mechanism must be invented to allow the whole of NATO to control and operate its own nuclear guarantee. The MLF and similar plans, such as the 1965 UK proposal for an **Atlantic Nuclear Force,** were the resulting suggestions, and the creation of such a force briefly became official US policy under the Kennedy administration. There was no great clarity about what the MLF should be, although the most common suggestion was a surface fleet carrying **Polaris** missiles, jointly staffed on an international basis by all NATO countries, and with command going to an American officer of equivalent status to **SACEUR,** but with authority residing in a council of all NATO members.

The problems, on close inspection, were tremendous, mainly because of the conflicting motives of the supporters of an MLF. These ranged from an American desire to dispense with the small **Independent Deterrent** forces under French and British control, to Western Europe's desire to make the US nuclear guarantee concrete. No plan made much sense, because any scheme would effectively produce a single nation veto. If the USA was unwilling to fire, it could lodge a veto. If Western Europe was unwilling to escalate, the huge US arsenal outside the MLF framework would make European control of the MLF weapons meaningless. In general it seemed that any scheme which satisfied the European desire to have some control over nuclear release decisions would merely create inertia, and thus reduce the **Credibility** of the force. Therefore the plans for the MLF never came to fruition.

Multiple Independently-Targeted Re-entry Vehicle (see MIRV)

Multiple-Launch Rocket System (MLRS)

The multiple-launch rocket system is a remote descendant of the multi-barrel rocket projectors used during the Second World War. However, the modern versions just coming into service with NATO armies are very highly sophisticated. Capable of firing their 12 rockets within a few seconds, they are used for two distinct purposes. The first is in support of conventional artillery tasks, along with the normal howitzers and cannons. Far more important is the second role, in which they are in the vanguard of the new generation of **Emerging Technology** weaponry. In this role they are capable of bombarding a target far beyond normal artillery range with about 7,000 grenade-size submunitions spread over an area equivalent to six football pitches. Thus any personnel and equipment not in a heavily-armoured cover will be destroyed from just one launch. The MLRS will become even more deadly in the near future, as the USA is co-operating with **European**

NATO members to develop 'terminally guided' or **Smart Bombs** to be fitted to these rockets, giving the weapon a very long-range anti-armour capacity in addition to its current predominantly anti-personnel function.

Multiple Re-entry Vehicle (MRV)

A multiple re-entry vehicle is a **Warhead** carried along with one or more others by a single missile. There are no MRV systems left in the American or Soviet strategic nuclear forces because of the development from MRV to multiple independently-targeted re-entry vehicle **(MIRV)** technology. The British and French **(SLBM)** forces, however, still rely on this generation of weapons. In the MRV system a missile carries several warheads, and releases them while outside the earth's atmosphere to complete individually the **Ballistic** trajectory so that the warheads hit the target area separately. However they cannot, unlike MIRV warheads, be aimed at separate targets, and instead detonate in a scattered pattern within a single area. The principal advantage of an MRV **Front End** over the older generation of single warhead missiles is that the several incoming warheads present a multiplicity of targets for any **Anti-Ballistic Missile** system to cope with. Furthermore, for certain targets like a very large city, the quite widely-separated detonation of three or more relatively small warheads can be far more destructive than the detonation of a single **Megaton**-level warhead (see **Megatonnage Equivalent**).

Multiple-Role Combat Aircraft (MRCA)

The ever-increasing cost of high-technology **Weapons Systems** has forced all nations to try to maximize the capacity of any such equipment which they procure. The price escalation is particularly fierce with aircraft because of the combined need for greater and greater performance from engines, and to provide pilots with very sophisticated computing and electronics gadgetry to enable them to fly the planes at all under prevailing military conditions. The consequence has been that each generation of combat aircraft costs, in real terms, nearly twice as much as its predecessor. Superpowers, although it strains even their budgets, can afford to develop and deploy several different models of aircraft, tailored for specific purposes, in part because the scale of production required to equip a large air force allows cost reductions. Second rank powers, even those as wealthy, and as prepared to spend money on defence as Britain, France, and West Germany can no longer afford to equip their air forces in this way.

The solution has been to build only one or two types of aircraft, and to design them to do all of the jobs that the USAF or the Soviet air force would assign to different design types. The multi-role combat aircraft concept is the result. The main example currently in operation is the

Tornado, flown by the British, West German and Italian air forces. It is intended to act as an army co-operation fighter to protect ground forces from air attack. An equally-important role is as a bomber to carry out the vital **Interdiction** role that modern NATO doctrine stresses so heavily in its **Follow on Forces Attack** strategy. Thirdly, in a slightly different variant, it is an **Air Superiority** fighter intended to protect against major Soviet bombing missions on, for example, the UK mainland, and in this role it has to be capable of taking on purpose-built Soviet fighters at long ranges.

The success of the Tornado is a very important example of the feasibility of arms construction co-operation by **European NATO** allies, a move vital to reducing costs and increasing **Interoperability,** as it was built jointly by a consortium of the three countries that fly it. It is also, however, a good example of the tremendous cost of competing with the superpowers in modern arms production. Britain's share of the cost of Tornado will total about £17,000 million (nearly twice the official government estimate for the cost of the **Trident** submarine programme), and each Tornado aircraft costs around £20 million.

There are obvious political problems about such programmes: an aircraft that has to carry out several tasks has to be a compromise between different service interests which, given **Inter-Service Rivalry,** can be hard to obtain. One that is to be built jointly by several countries involves an even greater degree of compromise, which sometimes proves politically impossible to obtain. This has been the case with the follow-on generation to Tornado, known as the European Fighter Aircraft (EFA). Originally France was to join with the builders of Tornado to form a new consortium, but after years of negotiating effort the different needs have proved irreconcilable. France has now withdrawn from the project, increasing the unit cost and the economic burdens for the remaining nations, even though Spain is also now involved with the EFA. (See also **Independent European Programme Group.**)

Multipolarity

Multipolarity is a term used to describe an international situation where there exist several nations of roughly equal power, rather than the two-superpower **Bipolarity** of today. It is, in fact, the basis of classic **Balance of Power** theories, in which individual states are protected by an automatic tendency for alliances to be formed against any one country that seems to be getting relatively more powerful than its partners in a multipolar system.

This classic multipolar international system certainly existed in the 18th and 19th centuries, when Britain, France, Russia, Prussia and Austria were either roughly equal in power or were otherwise prevented from establishing individual supremacy. There are those who wish to interpret current world politics as essentially multipolar because China has, at least potentially, enough economic and military power to offset either the USA or USSR. Others argue that Western Europe, were it only to achieve some real degree of effective political unity, would be superior economically to either of the superpowers. Neither Western Europe nor China has, however, realized its potential, and the current world system may be better described as **Bi-Multipolarity.** This, a more arcane part of balance of power theory, suggests that the rivalry between two nations in a bipolar relationship can be constrained by the secondary, but significant, power of their allies providing a multipolar second level of interaction.

Mutual and Balanced Force Reductions (MBFR)

The MBFR talks have been conducted between NATO and the Warsaw Pact since October 1973, with no substantial successes at all. By the end of the 1986 negotiating period there was increasing belief in Western **Arms Control** circles that the time might have come to close down the talks and try an entirely new forum. The Vienna-based MBFR negotiations are unlike, for example, the **SALT** or **START** talks, in that they directly involve most members of the two alliances rather than just the two superpowers. All of the Warsaw Pact countries participate, as do all NATO members, with the exceptions of France, Iceland, Portugal and Spain. This in itself leads to problems, because the NATO Council, in constructing the negotiating position for its diplomats to present, has to satisfy too many divergent internal views, while the USSR completely dominates the position adopted by the Warsaw Pact.

The aim of the MBFR talks is to contribute to stability and reduce armament expenses by limiting the land-based troops on the **Central Front** to a maximum of 700,000 army, and 900,000 army and air force combined, on each side. This would be so much more in NATO's interest that it is difficult at times to see why the Warsaw Pact should be expected to co-operate. Such an agreement would not just remove their traditional advantage in conventional forces. Until recently it would also have left the Warsaw Pact forces vulnerable to the NATO superiority in **Theatre Nuclear Forces.** However, as both alliances claim to be purely defensive there is no good reason for *not* accepting an equal numbers agreement: an invader would require a superiority of perhaps two or three to one before launching an attack with confidence, so both sides could feel safe if parity was established (see **Force Ratios**).

There have been two main stumbling blocks in the MBFR process. One is the familiar problem of **Verification** that occurs in most arms control negotiations. As with the problem of verification for a **Comprehensive Test Ban Treaty** or the **Chemical Warfare Convention,** NATO insists on the need for on-site inspections which the USSR have been unwilling to grant. **National Technical Means** are inadequate for these purposes. The second problem is often referred to as the 'methodological' problem.

Clearly, before a series of troop reductions that will lead to conventional parity can be introduced, there has to be agreement on the personnel levels which each side has deployed in Europe. Unfortunately the NATO figures for Warsaw Pact troop levels are considerably higher than those admitted to by the Soviet Union and its allies. This problem became so much of an obstruction that, in 1985, NATO offered to begin by agreed cuts in both sides *without* waiting to come to agreement on the existing levels. The Warsaw Pact had not accepted this proposal by mid-1987.

Although the MBFR process has been accused of being political time-wasting by both sides, it may in fact become of more urgent importance for the West. There are two serious problems facing NATO. One is the likelihood that the USA will need to reduce its **Forward Based** troops because of budgetary pressure. The other is that West Germany, with the biggest single permanently-deployed army, is facing a severe demographic problem and will not be able, even with **Conscription,** to maintain its troop numbers over the next decade. Therefore any agreement that can be gained from the MBFR talks to reduce Warsaw Pact troops, even if the Eastern bloc retains an advantage, would be of great benefit to NATO.

Mutual Assured Destruction (MAD)

Mutual assured destruction, usually known by its accidentally ironic acronym MAD, was the essence of Robert **McNamara's** new strategic policy for the USA, and has remained until recently the working, if not the official, doctrine. MAD involves both superpowers having a secure **Second Strike** capacity at a level guaranteed to utterly destroy the opponent in retaliation for a **First Strike.** Given such a situation it was argued that nuclear war was impossible, because neither superpower could ever hope to evade destruction, and consequently any major use of nuclear weapons would be effectively suicidal.

Obviously there is no easy definitive answer to what would constitute the **Assured Destruction** of a society, unless it is taken literally to mean the extinction of life in that country. McNamara set rather arbitrary limits, which varied from time to time. All the versions, however, were in the order of the killing of approximately one-third of the enemy's population and the destruction of two-thirds of its industrial capacity. These figures in fact derive from an accidental phenomenon of the industrial and urbanization structure of the Soviet Union. It so happens that population distribution and density is such that a graph plotting megatonnage detonated over cities against these two variables, civilian deaths and industrial capacity destroyed, flattens very rapidly at the 400 **Equivalent Megatonnage (MTE)** level. The delivery of this level of **Yield** would kill about one-third of the population immediately, but even doubling the megatonnage delivered would not significantly increase the death rate or industrial destruction. As 400 MTE was well within the capacity of any one of the three legs of the US nuclear **Triad,** it served as a convenient measure. As it happens, McNamara's real motive for establishing such targets had as much to do with controlling the **Strategic Air Command's** insatiable desire for more weapons as with deterring the Soviet Union from launching a first strike.

The idea of this form of **Counter-Value** strike has long been abandoned in US nuclear planning, if indeed it ever was a real goal, and replaced with what is known as counter-**Recovery Capacity** targeting. The USA officially no longer targets population *per se,* although it is, of course, impossible to attack recovery capacity, which effectively means major industry, without killing millions of people. As McNamara's own preference was for the **No Cities Doctrine,** and as NATO then and now rests on an **Escalation** ladder in its concept of **Flexible Response** which has **Central Strategic Warfare** as its ultimate step, mutual assured destruction may never have been more than a useful way of expressing a **Deterrence** position.

The logic behind the development of the **Strategic Defense Initiative (SDI)** does imply the abandonment of MAD, because the mutuality of the doctrine was all-important. It was felt that it was necessary for the USA to make itself vulnerable to an equivalent destructive threat which it held over the USSR in order to stabilize the **Arms Race,** which was one reason for America giving up its first attempt to create a **Ballistic Missile Defence.** Clearly some versions of the SDI theory, those that claim to protect US cities from nuclear attack, would remove that vulnerability. This may be less destabilizing than opponents of SDI like to claim, however, because there is no evidence that the USSR ever accepted MAD as its own doctrine. Instead, by investing in **Civil Defence** and developing a strategic doctrine of nuclear **War-Fighting,** they sought to make their own security depend on something other than a shared vulnerability.

MX

MX stands for 'Missile Experimental', the new generation of US **ICBMs** officially christened 'Peacemaker' by President Reagan. The MX missile is supposed to close a gap between Soviet and American missile forces that has led to the fear of a **Window of Vulnerability** in the USA. This fear has come about because the USSR has deployed a large force of very heavy and fairly-accurate ICBMs (see **SS-18**) capable of destroying **Hard Targets** in America, and possibly capable of carrying out a **First Strike.** The existing US ICBM force of **Minuteman** missiles are neither accurate enough, nor carry sufficient number or weight of warheads, to have an equivalent capacity. MX was first planned by the 1976 – 80 Carter administration, but has run into continued trouble with Congress on the grounds of both **Arms Control** and cost. President Reagan fought hard for funds to build the missile, and 50 have been authorized and are due to be deployed. The MX has a much higher **Throw-Weight,** of about 3,600 kg, than Minuteman III, and a similar range, at 13,000 km. Most importantly it will carry 10 **MIRV** warheads, against Minuteman III's three, each with the same yield of some 335 kilotons, but with far greater accuracy, expected to be measured at a **Circular Error Probable** of not more than 100 metres.

Clearly MX is a major enhancement of the USA's hard target strike capability. However, if deployment never exceeds the currently-planned 50 missiles it will not constitute a first strike weapon, because of the large number of targets it would leave undestroyed. The major disadvantage with the MX is that, as no one has yet solved the **Basing Mode** problem, it will remain as prone to problems of **Vulnerability** as the Minuteman, into the silos of which it is being placed.

N

National Command Authorities (NCA)

A major problem in designing a strategic nuclear system, especially one based on the **Fail-Safe** concept of protection against **Accidental War,** is that of establishing who can give the order to launch a nuclear attack and under what conditions. Ideally the authority to order nuclear release is restricted to the highest political level possible, in all countries. The US and French Presidents, the British Prime Minister, and the Chairman of the Politburo of the USSR are normally considered the only holders of this authority among the NATO and Warsaw Pact powers. However, it is manifest that a surprise attack could easily kill such individuals, or cut them off from communication, regardless of any precautions that might be taken. The phrase National Command Authorities is the inevitably vague way of describing the list of those on whom such authority might devolve under conditions of nuclear war.

Although the first few steps of this chain are political in all countries, it is less widely realized that military officers, possibly not of the highest seniority, could quite conceivably end up with national command authority. If the US President and cabinet are killed or prevented from exercising the authority, for example, responsibility would be transferred to the Commander of the **Strategic Air Command.** But officers of this rank may themselves be killed in a major attack aimed at destroying the command and control bunkers (see **C^{3}I**). The stand-by position in the USA, and almost certainly elsewhere, is to have an officer some ranks lower permanently airborne or on alert for immediate take-off. This officer would be empowered, under specific conditions, to authorize nuclear attacks, and many credible **Scenarios** involve such military personnel being the only surviving authorities.

National Security Agency (NSA)

The National Security Agency is the US equivalent to the British Government Communications Headquarters (GCHQ). It is much less well known than the **Central Intelligence Agency (CIA),** but arguably much more important. It handles the interception, decoding and interpretation of virtually all **Signals Intelligence (SIGINT),** and indeed most of the other forms of **Electronic Intelligence (ELINT),** for the USA. Unlike the CIA it is not fully independent, but comes loosely under the control of the **Department of Defense,** although most of its employees are civilians.

It is a huge institution, employing several thousand people with a vast headquarters north of Washington (DC), and with outstations for electronic eavesdropping throughout the world. In its first 20 years of existence after the Second World War it had a series of spy scandals, as a result of which many of its activities became public knowledge. These include flying ELINT aircraft near and sometimes inside the borders of the USSR in order to trigger their radar procedures, and intercepting diplomatic and military communications from almost every country in the world, including America's allies. Since the Congressional reform of the intelligence services during the 1970s it has come rather more under political control than previously. However, it remains very secretive and is probably America's single most important intelligence source. It is also known to have very close relations with Britain's GCHQ, with which it co-operates and shares information.

National Security Council (NSC)

The National Security Council is the US President's principal source of military and strategic advice because, unlike the **Department of Defense (DOD)** or the Department of State, which are independent government departments represented by cabinet secretaries fighting for their own departmental interests, it is directly under presidential control. The NSC is headed by the President's Adviser for National Security Affairs, which is one of the few posts not given to a leading politician or personal associate. Presidents take great care in their appointment to this post, often choosing either someone from the academic defence analysis world or a military professional. Such officials can, of course, go on to political careers—Dr Henry Kissinger was National Security Adviser before becoming Secretary of State.

The National Security Council is relatively small, and the staff, often young, tend to serve there only for one Presidential term, being replaced when a new incumbent enters the White House. As a source of intelligence evaluation, as well as representing the President's own **Think-Tank** for developing policy in all military, strategic and foreign affairs matters, the

NSC is a natural rival to the huge bureaucratic institutions like the Department of State and the **Pentagon.** Very few policies have any chance of presidential support unless the National Security Adviser and the NSC staff can be persuaded to agree to them.

Technically the NSC is, as its title suggests, a council, established by Congressional statute, which the President is bound to consult, and which includes the Secretaries of State and Defense, and the Director of Central Intelligence as well as the National Security Adviser. It was set up as part of the overall post-war reorganization which also created the DOD and the **Central Intelligence Agency.** In fact, what people almost invariably mean when they refer to the NSC is not the council at all, but the staff of the council, whose advice goes directly to the President who may or may not share it with the council.

National Service

National Service was the official name of the military **Conscription** system operated in Britain from early in 1939, when war was already feared, until the early 1960s. Once the Second World War was over conscription continued, with virtually every male over 18 serving in one of the services, usually the army. Unlike other systems, notably the American **Selective Service** system, there were no exemptions except on grounds of health and, for a very few, as conscientious objectors who could demonstrate the sincerity of their pacifist beliefs.

Politically, National Service was necessary because of the extensive imperial commitments Britain retained until the end of the 1950s, although it was never popular with the senior military commanders. The very large conscript army required an enormous investment in training, and inevitably led to a lower degree of military professionalism than in the all-volunteer army which replaced it after 1962. As soon as the colonial tasks were reduced, in the wake of successful independence movements, the Conservative government, in 1957, planned to abolish National Service, although various emergencies kept it in being for several years more.

It would probably be politically impossible to reintroduce conscription in Britain, although there are advocates of such a policy, impressed by the difficulty of matching Warsaw Pact troop numbers. Those who do urge the return of National Service, and of **Selective Service** in the USA, point out that the USA and the UK are the only major NATO countries not to require military service from their citizens; indeed, most of non-NATO Europe also has conscription. The loss particularly bemoaned is not just the active duty personnel, but the absence of the very large ready reserve of those who have had military training enjoyed by countries which do have conscription. As greater reliance on reserves is often cited as a way round the problem of troop shortage for NATO, the UK is clearly at a disadvantage. Whether the advantage of an all-volunteer military service overrides this is unclear.

National Technical Means (NTMs)

National technical means is a term which refers to methods of **Verification** in **Arms Control** agreements. NTMs are principally satellites which can relay sufficiently-detailed photographs to reveal whether a country is building missile silos, concentrating forces, or engaging in any other activity in contravention of an arms control agreement. However, developments in **Weapons Systems** are increasingly making NTMs less effective. A major example is the introduction by both the USSR and the USA of mobile **ICBMs** which, because they can be hidden or camouflaged, are less easily verifiable. Some crucial aspects of desired arms control agreements cannot, even in principle, be verified by NTMs: a satellite cannot, for example, show what is going on inside a civilian chemical plant that another country fears is being used illegally to make weapons for **Chemical Warfare.**

There is a variety of other technical means by which one country can check on another's compliance with an agreement without having to demand politically sensitive on-site inspection. For example, geological measuring techniques can check on nuclear testing, and **Electronic Intelligence (ELINT)** can monitor radar stations.

NATO

The North Atlantic Treaty Organization was set up by its original twelve members when the Treaty was signed in Washington (DC) on 4 April 1949. These members were: Belgium, Canada, Denmark, France, Iceland (which has no military forces of its own, but provides bases for other countries), Italy, Luxembourg, the Netherlands, Norway, Portugal, the UK and the USA. Four other countries joined subsequently, the most important by far being the Federal Republic of Germany in 1955. The other three were all on NATO's **Southern Flank:** Greece and Turkey in 1952, and Spain in 1982. Although there is no membership requirement that a country be democratic, it was because Spain was politically unacceptable that it did not become a member earlier. However, the USA did have NATO-related bases there from the early-1950s. France partially withdrew from NATO in 1966. While it has withdrawn only from the integrated military structure, it is not entirely clear that this is legally possible, given Article 5 of the Treaty. Nevertheless, France has remained a member of other NATO institutions and, in practice, has maintained a high degree of military co-operation and liaison with the other members on the **Central Front.**

The core of the North Atlantic Treaty is Article 5, which binds all members to regard an attack on any one of them, on either side of the Atlantic Ocean, as an attack on all of them. Although the Treaty does not specify how the various members shall come to the aid of one of its members under attack, coming to its aid in some way or other is not voluntary. Another vital clause, which has come to be more politically sensitive than was originally expected, is Article 6, which limits the area in which treaty support of a member state is required. The simplest definition of this area is that it applies to attacks on the territories of the member states in Europe or North America, and their ships and aircraft when north of the Tropic of Cancer. Thus it excludes areas such as the Indian Ocean and the Persian Gulf, where the USA nowadays tends to argue that vital NATO interests lie, and where it would like **European NATO** members to assist it in its 'world policeman' activities (see **Global Strategy**).

The most significant development of NATO, without which it would probably not have continued to be important, was the entry of West Germany. As soon as the idea of creating NATO to offset the power of the Soviet Union was thought of, West German membership became essential. In many ways NATO was, and remains, a pact with the primary aim of defending West Germany, not only out of consideration for that country, but because it is the barrier to any Soviet incursion into Western Europe in general. It made no sense to establish a huge, expensive multi-nation defence alliance without including the country whose defence was central to the whole enterprise. Furthermore, the troop numbers required could not be raised without the use of West Germany's population. Despite this the opposition in Europe, and particularly in France, to the rearmament of West Germany so soon after the Second World War was intense. Only after the failure of the **European Defence Community,** which would have integrated West German troops into a European army, but with them under non-German control, to become a reality, in 1954, was the way left open for the full rehabilitation of the Federal Republic. This was achieved through the Paris Agreements and West German accession to the **Western European Union,** leading to the acceptance of the Federal Republic into NATO, when it was allowed to raise an army entirely dedicated to NATO purposes.

The internal history of NATO has been one of continued tension, particularly between the USA and the European members. Europe tends to resent American leadership of the alliance, and America resents what it sees as European reluctance to pay properly for its own defence (see **Burden Sharing**). The former concern led de Gaulle to remove France from full participation in 1966 (see **Gaullism**). The latter objection has, several times, seen the US Senate on the verge of withdrawing or reducing the size of the US NATO contingent in Europe. There is validity in both arguments, but little chance of change. On the one hand the USA can hardly fail to lead when the official policy of NATO, **Flexible Response,** requires the use of nuclear weapons, in which the USA has a near monopoly. On the other hand, although the USA does spend about 60% of its defence budget on NATO, this is roughly its proportional share of total NATO wealth, in terms of Gross National Product (GNP) per caput, rather than just total GNP.

Despite the tensions NATO is a robust organization which seems capable of endless readjustment, and has weathered all problems so far. No other defensive alliance of its complexity has lasted so long, at such a state of readiness, during peacetime and remained, at least in broad terms, a democratic partnership of equals. The structure of NATO is extremely complex. This is because it is simultaneously both a high-technology military organization and a pluralistic political alliance. The political control and decision-making of the alliance is in the hands of the North Atlantic Council (NAC), where Ministers of Foreign Affairs from each of the 16 nations consult and reach binding decisions; such meetings take place at least twice a year. However, rather than being a fixed body of people the NAC is a functional body; meetings may also be held at the Heads of State or Government level, and the Council holds sessions at least weekly at the Permanent Representative level. Only slightly different from the Council is its Defence Planning Committee (DPC), which contains all the members except France, and exists in permanent session of Ambassadors with twice-yearly meetings of Defence Ministers. The DPC deals with the more strictly military and technical decisions that have to be made. Specifically nuclear matters are dealt with by the same people as the DPC, minus the Icelandic representative, at both levels, under the title of the **Nuclear Planning Group.** Technical military advice is given to these committees by the **NATO Military Committee,** assisted by an integrated International Military Staff.

The most important routine work of these committees and staffs is the preparation of force plans on a rolling two-year basis. A 'Ministerial Guidance' document is passed by the DPC, consisting of guidelines and political targets for defence planning both by NATO as such and by the individual members. On the basis of this a set of 'Force Goals' is prepared, delineating specific planning targets for each country for the next six years. Against the background of these documents there is an annual review of the individual national defence policy actions over the previous year, which allows, finally, the writing of a comprehensive NATO force plan for the next five years. This allows **SACEUR** to make strategic and tactical plans, with a good degree of confidence, to cover the five years ahead.

There are innumerable other agencies and institutions associated with NATO, some formally, and some less formally. One of increasing importance is the EuroGroup (short for Informal Group of NATO European Defence Ministers), which gathers to discuss and promote initiatives on purely European interests of the members. An offshoot of this, the **Independent European Programme Group,** has been particularly important in developing arms co-operation programmes, such as the Tornado aircraft, to reduce the heavy preponderance of US-manufactured weapons used by the alliance. There is also a body, technically independent of NATO, the North Atlantic Assembly, which functions as a sort of parliament of the alliance, but with absolutely no decision-making powers. It is not directly elected, but consists of delegations from the parliaments of the member states, and is not thought to have more than the most minor influence on the development of NATO policy.

NATO Military Committee

The Military Committee is one of the organs of the very complex **NATO** policy-making structure. It is the highest purely military body in NATO, although it has no command functions, these being handled by **SACEUR** and the equivalent supreme commanders for the naval roles (SACLANT and CINCHAN). The Military Committee consists formally of the **Chiefs of Staff** of all the member countries (except France which, despite not having belonged to the integrated military structure of NATO since 1966, is represented by a Chief of Mission, and Iceland, which has no military forces of its own). The Committee has the responsibility of making recommendations on defence policy to the higher civilian organs, and of giving guidance to the major NATO commanders. If, for example, SACEUR wants to adopt a new tactical doctrine, such as **Follow on Forces Attack,** or a new **Weapons System,** it would be necessary to persuade the Military Committee to make an official recommendation to the North Atlantic Council.

The regular work of the Committee is done mainly by the national Permanent Military Representatives, although the Chiefs of Staff do meet at least twice a year. The presidency of the Military Committee rotates annually among the member nations in alphabetical order.

Naval Diplomacy

Naval diplomacy has been practised at one time or another by all major naval powers. It consists of sailing ships or fleets close to the coast of smaller powers to intimidate them by a reminder of the force that could be exerted should the major power's interests be damaged unacceptably. It is related to the old British concept of gunboat diplomacy, where a minor naval bombardment of a small country would follow any hostile actions against British citizens or interests. A classic recent case of naval diplomacy was the American dispatch, in March 1986, of a **Carrier Battle Group** to contested waters off the Libyan coast in response to terrorism allegedly backed by the Libyan government. In April President Reagan ordered selective bombing attacks on government and military targets in Tripoli and Benghazi, after it was claimed that Libya had been involved in a terrorist attack on US military personnel in West Berlin. The Soviet Union has not, until recently, had the sort of **Blue Water Navy** required for such diplomacy, but will doubtless develop the tactic.

Need to Know

Security in the military, diplomatic and intelligence circles of most countries operates in two distinct and overlapping ways. People with access to sensitive material are given general security clearances at various levels, the most usual being 'Secret' and 'Top Secret'. Higher classifications also exist; the USA, for example, has a very high level of security clearance labelled 'Cosmic'. These clearance levels, however, are minimum qualifications for access to material with the corresponding classification. While it is necessary to have, for example, a Top Secret clearance to see a particular document, this does not mean that any individual with such a designation can see any document classified at that level. Here a second criterion comes into play—the so called 'need to know' rule. Any particular clearance holder may only see information that is necessary for his or her particular task or department, and would, indeed, not have access to information of a lower security classification unless there was such a need.

While the system is a sensible one, it has drawbacks, principally in over-regimenting the organization of data so that information which might be very useful, although apparently unconnected, to a particular officer's task will be withheld. Like all security systems it fails often because of an unavoidable problem. This is that the sheer mechanical workings of secretarial and clerical activity in typing out, photocopying and delivering the most highly-confidential material has to be carried out by those who have no need to know at all. Thus many espionage successes, for example even inside the ultra-secure US **National Security Agency (NSA),** have come about by corrupting messengers and clerks who have an adequately-high general security clearance, combined with an unlimited access to all information in their department which would be denied to senior officers on the need to know basis.

Nerve Gas (see Chemical Warfare)

Neutrality

Neutrality is often thought of as a rather vague state of non-involvement in international conflict, but it has, in fact, a fairly precise meaning in international law. If a state wishes to adopt a position of neutrality between two or more others who are at war, it has an obligation under international law to refrain from aiding any war-making party, or from allowing them to use its territory for any warlike purpose at all. In return for this it is to be allowed to continue trading with any or all of the war-making powers, except that the latter have the right to **Blockade** and prevent any prohibited trading, although they must exercise care to protect the nationals and ships of the neutral country. To remain within international law, no war-making party may attack the neutral state.

Although the idea of neutrality was at one time, during the era of **Limited War,** perfectly sensible and minimized the impact of war on the international community, it has not, in the last two World Wars, made a great deal of sense. In both wars, for example, protestations of neutrality did not save Belgium from invasion. In the First World War it was, to a large extent, Germany's refusal to avoid attacking neutral American merchant ships that brought the USA into the war on the side of Britain and France. Only Switzerland, which has been recognized internationally as permanently neutral since the early 19th century, was fully able to avoid favouring or being used by one side or the other. There are good reasons of mutual interest, even among the most bitterly hostile of enemies, to maintain a genuinely neutral intermediary to deal with matters such as negotiations over prisoners of war. Legally, in fact, not all nations even have the right to announce a general neutrality. All members of the United Nations, for example, share a common duty to defend each other and to assist in the punishment of an aggressor under certain conditions, and could not claim that their neutrality required or allowed them to be impartial between two parties if one had the sanction of the United Nations. In practice the only effective neutrality is what has come to be known as 'armed neutrality'. This state of affairs, and modern Sweden may be the best example, involves not just the general intention not to be involved in any war, but a manifest ability to defend its own frontiers effectively. The Swedes in fact have an efficient armaments industry and a very effective military capacity based on a large reserve and more or less total liability to **Conscription** for military training. Being able to defend oneself actually comes close to a legal definition of neutrality, because it is always open to a combatant nation to claim the need to occupy a neutral state to prevent its enemy from so doing, if it cannot trust the neutral state itself to be able to honour its legal obligation not to allow any other party to benefit from its weakness.

Considering the readiness of aggressors to invade neutral countries in the major wars of this century, the notion of neutrality in any third world war is largely imaginary. Not only is a **Central Front** war inherently likely to be nuclear, but the strategic position of a country such as Sweden would make it extremely difficult for either the Warsaw Pact or NATO to respect the neutrality of at least its airspace. Neutrality is, of course, entirely possible in limited and small wars not involving the superpowers or the major alliances, but this is largely the neutrality of those who do not care to be involved, rather than that of a small nation which *fears* to be involved. Furthermore, there are very few potential conflicts that are not at least on the fringes of the interests of the superpowers, and neutrality in the full sense of giving no aid or preference at all has not been practised by either the Soviet Union or the USA in any important post-war conflict.

Neutron Bomb (see Enhanced Radiation Weapons)

New Look

Unsurprisingly, the temptation to label strategic theories as 'new', rather as if they were brands of toothpaste or dress fashions, is seldom avoided, even though most analysts admit that there has been surprising consistency in US strategy, at the employment level if not at the more public level of **Declaratory Policy.** There are at least two distinct strategic postures that have received the nomenclature of the 'new look'. The most usual application is to the shift from conventional defence against a possible Soviet invasion of Western Europe to an almost entirely nuclear response, associated with the Eisenhower administration's **Massive Retaliation** doctrine. This plan had its inception in a vital **National Security Council (NSC)** policy paper (NSC-162/2, entitled *Basic National Security*) written in October 1953. This formal beginning of the new look combined the need for the USA to follow the **Truman Doctrine** guarantee to US allies with the need to protect the US economy from the burden of expensive rearmament. A clear-cut nuclear deterrent was advocated as the optimum solution to both goals.

The other development in national security policy sometimes called the 'new look' is the series of decisions and statements associated with US Secretary of Defense James Schlesinger's advocacy of **Limited Nuclear Options** in 1974 (see **Schlesinger Doctrine**), through President Carter's **PD-59,** to the Reagan administration's ambivalent support for the doctrine of nuclear **War-Fighting.** While very different in strategic context, these two 'new looks' share an opposition to the **McNamara** doctrine of **Mutual Assured Destruction** which was essentially a question of *never* threatening to use nuclear weapons except in retaliation to a major nuclear attack.

Nike-Hercules (see Anti-Ballistic Missile and Anti-Ballistic Missile Treaty)

No Cities Doctrine

When Robert **McNamara** became President Kennedy's Secretary of Defense in 1961 he inherited Eisenhower's **Massive Retaliation** doctrine for the employment of US **Strategic** nuclear weapons. Initially, in an effort to increase the **Credibility** of this threat by making it more likely that it would in fact be carried out, he adopted the 'no cities doctrine'. This was essentially a statement that nuclear weapons would be used, as far as possible, in the traditional way of all military power. In effect he was abandoning the threat to destroy the civilian population of the USSR for what would now be called a **Counter-Force** strategy.

This new policy was attacked immediately from many directions. To conservatives in the USA itself it seemed a weakening of resolve, in line with their general impression that the Kennedy administration was too 'liberal'. More important was the fierce resistance of America's NATO allies in Europe. From the European perspective it amounted to a partial withdrawal of the US **Nuclear Umbrella,** and an invitation to Western Europe to join in a prohibitively expensive conventional **Arms Race** against the Warsaw Pact. The Western European fear was that, by abjuring the threat of strikes against the urban structure of the USSR, McNamara's new doctrine actually made war more likely. It was seen, as indeed it was, as an attempt to undo the Eisenhower administration's attempt to replace **Defence** with **Deterrence.** Opposition to the policy was so intense that McNamara had to withdraw it and replace it with the idea of **Mutual Assured Destruction,** which remains the basis of NATO's **Flexible Response** strategy.

No First Use

No first use usually refers to the policy of guaranteeing that a power will not initiate the use of nuclear weapons. Demands that NATO in general, or the USA in particular, should formally adopt such a policy have been made with increasing frequency, and from more and more different circles in the last few years. The most famous call for no first use was published in the influential American foreign policy journal *Foreign Affairs* in 1983, in an article signed, among others, by Robert **McNamara** and McGeorge Bundy. The significance was that, far from being well-known peace activists, these men had been of vital importance in developing the US nuclear arsenal. Their call was mainly addressed to the first use of **Battlefield Nuclear Weapons** or **Intermediate Nuclear Forces** in the European theatre. The argument was that such weapons served no useful military function whatsoever. Consequently possession of them serves only to deter the Soviet Union from making a theatre nuclear attack. If NATO was publicly to promise never to initiate such an attack, it was felt that **Arms Control** progress would be facilitated.

Any such no first use declaration would contradict the official NATO policy of **Flexible Response,** which is deliberately ambivalent about when any particular level of **Escalation** will be reached. The USSR has itself promulgated a no first use declaration, but this is not taken very seriously by Western military planners, and there is no reason to assume that the Warsaw Pact would be any more impressed by a similar NATO statement.

Non-Proliferation Treaty (NPT)

The Non-Proliferation Treaty was signed in 1968 by three of the **Nuclear Club** of the time—the UK, the USA and the USSR—and by a large number of non-nuclear powers. It binds the nuclear powers not to aid any other country to develop nuclear weapons, and the non-nuclear signatories not to try to acquire them. Unfortunately the NPT is of very little use. The three nuclear members had not the slightest intention of increasing world instability by letting other smaller powers acquire such weapons. The non-nuclear powers of the day who had any desire to build such weapons simply refused to sign it. Thus, for example, Canada is a signatory as it had long been Canadian policy to renounce the ownership of nuclear weapons, while India, South Africa and Israel, all of whom are now known or believed to have such weapons, or at least the capacity to build them, all refused to sign.

The more curious omissions from the list of signatories were the other two countries which *did* have nuclear weapons in 1968, France and the People's Republic of China. In both cases it was more a matter of status than of politics. France was still in its mood of **Gaullism** in which it resented America's nuclear leadership, and China was at the height of its ideological split with the Soviet Union. Neither country would find it any more in its interest to foster nuclear proliferation than would the UK, USA or USSR.

North Atlantic Treaty Organization (see NATO)

NORTHAG

NORTHAG stands for Northern Army Group, Central Europe. It is one of two **Central Front** land forces under NATO's Allied Forces Central Europe (AFCENT) command, which is itself responsible to **SACEUR.** It is often commanded by a British general, because it is the organization to which the **British Army of the Rhine** is committed, and defends the North German Plains, one of the two historic invasion routes into Germany. The southern route is covered by CENTAG (Central Army Group, Central Europe), and is dominated by US troops. AFCENT also commands **Tactical Air Forces,** to which RAF Germany is a major contributor.

Northern Flank

The northern flank of NATO is usually taken to mean Norway, and particularly northern Norway where Soviet territory and Soviet troops are nearer to vital NATO interests than anywhere except on the **Central Front.** The particular importance of northern Norway is that aircraft based there could command the sea approaches to the Kola inlet, which is vital for Soviet naval deployment. While there is no other intrinsic value to this largely desolate area, and even though it does not provide a useful avenue of invasion to the rest of Western Europe, northern Norway is likely to be an immediate target for Soviet forces should a central front war start.

As such it has a separate NATO command of its own, Allied Forces Northern Europe (AFNORTH), and is also the primary commitment of the **Allied Command Europe (ACE)** Mobile Force, both of which are responsible to **SACEUR.** The northern flank is of major importance to the UK and the USA because of their naval role in controlling the approaches to the Atlantic, and both the Royal Marines and elements of the US Marines are designated for rapid reinforcement of the area (along with Dutch and Canadian troops). It is because of this vital role that considerable anxiety is raised whenever, as has often happened in the last few years, a British government seems hesitant about retaining the Royal Marines' amphibious warfare capacity.

NSC-68

National Security Council (NSC) memorandum Number 68, which was approved by President Truman in September 1950, was the first very formal and high-level decision that the post-war strategy of the USA would have to be based on nuclear weapons. Ironically it was mainly prepared by Paul Nitze who, much later, was to become the 'Dean' of America's **Arms Control** fraternity. NSC-68, unlike its successor document which established the **New Look,** did not argue that the USA had a special advantage in **Atomic Bombs.** On the contrary it was much less sanguine about the lasting American monopoly, predicting a major Soviet atomic bomber force by 1954. It also stressed that the natural way to wage a nuclear war was by surprise attack, and felt that the USA would always be more vulnerable than the USSR to such a strategy. Nevertheless, the authors of NSC-68 saw no alternative to US reliance on nuclear force until the weakness in conventional arms, compared to the USSR, could be rectified. As it never *was* rectified the memorandum was effectively the first step to a permanent reliance on possible first use of nuclear weapons (see **No First Use**), although it was not until later that it was openly welcomed as an efficient long-term defence plan.

Nth Country Problem

The Nth country problem was a fashionable way of talking about what is now known as the problem of nuclear **Proliferation,** and was in common usage during the late 1950s and early 1960s. The real concern of those, mainly Americans, who worried about the problem was their distaste for the British and French **Independent Deterrents,** and the nomenclature of 'Nth country' was a feeble attempt to disguise this fact. There were two distinct aspects of the problem. One was a purely American concern that NATO's power, particularly its nuclear power, should be as much under their control as possible. To this end officials and consultants from both the US **Department of Defense,** especially under **McNamara,** and the US Department of State openly opposed American aid to the UK for developing its replacement for the **V-Bomber** force, and did what little they could to dissuade the independent efforts of the French to build the **Force de Frappe.**

The second subject of concern was more genuinely indifferent to which country might become 'the Nth' to join the **Nuclear Club.** This, a common worry on the part of the more academic strategic theorists, was largely the fear of **Escalation** into a **Catalytic War** between the superpowers as a result of minor nuclear powers using their weapons against other countries which were linked to a superpower by alliance obligations.

Nuclear, Biological and Chemical Warfare (NBC)

NBC is a portmanteau concept referring to the three forms of weapons that fall outside the definition of **Conventional** war. Of the three, **Biological Warfare** is the one least understood and least clearly part of the arsenals of modern states. The grouping together of these three forms of attack comes about because, while scientifically very different, the three threats require essentially the same defensive response. While, of course, there is no defence against the direct blast and heat of a nuclear explosion, the problems of **Radiation** and **Fall-Out** necessitate the construction of airtight compartments in ships, aircraft and tanks, and the wearing of protective suiting by soldiers. Exactly the same sort of protection, and the provision of a filtered air supply, are needed to protect against the short- and possibly longer-term effects of both biological and **Chemical Warfare.** Regular exercises are carried out in all branches of the military to practise both the cleansing of personnel and equipment after an attack, and the carrying out of routine military tasks while under such an attack. It makes relatively little difference to these exercises which of the three forms of NBC warfare is being considered.

Nuclear Club

The term 'nuclear club' usually refers to those states which had gained nuclear weapons by the early 1960s and which, openly admitting to owning them, base their defence policy to a greater or lesser extent around the use of such weapons. Significantly, they are the traditional five major powers of the UN Security Council—the USA, the USSR, France, the UK and the People's Republic of China. Any more recent nuclear powers are probably not to be considered members of the nuclear club for several reasons. Firstly, there is no other power which is absolutely open about owning nuclear weapons. The only other country which is generally acknowledged to possess nuclear weapons beyond any doubt is Israel, even though the Israeli government still does not publicly admit this. India is known to have tested, but not known to have built, a nuclear weapon. With a further moderate-sized group of countries evidence mainly points to a capacity rapidly to test and build, South Africa being a prime example, but their nuclear status remains one of potential rather than actuality.

Secondly, even if several Third World countries did build nuclear weapons, it is improbable that their defence strategies would ever be based on these, rather than on conventional forces, in part because they are not locked into a complex **Deterrence** system with other nuclear powers. An exception here might be the potential for India and Pakistan both to develop such weapons and end up in a nuclear deterrence deadlock. Finally, the 'club' is to some extent a real club, where entry is dependent on the existing members letting new powers in. The common interests in holding a collective nuclear monopoly, and in trying to ensure nuclear non-proliferation (see **Non-Proliferation Treaty**) that the current five members have, especially now that nuclear **Arms Control** is a serious objective, makes it unlikely that they would ever grant a status of equality to a new nuclear power.

The remaining question of interest is why the club was so small in the first place. If two secondary powers like France and Britain could develop serious nuclear forces only a few years after the superpowers, why did more western nations not follow the same path? There is no single answer, of course. Partly it is an ethical matter: many of these countries, Canada being one, simply did not believe that they should have such weapons. More generally it is a matter of a lack of military utility. There are few **Scenarios** where it makes any sense to contemplate using nuclear weapons, and many potential developers, Sweden or Norway for example, would never need to use them. Finally there is a historical explanation. The American **Nuclear Umbrella** was already in place over Western Europe by the 1950s, and was quite sufficient. Britain and France built their forces, originally, more for reasons of prestige than out of military necessity; that drive to retain the status of a great power is not felt in many other countries.

Nuclear Depth Bomb (NDB)

A nuclear depth bomb is a depth charge to be used against submarines which is fitted with a low-**Yield** nuclear warhead. These would be dropped by aircraft, or possibly ship-borne helicopters, during **Anti-Submarine Warfare (ASW)** operations. They are usually of relatively limited power, within the range of one to 20 kilotons, and are capable of being used by most aircraft and helicopters in the naval aviation forces of members of the **Nuclear Club.** For obvious reasons orthodox depth charges just fired at short range from ASW **Destroyers** and **Frigates** cannot be nuclear-armed because of the risk to the ship. However, the US navy, and probably the Soviet navy, is equipped with short-range nuclear missiles that can be fired from a ship against a submarine as near as 10 kilometres, and not more than 65 km, away. The most sophisticated weapon in this category is probably the US SUBROC, which can be fired through the torpedo tubes of a submarine. It rises to the surface, flies through the air for up to 55 km, and then descends to drop a nuclear depth bomb. The increasing toughness of submarine design, especially the twin-hulled Soviet submarines, is likely to make them extremely difficult to sink with conventional depth bombs. Many **Scenarios** suggest that nuclear engagement in the sea war is much more likely than in land combat at the **Central Front.**

Nuclear Fission

Nuclear fission is the physical process harnessed to produce **Atomic Bombs.** It is dependent on the fact that atoms of some of the heavy elements are unstable and, if split by bombardment with sub-atomic particles, release very large amounts of energy. A heavy element is one with a large number of neutrons and protons in its atomic nucleus. The neutrons, being electrically neutral, can be detached from the nucleus when struck by a free neutron, which acts as a sub-microscopic bullet. With naturally unstable elements there is a continual process of particle loss which, over time, leads to the transformation of the element from one variant (isotope) to another. This process was first understood during the inter-war years, and it became apparent that there was a theoretical possibility of a chain reaction being artificially created. Such a chain reaction would occur if the impact of one neutron 'bullet' was to force the expulsion from a nucleus of two or more further neutrons which would split more nuclei, and so on with increasing rapidity.

Some heavy isotopes (such as Uranium 233 or 235, and Plutonium 239) were known to be particularly prone to this natural neutron release. Amounts of one of these heavy elements too small to give way to a chain reaction are referred to as 'sub-critical'. Bringing together two samples of a suitable element which together formed a critical mass would cause an immediate and spontaneous release of huge amounts of energy. The chain reaction requires containment of the free-flying neutrons so that they impact on other uranium, or alternative radioactive element, atoms, rather than escape into the atmosphere. The technical problem is that the neutron emission/nucleus splitting process is very fast indeed. If the two sub-critical masses were brought together too slowly they would produce enough energy to throw themselves apart, or to melt their container, before the full chain reaction could start. It was the solution to this problem of keeping the two radioactive samples sufficiently apart to prevent premature energy release, and then bringing them together fast enough to force the chain reaction, that was finally solved by engineers working on the **Manhattan Project.** Two successful designs were found to be the **Gun-Type Bomb** and the implosion bomb. The other major problem was the refining of suitable nuclear material. Although uranium occurs naturally it has to be 'enriched' to get enough of the relevant isotopes; weapons-grade plutonium is itself the product of nuclear energy release in an atomic reactor. Much of the engineering work and financial investment of the atomic bomb project went simply into creating enough material for some half-dozen test devices and two bombs.

Nuclear-Free Zone (NFZ)

There are two different types of nuclear-free zone. Under one meaning of the term many local governments throughout Western Europe, and a few in the USA, have declared themselves to be nuclear-free zones. They claim that they will not participate in any effort to store, deploy or transport nuclear weapons. Some authorities, particularly in the UK, have taken this a step further by refusing even to co-operate with **Civil Defence** planning for nuclear war, either on the grounds that such preparations are futile, or that they actually encourage war.

A second category of nuclear-free zone is that where a political demand is made for nuclear deployment in a certain area of the globe to be banned by treaty. Examples include the New Zealand government's demand for a Pacific NFZ and the suggestions of a nuclear-free corridor on either side of the border between East and West Germany. A more ambitious version of the latter is the demand made by the peace movement **European Nuclear Disarmament (END)** for the whole of Europe to be declared an NFZ.

The local nuclear-free zones mean nothing at all in terms of national and international politics. They are largely an invention of the anti-nuclear movements of Western Europe, and central governments are not likely to be heavily influenced by such political gestures if their general policy is to facilitate nuclear deployment. Even the more general demands for treaty-based denuclearization of whole areas are less powerful than they may seem. Given the range of most missile- or aircraft-delivered nuclear weapons, both NATO and the Warsaw Pact could accept a broad range of zones in which such weapons could not be deployed. Targets within, for example, a continental European NFZ could easily be hit by missiles based in the USSR or the UK. Treaties forbidding the *use* of nuclear weapons in specified zones would not, in the appalling context of a nuclear, or potentially nuclear, war, be worth the paper they were written on.

Nuclear Fusion

Fusion energy is the process underlying what are referred to as **Hydrogen Bombs** or **Thermonuclear** bombs, in which two light atoms combine together to form a heavier element. In the process a huge amount of energy is released. The lightest elements which can be used for this process within the capacity of modern physics are isotopes of hydrogen, hence the name. (An isotope is a variant of an element. Each element has a fixed number of protons in its nucleus, but can have more than one form, depending on the number of neutrons. The isotope number is the sum of protons and neutrons.) The two isotopes of hydrogen that are usable for nuclear weapons purposes are deuterium and tritium. The problem with nuclear transformations of a light element into a heavier one is that while an enormous amount of energy is released, very high temperatures are required to start the process. A vivid illustration of this can be seen if it is remembered that the energy release phenomenon that powers a star is nuclear fusion of helium and hydrogen into heavier elements. It is the extremely high gravitational pressure generated in the core of such immense objects as stars that creates the temperature to sustain fusion energy release. The only way it is possible to create such temperatures by human design is to use the heat of a **Nuclear Fission** chain reaction. Thus the trigger for a hydrogen bomb is itself an atomic device. It was inevitable, in other words, that the **Atomic Bomb** programme would come before the development of fusion weapons. The reason for the second name, thermonuclear, is now clear—it is an explosion generated by the heat of a nuclear explosion.

Unlike atomic bombs, where critical mass considerations limit the possible **Yield,** there is no theoretical limit to the size of an H-bomb. With the provision of a fission trigger, the amount of deuterium or tritium present is the determining factor. Nuclear fuels for hydrogen weapons are not as cheap and abundant as it might seem, given that hydrogen is the second most plentiful element in the universe. Neither deuterium nor tritium occur naturally, and have to be made as by-products of chain reactions in the uranium used in nuclear reactors. The most limiting fuel problem, however, is the creation of enriched or 'weapons-grade' uranium, and even more so plutonium, to provide the trigger devices. As a rough example of this, in 1951 the UK had two reactors working from which they managed to extract a total of only 24 kg of plutonium over a whole year. As it takes about 3 – 4 kg of plutonium to make the trigger for a single hydrogen bomb, some idea of the considerable investment involved is easily obtained.

Nuclear Hostages

The idea of nuclear hostages has been present since the early days of nuclear strategic thinking, and gained prominence during US Secretary of Defense Robert **McNamara's** attempt to move away from the doctrine of **Massive Retaliation.** It has re-emerged as an important concept with the development of more sophisticated approaches to nuclear **War-Fighting** since the early 1970s. A nuclear hostage is some target of presumed great importance to the enemy that one does *not* attack in a first (or subsequent) strike, but which one clearly retains the capacity to destroy. The usual version is that by attacking the enemy's nuclear forces, or **Other Military Targets,** but leaving its cities undamaged, the latter are effectively held hostage (see **No Cities Doctrine**). The enemy is forced to consider whether it wishes to retaliate to the 'limited' strike, knowing that doing so carries a great risk of incurring a **Second Strike** which would do very much more damage to the social and economic structure of the nation than the inevitable **Collateral Damage** suffered in the original **Counter-Force** attack. The enemy's civilians are thus held hostage against the moderation of their own leaders.

Setting aside the unreal macabre precision and rationality of the theory, which is found in all nuclear thinking, the assumption that the Soviet Union shares the same sense of nuclear policy as the West does present a serious problem in considering the concept of nuclear hostages. While there is good evidence that Soviet targeting is, if anything, *less* likely to involve the deliberate destruction of urban targets than does that of the Western nuclear powers, it is also true that they have much less faith in either **Limited Nuclear Options** or **Escalation Control.** The Soviet reaction to a 'non-city' strike by the USA which itself avoided hitting American cities would probably depend much more on whether they saw any point in targeting US population centres than on the fact that their own had been left alone. A further problem is that the doctrine is, in any case, dependent on the viability of a **Surgical Strike** which could avoid killing millions of the enemy's citizens. No one but the country attacked can know when they would see the citizen death rate as 'merely' collateral damage, and when it would be taken as virtually a deliberate **Counter-Value** strike. Consequently it would be very difficult to be sure that the attack planned was, indeed, going to be seen by the enemy as having left its vital concerns safe and available to be considered as nuclear hostages. (See also **Intra-War Bargaining.**)

Nuclear Planning Group (NPG)

The Nuclear Planning Group is an *ad hoc* committee of **NATO** set up in 1967 to provide a forum in which the USA, as the main provider of nuclear force for the alliance, could consult with its partners. It was set up in response to the discontent felt by many European members about the effective US monopoly of decision-making about the nuclear guarantee, which has always underlain NATO's strategy whether in the days of **Massive Retaliation** or of the subsequent strategy of **Flexible Response.** The immediate reason for its creation was the failure to be realized of plans for a joint NATO nuclear **Multilateral Force.**

Originally the Americans had hoped to have a small committee of representatives from not more than five or six major NATO countries, but it has now grown to include 14 of the 16 members, excluding only France, which is not a full member of NATO's integrated military structure, and Iceland, which has no military forces of its own. There has never been any chance that the NPG could exercise any control over the US nuclear force itself, but it has been useful in managing NATO-wide concerns, as

with the Intermediate Nuclear Forces (INF) modernization during the **Euromissile** crisis. Like all consultative bodies, it is seen as being too weak by those who are only consulted, and the USA tends totally to ignore it when developing plans of their own. Neither the announcement of the **Strategic Defense Initiative** plan, nor the surprise offer by President Reagan to abolish all INF forces in Europe made at the Reykjavik Summit in October 1986, were even mentioned at the NPG meetings that had taken place shortly before.

Nuclear Sanctuary (see Sanctuarization)

Nuclear Sufficiency (see Effective Parity)

Nuclear Threshold

Strategists often talk about the need to 'raise the nuclear threshold', or about their fear that some inadequacy in conventional armaments might 'lower the nuclear threshold'. The nuclear threshold is the point during a war at which nuclear weapons, of any sort, are introduced, in addition to or instead of conventional arms. As such it is one of the **Fire-Breaks** discussed in the theory of **Escalation,** and is of particular importance to NATO planners. The most common use of the concept is in the context of NATO's **Flexible Response** strategy for defeating a Warsaw Pact attack. It is assumed generally that NATO would be able to sustain a purely conventional defence effort for only a limited time, often put at as little as 10 days, and certainly no longer than a few weeks. If the Soviet Union was still able to mount a forceful conventional attack after this time, NATO would be forced to 'go nuclear', to cross the threshold and use at least **Battlefield Nuclear Weapons,** and possibly its **Theatre Nuclear Forces.** Thus political initiatives to improve NATO's conventional defence strength, such as the **Long-Term Defence Programme** or the **Conventional Defence Initiative,** are described as raising the nuclear threshold because they will defer the stage at which NATO might have to use nuclear weapons.

Nuclear Umbrella

The nuclear umbrella is an informal and much-used way of describing the protection extended to Western Europe by what Americans refer to, more technically, as **Extended Deterrence.** This nuclear guarantee has underpinned NATO policy from the days of **Massive Retaliation** to the current doctrine of **Flexible Response.** The 'umbrella' is provided, at its strongest, by US intercontinental missiles. If necessary, the USA would launch **Central Strategic Warfare** against the Soviet Union to prevent the latter from capturing a significant part of Western Europe, or as retaliation for a nuclear strike in Europe.

The umbrella has become increasingly 'leaky' for two different reasons. The first, relatively technical reason, is that it is unclear quite what sort of nuclear response to which particular Warsaw Pact incursion is actually contemplated. Would US presidential permission for **SACEUR** to use **Battlefield Nuclear Weapons** be enough to fulfil the nuclear guarantee? Would a Soviet nuclear strike with, for example, an **SS-20,** against a French town demand that an American **ICBM** be launched against a Soviet city? How should the USA react to Warsaw Pact **Salami Tactics** in which they halted an advance after taking some hundreds of square miles of West German territory? The second, and more important, reason for doubting the US nuclear umbrella is the essential lack of **Credibility** that it carries in a period of nuclear **Parity.** When the doctrine was developed, and even until the early 1970s, the USA had such superiority in intercontinental nuclear weapons that it could be plausible for a US President to threaten to strike at the heartland of the USSR to prevent even a conventional defeat of NATO. However, in a situation where a broad nuclear parity exists, it is inevitable that the question of whether a US President would, for example, risk Detroit to avenge Düsseldorf is raised. The doubt implied in this question is not confined to Europeans, but is shared by many Americans, among them even leading Senators.

It is partly to allay such fears that NATO has put so much stress on the deployment of US missiles in Europe. Thus, it is believed, the 'coupling' of US nuclear weapons to the defence of NATO's interests in Europe will be ensured (see **Decoupling**). Similarly, a major reason why European governments fear the withdrawal of US troops from Western Europe is that a substantial American presence makes it more likely that, if only to protect them from being overrun by Soviet armies, the USA would be forced to use its nuclear arsenal. This argument is, to some extent, a re-creation of the **Tripwire Thesis.**

Nuclear Winter

The nuclear winter hypothesis was first publicized in 1983 by a group of American scientists including Carl Sagan, the noted astronomer. Since then there have been a series of scientific reports subject to varying and contradictory interpretations, and the hypothesis is still much in doubt. Shorn of technical detail the argument is this: the result of nuclear explosions, especially in cities, will be to start fires that will send as much as 180 million tons of smoke and fine soot particles into the upper atmosphere. There is some chance that these will stay aloft, circulating all round the world, for months, and perhaps even years. The consequence would be that a major part of the earth's normal quota of sunlight would be blocked out, and that average temperatures would thus be reduced in all parts of the earth. If the temperature drop was large enough there could be a partial or nearly total failure of world agriculture, with a human death rate that could run into thousands of millions.

Everything in a theory like this depends on the precise assumptions fed into the computer models. No one doubts that *some* smoke would reach the upper atmosphere. No one doubts that very large amounts of smoke would produce *some* temperature drop, and no one doubts that large temperature reductions would have *some* impact on agriculture. Whether an agreed model can be produced is improbable, because there can be little solid evidence for any set of assumptions on something as untested as **Central Strategic Warfare.** At the moment apparently equally-valid models suggest a temperature drop of only a few degrees, producing September temperatures for July in the northern hemisphere, with little attendant long-term danger, or a drop of as much as 20° Centigrade, with potential near extinction of the human race. All that can be said with certainty is that the nuclear winter hypothesis does raise a problem that strategists had never bothered with for the first three decades of the nuclear age.

O

Office of the Secretary of Defense (OSD)

When people talk loosely of the **Pentagon,** or the **Department of Defense (DOD),** in the USA they are most often actually referring to one specific part of a very complex institution, known as the Office of the Secretary of Defense. The OSD is the central controlling and integrating department of government which presides over quite separate service departments, with their own civilian bureaucracies and Secretaries. It is the OSD, rather than the DOD in general, which is responsible for presenting budgets to Congress, approving procurement, and setting defence goals. Despite this it is actually a relatively small unit compared with the enormous civil and military administrations of the individual services. Consequently much that the OSD might prefer to see done, or not done, can never completely be imposed on the services themselves.

Neither is the OSD an operational command; any control that the Secretary of Defense can exert over the services in an operational context has to be channelled through the **Joint Chiefs of Staff** Committee. That, in turn, exists in a completely separate chain of authority from *either* the OSD *or* the service Secretaries. For these reasons there have, for some time, been serious attempts in Congress to increase the power of the OSD in relation to other parts of the DOD complex.

Operational Level

Traditionally, military activities have been divided into two levels, **Strategy** and **Tactics.** Strategy tended to concern the planning and movements of whole armies, while tactics was understood to deal with the immediate combat level. The latter covered procedures for manoeuvring units up to the level of a **Division,** but also as small as the **Infantry** rifle group. In slightly different ways, during the Second World War, both the Germans and the Soviets began to think about an intermediate range of activities. This went beyond the relatively mechanical business of bringing small units into contact with the enemy, but did not touch on the major, often political, questions of where a nation's overall effort should be directed, and what its war aims should be.

This intermediate range of activities has subsequently come to be called the 'operational level'. It forms part of the doctrine and war planning of many armies, including not only its originators in Germany and the USSR, but also armies as different as those of Israel and the USA. No two armies define the operational level (the technique is sometimes called 'operational art') in quite the same way. The easiest way of thinking about it is to see operational activities as covering the whole of a **Theatre of War,** for example the **Central Front,** and necessarily involving all services and arms, in the pursuit of very broad definitions of victory in that area. It differs from strategy in that decisions on major resource allocation and general war objectives, such as whether to aim for unconditional surrender, armistice or other final outcome, will already have been made. However, commanders will retain a considerable degree of discretion within the operational level, and their task will be much greater and more complex than simply to bring the enemy to battle in favourable circumstances.

Operational Manoeuvre Group (OMG)

Operational manoeuvre groups form part of the modern Warsaw Pact strategy for a **Central Front** war. The general doctrine is based on the fear of **Battlefield Nuclear Weapons;** a wide distribution of troop formations is essential to avoid presenting easy targets for NATO's nuclear forces. As a consequence the Warsaw Pact is committed to a plan in which its armies would approach the battle area in a series of waves (see **Echeloned Attack.**) This means that the whole military force is not assembled for a breakthrough at any one or two points and does not, therefore, present such a suitable target for nuclear attack.

The main problem with this is that such a deployment deprives the commanders of the weight of troops with which to exploit any breakthrough that does occur (see **Concentration of Force**). To deal with this the Warsaw Pact would combine the echeloned forces with a series of compact, mobile formations called operational manoeuvre groups. These would be of roughly divisional size (of perhaps 10,000 troops, in armoured and armoured infantry units), too small and fast-moving to make a particularly attractive target, but capable of exploiting weak spots exposed by the linear attack of the first or second echelon. They would not fight pitched battles against strong reserves, but bypass any fortified areas and penetrate as far as possible behind NATO front lines in order to cut communications and paralyse efforts to reorganize.

The combination of echeloned attack and OMGs is intended to maximize the problems of NATO's defensive strategy by forcing the latter to fight a mixture of concentrated and dispersed forces. Pressure to improve NATO's **Deep Strike** and **Manoeuvre Warfare** capacities is a reaction to the Warsaw Pact's conventional superiority and, in particular, to the threat posed by OMGs.

Order of Battle

An order of battle is simply a list of the units in a military force deployed and intended to take part in any imminent or ongoing conflict. To discover the enemy's order of battle is a prime task of military intelligence, because knowing exactly which units, with their special capabilities, are scheduled for combat in any theatre can tell an intelligence analyst a good deal about the **Tactics** and **Strategy** which the enemy has planned. In past wars the order of battle was a highly-prized secret, but the development of **Signals Intelligence (SIGINT)** and other technical means of intelligence gathering, such as satellite observation, has made it much less possible to disguise the order of battle.

Other Military Target (OMT)

There is often confusion in discussions of nuclear targeting policy over the meaning of **Counter-Force** targeting, which is sometimes assumed to mean anything that does not form a purely civilian target, such as a city or a factory complex. In fact, the meaning of counter-force is restricted to cover the enemy's own nuclear forces and associated facilities, especially the **C^3I** bunkers. The remaining military targets—armaments depots, logistic **Bottlenecks,** troop concentrations, and air-bases and ports not involved with the nuclear forces—are collectively known as 'other military targets'. While these are not **Hard Targets,** they still require a precision of attack, particularly if **Collateral Damage** to civilians is to be minimized, that makes them unsuitable targets for many nuclear weapons. Consequently, the 'OMT' category has always featured in the **Single Integrated Operations Plan (SIOP)** as a major drain on **Strategic** forces, and would probably require attack by **ICBM** rather than **SLBM** forces, unless they came into **SACEUR's** ambit as theatre targets. One account suggests that of the 8,000 or more targets which could be specified in a single all out strike model of SIOP about 1,600 would be in the OMT category, against 2,000 nuclear targets, 700 C^3I targets and over 4,000 economic industrial targets.

Out of Area (see Out of Theatre)

Out of Theatre

The term out of theatre (also referred to as out of area) highlights the fact that **NATO** is technically a limited regional defence pact, and not a general

'Western Alliance'. NATO's treaties bind members to come to the aid of another member whose national territory, or whose aircraft and shipping, is attacked—within well-defined geographical boundaries. It does *not* exist to protect the vital interests of any one or more members in any other way. To take two of the clearest examples, there was no obligation for any NATO member to aid the UK in its military campaign to recover the Falkland Islands, nor to give any assistance to the USA in its bombing of Libya in 1986. Inevitably, ambiguous situations could arise. For example, there is a *prima facie* case for saying that if a ship of the US navy was attacked by the Soviet navy in the Mediterranean, this would be covered by the terms of the treaty. However, would the same apply if the incident had arisen because the USA, in support of its non-NATO ally, Israel, had deployed an intervention force in the Middle East which came into conflict with a similar force which the USSR had dispatched to aid its non-Warsaw Pact ally, Syria?

Over a period of time the USA has become increasingly impatient with what it sees as the refusal of its **European NATO** allies to participate in its self-imposed 'world policeman' role outside the NATO area (see **Global Strategy**). It claims, for example, that a NATO presence is vital in the Indian Ocean and the Persian Gulf to protect the oil supplies on which Western Europe is highly dependent. Such demands are for 'out of theatre' co-operation. With the exceptions of France and the UK, the demands are ignored, in part because the other NATO members lack any general **Force Projection** capacity, and in part because of ideological and practical arguments that NATO ought not to be drawn away from its major security policy role. The attitude of the UK fluctuates, depending partly on which party is in power, but increasingly lacks the military capacity, or the defence funding, to engage in out of area operations, while France has its own concerns, particularly in Africa. The issue is one of those contributing to the increasing strains in US – European relations within NATO. The irony is that the tight geographical definition of NATO obligations was actually included at the insistence of the Americans who, in the late 1940s, feared getting embroiled in French and British colonial problems.

Overkill

Overkill is a concept to be found more in the language of nuclear disarmers than strategic theorists, and has no very precise meaning. In general, those who believe that overkill has been reached in nuclear weapons argue that the total of the nuclear arsenals of members of the **Nuclear Club** is so high that enough destructive power has been accumulated to kill every inhabitant of the world, and to leave some to spare. An alternative and weaker version simply holds that the USA and USSR each have more weapons than they need to destroy the other nation completely.

It is unclear quite what the implications of this state, if it exists, are. It is entirely possible to have a combination of weapons of different types such that the use of all of them exceeds any possible need. But, of course, there is no plan to use all existing weapons, given that they are tailored to different purposes. For example, an all out **Counter-Value** strike against the USA by the USSR, which brought forth a maximum retaliatory strike, might leave significant parts of the arsenals of both countries unused. There would be little point at all in using the relatively low-**Yield** and highly-accurate **ICBMs,** such as the **MX** missile, designed for **Counter-Force** strikes against **Hard Targets** and probably as a **Limited Nuclear Option.** Similarly, a superpower could conceivably run out of stocks of **War-Fighting** missiles, and 'lose the war' as a result, without ever employing their 'city busting' weapons.

As an example of how overkill probably does exist, but is irrelevant, the following figures are useful. If the whole of the planned US **Trident II** fleet fired its entire broadside, it would deliver perhaps 1,800 MTE (see **Equivalent Megatonnage**) over Soviet cities. This would probably kill, outright or within a short span of time, 80 – 90% of the population of the USSR. Yet this would still leave over 1,100 MTE in the **Minuteman** force, as it is currently structured. There would be no useful targets, however, at which to aim these; the only plausible **Scenario** under which either superpower destroys the population of the other is in response to a **First Strike.** In any such first strike the missiles against which the Minuteman force are targeted would already be used up.

P

Parity

Parity is an idea of increasing importance to **Arms Control** analysis. It is the basic goal of both sides in arms control negotiations, because neither the USSR nor the USA can expect the other side to accept a situation of strategic inferiority as legally acceptable in a treaty, although they might in practice settle for such a situation as a *de facto* condition. Equality of nuclear weapons as such is a largely meaningless concept because of the large number of variables involved, and the different needs each side may have both because of its strategic posture and because of the nature of the opponent as a target. Parity tends to be defined in terms of a situation where neither side would believe itself to have an adequate superiority to risk an attack. However, this essentially subjective element leads to problems where one superpower insists that it is in an inferior position, while its opponent has a quite different interpretation.

Consequently parity situations may look numerically very unequal. There are a series of basic variables that enter into such an analysis. The sheer numbers, both of warheads and launchers (see **Delivery Vehicles**), are obviously important, but so too are the total **Throw-Weight** which each side is capable of delivering and the total megatonnage, which might be different still. A further complication is the accounting problems which defy anything but arbitrary definition. For example, does a piloted intercontinental bomber count as one launcher, or should each bomb or each **Air-Launched Cruise Missile** it can carry be counted? Or, given that the same missile can have varying numbers of warheads, does one assess it according to what its owner says it is loaded with, or to its maximum capacity? Should the accuracy of a missile/warhead combination be taken into account (see **Circular Error Probable**)? If so, is a very inaccurate one-**Megaton** warhead equivalent to more or less than a highly accurate 200-kt warhead? In terms of attacking a **Hard Target** they might be equivalent, but clearly they are not in terms of a **Counter-Value** strike against cities.

It is because of such assessment problems that arms control negotiations are so difficult, and why what often seems to a lay person a generous offer by one side is unaccountably rejected by the other. A simple example might be an offer by the USSR to cut its total number of intercontinental missiles by 50% in return for the USA doing the same. Is this to mean that each category of missiles, **SLBMs** and **ICBMs,** are to be cut by 50%, or the total number only? If the latter, the USSR might choose to abolish all of its SLBMs, hoping that the Americans would cut each of their categories in half. Were that to happen the current USSR superiority of around 2.5:1 in heavy land-based missiles would grow to over 3:1, while the USA might, or might not, feel that their resultant monopoly of submarine-launched missiles made up for it. Such arms control agreements as have been reached, notably **SALT I** and **SALT II,** have been quite crude in their counting rules, but also very complex, with different maximum sub-totals allowed in different categories in order to accommodate the conflicting strategic preferences of the two superpowers. All missiles are rated at the maximum number of warheads they can carry, for example, and no attempt is made to take account of **Yield.** Indeed, SALT I, is often characterized as based simply on 'counting holes', that is, silos, because these at least are obvious and undeniable, although the number of silos a nation has says very little about its nuclear arsenal. This illustrates the general problem of **Verification** which characterizes any quest for parity.

The problems of assessing parity have led some to argue for a concept of **Effective Parity.** This is based on much the same logic as the **Mutual Assured Destruction** thesis—when both superpowers have a secure **Second Strike** capacity at the level of **Assured Destruction** of the enemy they are in a state of parity, because there is nothing more that can be done. An alternative conception of parity, which puts much more **Arms Race** pressure into the international system, is known as **Essential Equivalence,** and requires parity in each leg of the nuclear **Triad.** It is this last interpretation of parity which has dominated US arms procurement policy since the mid-1960s.

Partial Test Ban Treaty (see Comprehensive Test Ban Treaty)

Particle Beam Weapons

Particle beam weapons are similar in general principle to **Laser Weapons.** Like lasers they involve the output of a narrowly-focused beam of very high-energy radiation. But while the energy output of a laser is electromagnetic, particle beams, as the name implies, consist of high-energy nuclear particles. These can be electrically charged particles, protons or electrons, or neutral particles like neutrons. They would be produced in a machine like the accelerators used in laboratories to study sub-atomic matter: to be able to design such a machine which was mobile and flexible enough to serve any military purpose seems highly unlikely. Nevertheless, particle beam projectors are among the many weapons under consideration in the US **Strategic Defense Initiative.**

Payload

The payload of a missile is the amount of ordnance which it can carry. The term does not necessarily have the same meaning as **'Warhead',** because the payload may also include penetration aids (see **Decoy**). An alternative use of the term includes the whole of the **Front End** of the missile. In the context of the **Throw-Weight** of a missile, which refers to the capacity to carry a load into the atmosphere, 'payload' usually has this alternative, broader meaning.

Pay-Off

A pay-off, technically, is the result to each player of a strategic interaction in **Game Theory.** It has come, in strategic as much as everyday language, to mean simply the result of any decision, action, or investment.

PD-59

Presidential Directive 59, signed by President Carter in 1980, was his administration's version of the type of **Limited Nuclear Option** strategy originally developed under President Nixon's Secretary of Defense, James Schlesinger (see **Schlesinger Doctrine**). As with all US **Declaratory Policy,** it was a matter of gradual adjustment of a targeting philosophy which had always contained more 'flexibility' than critics had assumed. If PD-59 was any different from its predecessors it was in stressing the need to be able to fight a prolonged but limited nuclear war (see **Limited War**). This required both a secure **Second Strike** capacity, also known as a 'secure strategic reserve', and enhanced **C³I** abilities.

PD-59 did make public a preference for targeting the Soviet leadership itself, as opposed to only its nuclear forces. It was argued that a highly-centralized State, such as that of the Soviet Union, might well be paralysed by hitting a few command bunkers, whereas the US **National Command Authorities** structure does not display such **Vulnerability.** PD-59 may have also contained an increased effort to define **Counter-Value,** but still 'limited', targets by concentrating on destroying the economic **Recovery Capacity** of the enemy, without widespread 'city-busting' attacks. The problem with the first of these new aims was that the intelligence information required to select from among the hundreds of possible emergency locations for the Soviet government, and to hit the relevant super-hardened bunkers (see **Hard Target**) with sufficient accuracy, would probably be impossible to collect. The second aim, targeting recovery capacity, has been investigated using very sophisticated economic models, most of which have tended to show that any strike which seriously seeks to avoid mass deaths among the civilian population (see **Collateral Damage**) will allow surprisingly rapid recovery.

Peace Research

Since the Second World War there has developed an academic discipline called, variously, peace research or peace studies, which is also practised under less obvious labels such as 'conflict studies'. Although the labels make it clear what the ideological emphasis of those working in the field is—that they believe war to be avoidable and that it should always be avoided, these titles can be misleading. The people working in this field are not innocent pacifists convinced that good intentions are all that is necessary. Their work is highly scholarly and analytically powerful, with full understanding of the sociological and technological forces driving defence policies. Some of the more distinguished peace research institutes, like the **Stockholm International Peace Research Institute (SIPRI)** and the Bradford University School of Peace Studies, are invaluable providers of detailed and often very technical information about armaments systems.

Peacemaker (see MX)

Penetration Aids (see Decoy)

Pentagon

The Pentagon, strictly speaking, is the enormous five-sided office building in Washington, DC, that is the main home of the US **Department of Defense (DOD).** Built just before the Second World War, it was until recently the biggest single office building in the world, with thousands of offices and literally hundreds of miles of corridors. The word 'Pentagon' is frequently used in a more general sense to describe either the High Command of the US military, especially the **Joint Chiefs of Staff,** the civilian authorities in the DOD, or both. In this sense the use is figurative, because important parts of both of these institutions are not in fact located in the Pentagon building at all. A measure of how important and huge the Pentagon is as a single site of military power is that the Washington underground rail network has *two* stations to serve it.

Permanently Operating Factors

'Permanently operating factors' is a phrase originally used by Stalin to organize a Marxist-Leninist theory of war. In fact the permanently operating factors have little to do with Marxism, and resemble many military historians' lists of the principles of war. It is unclear how important they really have been in the development of **Soviet Doctrine.** Much lip-service was, and some still is, paid to them, but it is characteristic of the presentation of Soviet thinking to use approved concepts and theories, even when the meaning is quite novel.

Nevertheless, there seems to be a historical continuity of Soviet military planning which is at least compatible, even if it does not derive from, Stalin's 'factors'. There are three characteristics which continue to dominate Warsaw Pact strategy. The first is the enormous importance attached in **Soviet Strategic Thought** to surprise, at least tactical surprise, in military operations. The second and third in fact tend to mitigate against surprise; they are the paramount importance of massive **Concentration of Force** and the crucial role of artillery in winning battles.

These 'factors' have a real impact, not just in Soviet, but even in NATO thinking. Much NATO planning effort goes into preparing for the worst case **Scenario,** in which the Warsaw Pact armies are able to launch an attack before NATO is fully mobilized. In the last few years the Warsaw Pact has been trying to improve its ability to fight from a standing start, with no major mobilization that would reduce the advantages of surprise. The importance of massed artillery is shown not only in the Warsaw Pact's considerable superiority in tube artillery, but also in the way that missile forces are highly regarded in the USSR. Finally, the problem of mounting overwhelming force on the battlefield while avoiding tactical nuclear weapons has led to the development of the **Echeloned Attack** and **Operational Manoeuvre Group** elements of Soviet strategy.

Pershing II

The Pershing II missile is one of the two **Intermediate Nuclear Force** weapons systems deployed by NATO in response to the Soviet deployment of the **SS-20** during the **Euromissile** crisis from 1977 onwards. (The other NATO weapons system is the **Ground-Launched Cruise Missile.**) Although there was an earlier model, the Pershing IA missile which was still in operation with the West German army in 1987, the Pershing II was specially designed as a new weapon when NATO decided that it needed to upgrade its **Theatre Nuclear Forces,** and is very different from its predecessor.

The new weapon is a **Ballistic** missile with a range of at least 1,800 km. There are strong claims from the Soviet Union, backed by many Western analysts, that its range is intentionally understated to make it possible to deny that Pershing II could reach **C³I** bunkers around Moscow, which would require a minimum range of 2,100 km. In either case the Pershing II's range certainly puts any likely Warsaw Pact military target at risk. The guidance technology on the missile is so accurate that it is thought to have a **Circular Error Probable** as low as 45 metres, which allows it to be loaded with a relatively low-**Yield** warhead of only some 50 kilotons. It is planned to deploy only 108 of these mobile missiles, all in West Germany, but in principle it would be possible to equip the batteries for multiple launches.

Phased Array Radar

Traditional radar systems rely on the familiar rotating radar disc to cover the whole sky, or any large portion of it beyond the arc that can be covered by a stationary aerial. These systems, while adequate for most purposes, are far too slow for some modern military needs. Obviously the rotating aerial cannot receive signals from an object from which it is turned away. Some military tasks, especially those dealing with **Ballistic Missile Defence,** cannot tolerate the delay involved in waiting for the aerial to come round for a second position check on an incoming object. The speed with which incoming **Re-entry Vehicles** move, and the tremendous pressure to observe, calculate and respond in fractions of a second, have created a requirement for rapid and continuous radar monitoring. Phased array radars provide the technology to meet this demand.

Phased array radars depend on complex signal processing and electronics. Such a radar station is immobile and relies on a very large number of transmitting and receiving circuits, set in a bank, with overlapping arcs of search. By very rapid switching between arrays of receptors the same area of coverage as a moving radar bowl is achieved with no, or only very short, interruptions. Both the USA and the USSR have built major phased array radar stations. The most recent development by the USSR, at Krasnoyarsk, has been alleged, by the USA, to be in breach of the 1972 **Anti-Ballistic Missile Treaty,** which bans radars intended to detect incoming missiles for the purpose of directing attack on them. This may well be the purpose of the Krasnoyarsk system, but phased array radars also have vital roles to play in ordinary anti-aircraft defence (as well as in space programmes), which present the same rapid **Target Acquisition** and continuous monitoring needs.

Planning, Programming and Budgeting System (see PPBS)

Point Defence

Point defence is mainly a concept applying to anti-aircraft or **Anti-Ballistic Missile (ABM)** systems. It is different from, but often complementary to, **Area Defence,** which consists of rapid-firing guns or missiles, attempts to track and attack incoming threats at some distance and over a broad area. Thus some specialist ships in any modern fleet will have the duty of trying to protect the entire fleet from airborne attack. Point defence is a capacity to protect only one specified target at relatively close range. Again examining the naval context, each ship in a fleet will have at least some weapons with which to shoot down attacking aircraft or missiles which will be unable to help even a nearby friendly ship, even if their general function is area defence.

A current vitally important example of the contrast between point defence and area defence occurs in ABM defences, particularly as incorporated in the US **Strategic Defense Initiative.** There are different versions of this plan, and the one that caught President Reagan's attention was a general area defence over the USA. However, by far the more likely and technically feasible option would simply involve providing point defences around **ICBM** sites. (See also **Layered Defence.**)

Polaris

The Polaris missile was the first **SLBM** to be deployed. Originally put to sea by the US navy in 1960 aboard the USS George Washington, a more advanced version, the Polaris A3, entered into service as the British nuclear deterrent in 1968. Until the **Trident** II (or D5) missiles replace Polaris, as they are scheduled to do so, in the early 1990s, the latter will remain in service with the Royal Navy, even though the US navy replaced it with the more advanced **Poseidon** C3 missile from 1970 onwards. This *itself* has already been replaced by the Trident I (or C4). It is thus clear that the British Polaris is a first generation weapon in a world where the *third* generation—Trident I—is already in existence. Two specific factors make it a more primitive SLBM than its successors: it has never had a multiple independently-targeted re-entry vehicle **(MIRV)** capacity and, until the **Chevaline** refit, it carried three warheads that all had to be aimed at the same target; and the inaccuracy of these 200-kiloton warheads, as measured by their **Circular Error Probable** of 900 metres, means that they cannot be used against any **Hard Target.** Furthermore, the range of the Polaris missile is only 4,600 kilometres, thus restricting the total sea area in which the submarines can operate, and increasing the chances of detection. The Trident C4 has an operating range of 7,400 km, and the Trident D5 one of 9,700 km, which greatly improves the safety of the submarines against **Anti-Submarine Warfare.**

POMCUS

POMCUS (Pre-positioning Of Material Configured to Unit Sets) is the US military's jargon for the equipment stored in Europe to be used by reinforcements flown in during a NATO mobilization. It is not just personal weapons and general stores, but also tanks, guns and vehicles for entire divisions which are pre-positioned ready for use. This preparation would allow the reinforcement of **Forward Based** troops to proceed much faster than if the equipment, as well as the personnel, required **Sea-Lift** to Europe. It is, of course, an enormously expensive process, not least because the divisions based in the USA need separate sets of equipment there for training purposes. There are serious drawbacks to the policy, the main one being that the POMCUS sites, well known to the Warsaw Pact forces, make obvious targets. If either a conventional or nuclear attack could destroy the equipment dumps, the incoming divisions would be rendered virtually useless. Furthermore, as the divisions, wherever they enter Europe, have to travel to these few fixed sites, they will themselves become obvious targets. At a lower level the US Marine Corps tries to operate a similar system by having advance equipment stores on ships deployed in the various theatres, particularly the Mediterranean, where they may be required to effect a rapid landing. The policies prospered with the temporary surge in military budgets under the Reagan administration. However, the long-term prospects for the development of POMCUS facilities are unlikely to be so good.

Poseidon

The Poseidon missile, also known as the C3, replaced the **Polaris** missile, which was the first American **SLBM,** from 1970 onwards. Poseidon differed from Polaris mainly in that it employed **MIRV** (multiple independently-targeted re-entry vehicle) technology. Polaris A3, the last model, although it carried three warheads, was only of **Multiple Re-entry Vehicle** capacity, meaning that the separate warheads all had to be fired at the same target. So although Poseidon was not superior in any other way—its range was much the same as Polaris at slightly under 5,000 km,—it effectively multiplied the effect of the fleet of 30 or so **SSBNs** many times over. Britain retained its original Polaris missiles, although there was brief discussion during the 1966 – 1970 Labour government of acquiring Poseidon, largely because it was thought that the shift to MIRV warheads would seem to give the UK excessive nuclear capacity. Subsequently another way, the **Chevaline** warhead modernization programme, was found to retain the guarantee of the **Moscow Option** against Soviet **Ballistic Missile Defence.** The USA began to emplace the next generation missile, the **Trident** I (C4), in the same vessels that had carried both Polaris and Poseidon, in the early 1980s. This, however, was only a stopgap measure until the old fleet, facing obsolescence, as is the British Polaris fleet, can be replaced by new, bigger submarines able to carry the Trident II, or D5, missile.

PPBS

PPBS (it is invariably known by its initials, which stand for Planning, Programming and Budgeting System) was the famous management system enforced on a reluctant **Pentagon** by Robert **McNamara** as Secretary of Defense from 1961 to 1968. It is an elaborate version of 'programme budgeting' which requires managers to identify programme objectives, develop methods of measuring programme output, calculate total long-term programme costs, prepare detailed multi-year programme and financial plans, and analyse the costs and benefits of alternative programmes through cost-benefit analysis. This may all sound so obviously sensible that it may be difficult to understand why it was so hated by the military leaders, and changed how the Pentagon operated so much. What had happened in the past, and still does to a large extent in weapons acquisition, bore little resemblance to a PPBS approach. PPBS would require an armed service to identify a specific military need, compare the efficiency of various weapon designs, and carry out the long-term budgeting; meanwhile they also have to consider whether or not the utility of satisfying that need even by the most efficient alternative is sufficiently high to offset the lost opportunity cost of not dealing with some other problem.

Nothing remotely like this happened during McNamara's period of office. Instead, a particular weapon would be procured for other reasons, often with no need identified, and it would be bought at whatever cost, and with financial planning so inadequate that the ultimate cost bore little relation to the initial estimates. A fighter aircraft might be procured because the **Tactical Air Force** was currently politically more powerful than the **Strategic Air Command;** even the design of the fighter might depend more on the dominance of an ideological school of pilots who despised modern missiles and wanted to retain traditional cannon-style weapons. A second reason for the unpopularity of PPBS was that it gave enormous influence to a group of young civilian systems analysts rather than to the military. PPBS still lives on, although in recent years the Reagan administration has relaxed civilian and centralized control and let the services have their own head much more—with, at times, some quite disastrous economic and planning consequences.

Precision-Guided Munitions (PGMs)

Precision-guided munitions are at the heart of the development of **Emerging Technology** in conventional warfare. The first PGMs were probably the **Smart Bombs** used by the US Air Force in Vietnam. These were laser-guided bombs which, when dropped from an aircraft onto a target that has been lit up by a laser beam, steer themselves towards the objective while falling. Similar bombs were used by, among others, the British in the Falkland Islands campaign and the Americans when they bombed Tripoli in retaliation for Libyan-inspired terrorism.

The technology of PGMs is now highly-developed and complex, but the enormous change which they are likely to bring to warfare is simple to comprehend. Previously, the accuracy of all weapons has been so low that a target, even if clearly identified and within range of a gun, cannon, rocket or bomb has had a good chance of surviving fire directed on it. Whether one thinks in terms of a handful of hits per million rifle bullets fired in the US Civil War, the millions of artillery shells stockpiled for a First World War battle or the inaccuracy of Second World War bombing campaigns, the picture is the same. The chance of hitting the target fired at is known as the **Single Shot Kill Probability (SSKP).** With PGMs, which can be artillery shells, **Multiple-Launch Rocket Systems,** short-range missiles or anti-tank rounds, as well as bombs and air-launched missiles, the SSKP can easily be over 80%—that is, only one or two rounds in 10 would ever miss.

The impact of PGM technology is furthered by the way that the techniques are being applied to sub-munitions, where a single missile warhead may carry dozens of individually precision-guided 'mini-bombs'. A major task in emerging technology development, for example, is to design a missile warhead that would explode over an enemy tank squadron, scattering dozens of PGM anti-tank bombs which would home in on the heat emissions of the tanks. Thus, according to the PGM enthusiasts, a single conventional missile could do the work of a low-**Yield** nuclear warhead. Although the technological problems of PGM development are great, they are probably soluble with the provision of enough money. What is less clear is whether the **C³I** capacities will indeed enable PGMs to have the impact on the battlefield that some hope.

Pre-emptive Strike

In terms of nuclear strategy a pre-emptive strike would be an attack on the nuclear weapons or related assets of an enemy which, it was believed, was highly likely to use them against one's own country. It is a particular application of the general concept of **Pre-emptive War.** The main point about pre-emption in nuclear conflict is that the opposite strategy, to 'ride out' a nuclear attack and then retaliate, is a much harder option to accept than the general case of waiting for an enemy's attack. On the other hand the near impossibility of successfully pre-empting, especially if the opponent has a secure **Second Strike** capacity, persuades most analysts that it is a suicidal option.

There is a version of nuclear pre-emption that *has* tempted strategists from time to time. This envisages an attack on a potential opponent just as it begins to develop a nuclear force. Thus some US planners may have urged a nuclear strike against the USSR's nuclear weapons stockpile and industry in the early 1950s. There has, for years, been a strong rumour that the Soviet Union contemplated, and invited the USA to join them, in an attack on the nuclear research institutions of the People's Republic of China to prevent them from developing a nuclear challenge. Similarly, the Israeli attack on the Iraqi Osirak nuclear reactor has been interpreted as an attempt to pre-empt Arab nuclear armament. In this case the threat was so distant in time, and so uncertain of ever being realized, that it could not be regarded as true pre-emption.

Pre-emptive Surrender

Pre-emptive surrender is a concept coined by strategic theorists mainly to stress the ways in which the speed, inflexibility and devastating capabilities of ballistic missiles make ordinary ideas of war and armed conflict redundant. Normally one thinks of an army surrendering after some period of fighting has shown that it cannot win. No such period of trial would exist in **Central Strategic Warfare,** leading to the suggestion that immediately one government had reason to believe its opponent was planning to attack, it ought to 'pre-empt' by surrendering. The argument is not meant to be taken as a literal prediction of a possible **Scenario,** but to make the point that surrender after a major nuclear exchange would arguably be too late to protect anything. It is obviously contradictory to the recently-fashionable **War-Fighting** theory of nuclear use, which stresses **Intra-War Bargaining.**

Pre-emptive War

A pre-emptive war is one where the country initiating hostilities does so not for an inherently aggressive motive, but because it is certain that it is about to be attacked. Those who try to defend such a strategy see a pre-emptive war essentially as launching a legitimate defence ahead of the attack. There obviously can be situations where a country is so vulnerable to an attack that it cannot hope to defend itself successfully if an aggressor is allowed to strike first. Similarly there can, in principle, be situations where an opponent's absolutely firm intention to attack can be known ahead of time. However, the situations under which these two necessary conditions would apply will be very rare indeed. More usually, a country which attacked first as a form of defence would be risking causing a war that might not otherwise have happened had it simply alerted its forces and mobilized to demonstrate its preparedness, thus adopting a posture of **Deterrence.** It is mainly to try to remove the fears that could cause pre-emption that **Arms Control** policies concentrate quite heavily on **Confidence-Building Measures.**

Presidential Directive 59 (see PD-59)

Prisoners' Dilemma (see Game Theory)

Proliferation

In strategic analysis and **Arms Control** circles 'proliferation' always refers to the spread of nuclear weapons and the technology to build them to countries outside the **Nuclear Club.** At least to the superpowers, there is a self-evident mutual interest in keeping to the minimum the number of countries with nuclear armaments, and this is one area of arms control in which there has been considerable co-operation.

The main basis for control over proliferation of nuclear weapons has been the 1968 **Non-Proliferation Treaty,** which has been signed by the USA, the USSR and the UK, along with over a hundred, at least officially, non-nuclear countries. The treaty obliged the nuclear states who signed it not to make their nuclear weapons technology available to any non-nuclear country. The non-nuclear signatories committed themselves not only not to build nuclear weapons, but also to submit their civilian nuclear industries to international inspection. The more economically-developed signatories were also supposed to be committed to making their nuclear reactor technology available to poorer countries.

It is unclear whether the treaty has achieved anything that would not otherwise have happened. Those countries which do not, but could have nuclear weapons, or at least do not admit to them, such as Israel, South Africa and India, have declined to sign the treaty which, because of the refusal of France and China to sign, does not even cover the whole of nuclear club. Nevertheless, with the possible exception of France's aid to Israel, the nuclear powers have not been tempted to aid their non-nuclear allies to develop nuclear weapons. The real problem of proliferation is that the scientific and technological capacity required to build at least simple nuclear weapons from scratch is within the scope of so many nations that superpower aid is probably not required. In this respect the inspection rights of the International Atomic Energy Agency are perhaps the most useful part of the treaty, because the easiest short cut to nuclear weapons acquisition is to use a civilian power-generating nuclear reactor to enrich a nuclear fuel like uranium to weapons grade level. But, of course, no country can be required to allow such inspection unless it has signed the treaty.

Proportional Deterrence

Proportional deterrence is a French nuclear strategic doctrine originally promulgated by General Charles de Gaulle, and subsequently developed by French strategic thinkers such as André Beaufre. The problem which they saw themselves facing was to explain how a relatively small nuclear force such as the early French **Force de Frappe** could have **Credibility** as a deterrent against a superpower like the USSR. The problem is that the maximum strike the French could launch would not inflict crippling damage on the USSR, whose missile forces would still be able to destroy the whole of France's urban population in retaliation. The problem applies equally forcefully to the UK, but the British strategic position has always been hidden by the assumption that Britain's nuclear forces are NATO-dedicated. As the introduction of the Force de Frappe coincided with France's withdrawal from NATO's military organization, there has always been a need for France to justify its small and **Independent Deterrent** force.

The theory of proportional deterrence holds that a country can be deterred from acting against another by the threat of incurring damage proportional to the benefit it would have gained by attacking. Put very simply, France's nuclear force might not do very much damage to the Soviet Union as compared to the impact of an American nuclear strike, but the damage would still be enough to ensure that the USSR was, overall, a 'loser' in the exchange even if it went on to conquer and occupy France. This theory was developed further, largely along the lines of 'the deterrence of the strong by the weak' to justify, by making credible, the threat from a small nuclear force.

The main flaw of the theory is that it over-rationalizes the process of nuclear strategic decision-making, and that it is based on a curious misperception of possible Soviet strategic intentions. It may well be the case, as advocates of **Minimum Deterrence** would argue, that the horror of *any* nuclear stike, however small, is enough to dissuade a rational nation from adventurism. This does not, however, deal with the fact that should France launch a nuclear strike on the USSR it would risk annihilation in retaliation: accepting a threat of national suicide is an option which is not easy to believe. Furthermore, the 'proportional' aspect of the argument is difficult to make much sense of. The point is that the conquest of France, which would have to be preceded by the conquest of West Germany, is not a lesser goal than others. Rather, controlling the whole of Western Europe is the maximum goal the USSR could conceivably have, and there would seem no reason to exclude France from this. Although there is perhaps no good reason to believe that the USSR would have any such strategic intention in the first place, the **Deterrence** problem is of a different kind than the much larger nuclear force of the USA, which does not exist to protect America from a Soviet invasion, which is unthinkable. Only France and Britain are in the position of possibly trying to deter conventional invasion and conquest by nuclear forces, so they represent the maximum prizes. The calculation really cannot be undertaken rationally, but it is far from clear that the relatively small damage that France could inflict on the Soviet Union in the early days of its strategic forces would have outweighed France's value as a prize.

Proportional deterrence is heard of less often today, as the French have shifted more to a doctrine of nuclear **Sanctuarization,** and as the modernization of their forces increases the level of damage which they could inflict. It was very much a facet, along with the **Trigger Thesis,** of a **Declaratory Policy** addressed as much to the French electorate and France's allies as to the supposed object of deterrence. (See also **Gaullism.**)

Proportionality

Proportionality is a doctrine in the **Just War Theory** which holds that the damage which any military action does can only be justified in terms of its proportionality to both the immediate military accomplishment it brings, and the original justification for going to war. For example, the use by the US military of 'free-fire zones' and 'search and destroy missions' in Vietnam may be judged as disproportionate because, while they prevented Vietcong activity, they did so at the cost of innocent lives when other tactics could have achieved the same result. More dramatically, it can be argued that a nuclear strike on an enemy would be so appalling that, even if it were the only possible response to a clear-cut aggression, the right of self-defence could never be seen as proportional to the evil inflicted.

Psychological Warfare (Psy Opps)

There has probably never been a time when military commanders have not put effort into psychological warfare, because it is directly related to **Morale,** the importance of which has always been obvious. Certainly propaganda efforts in the First World War were early attempts to organize psychological warfare, or 'PsyOpps' as the approach is often known. In this case, however, the major efforts went into boosting the morale of one's own army. The early days of the Second World War saw a deliberate attempt by Britain to carry out psychological warfare by mounting bombing raids over Germany in which only propaganda leaflets were dropped, in an attempt to destroy civilian backing for the war. Later, most forces in most theatres attempted to weaken the morale of the fighting troops by dropping leaflets guaranteeing good treatment and safety if a surrendering enemy soldier presented the leaflet to the opposing troops.

The range of techniques available is enormous, but they fall into two tactical categories: increasing the terror felt by the enemy, or trying to demoralize the enemy by removing the psychological props, such as faith in the cause being fought for, that all soldiers need to face the ordeal of battle. All armies of any importance have separate psychological warfare departments, but there is very little clear evidence that any of them have achieved much more than earlier primitive techniques such as training soldiers to utter terrifying battle calls when charging.

R

Radar Signature

All objects which can be detected by radar (that is, primarily metallic or otherwise highly-reflecting surfaces), have a characteristic 'signature'. Considerable effort is taken by intelligence services in countries with advanced defence systems to obtain clear radar records of known enemy ships, aircraft and weapons systems so that, in times of crisis or war, these signatures can be identified. In this way defending forces can recognize the type of vehicle or missile which is approaching, as well as how many of them. The key point about radar signatures is that they are largely determined by the shape of, for example, an aircraft. The more care that goes into designing a very smooth shape, with few protuberances and minimal cross-section, the smaller the radar echo returned will be. It is this design criterion which lies at the heart of **Stealth Bombers**—aircraft designed to be as close to invisible as possible to defensive radar. (See also **Electronic Counter-Measures** and **Electronic Warfare.**)

Radiation

Upwards of 50% of the energy released by a nuclear detonation is in the form of 'invisible' radiation, ranging from X-rays, neutrons and gamma rays to alpha rays and low-energy beta rays. Even a short exposure to a low dose of radiation will do some physical damage to the human body and, as is well known, the exposure from a major nuclear explosion far exceeds such levels. There are two distinct ways in which radiation damage can occur. The more energetic particles, X-rays, neutrons and gamma rays, can directly penetrate the body and are the source of most short-term radiation illness. Others can only cause harm when taken into the body, through breathing contaminated dust particles or drinking and eating contaminated foodstuffs.

There is little direct evidence of the amount of radiation illness nuclear detonations of varying sizes will cause; Hiroshima and Nagasaki are, in many ways, inadequate guides because both explosions were of relatively low **Yield** compared with modern weapons. Even in these cases some degree of radiation damage was found in between 40% and 50% of survivors. It was estimated that 30% of short-term deaths among those who survived for longer than 24 hours at Hiroshima were caused by radiation. The long-term effects of radiation are in some ways even more frightening, covering as they do most forms of cancer and with potentially very long periods before the results appear. In the period 1950 to 1972 there were 141 cases of leukaemia, for example, among the 82,000 survivors who were followed up. (See also **Fall-Out, Nuclear Fission** and **Nuclear Fusion.**)

Raison de Guerre

Raison de guerre is a doctrine, akin to the concept of raison d'état, justifying certain actions in war. Raison d'état argues that at times when the whole safety of a state is at stake, the state can do anything necessary to protect itself, regardless of normal moral or legal restrictions. Raison de guerre, by analogy, is the doctrine that the vital necessity of winning a battle may, at times, justify behaviour which would not normally be acceptable even in the context of the already less-constrained standards of morality which exist in war. While raison d'état has long been recognized in international law, raison de guerre is much less well established—otherwise allegations of 'war crimes' could hardly be regarded as holding any validity. (See also **Just War Theory.**)

RAND

RAND (the name is an acronym from the full title 'Research and Development Corporation', which is never used) was the first of the big American defence **Think-Tanks.** It was set up by the US Air Force in 1948 to help with the development of both doctrine and technical plans, and gave the USAF a considerable edge both against the other services, and against central control from the **Office of the Secretary of Defense** through the sophistication of the analyses which were produced. It is, legally, an independent corporation, although the great bulk of its funding came initially, and much of it still does, from USAF research grants. One area of particular importance in RAND's work was the crucial role it played in developing strategic nuclear doctrine, and the detailed planning for the **Strategic Air Command's** early war plans. Most of the famous thinkers of the early days of strategic theory worked at RAND at one time or another. Although it is still very important, its relative influence has declined as the other services have developed their own think-tanks, and as Congress has developed powerful analytic agencies of its own. (See also **Center for Naval Analyses** and **Institute for Defense Analyses.**)

Rapid Deployment Force (see CENCOM)

Readiness

Readiness means exactly what the word suggests (which is by no means always the case with military terminology!): it is a general measure of how prepared a country's military services are to go to war. Readiness covers all the relevant elements—ammunition and fuel stocks, training levels, mobilization plans and, sometimes, the availability of necessary reserves in a hurry. It has come to be exceptionally important because of the likely nature of any **Central Front** war between NATO and the Warsaw Pact. Such a war will almost inevitably be short, whether or not it 'goes nuclear'. This means that no inadequacies in equipment, war stocks or training can be made up for during the course of the war. As a graphic American phrase puts it, the war will be a 'come as you are party'. Unfortunately it is precisely in the area of readiness that NATO is least well prepared. The stress on weapons acquisition, always more glamorous and professionally interesting for military officers, combined with restricted defence budgets has led to underinvestment in all aspects of readiness. Air forces, for example, would always prefer to buy more aircraft than to allow their pilots more flying time to train on the aircraft which they already have; armies let their ammunition stocks run down, and restrict the number of training rounds tank crews can fire in order to procure more tanks. Some US studies have suggested that the entire national inventory of certain crucial armaments, for example air-to-air missiles, would run out before a major war was 10 days old. (See also **Logistics.**)

Real Time Information

Throughout most of the history of warfare, commanders at anything above the lowest unit levels have not been able to exercise detailed control of their armies, navies or (more recently) air forces. The problem has been one of communication time and range of observation. Before this century, for example, an army commander could be utterly ignorant of the arrival of an enemy who have been steadily advancing for days. Even if the information-gathering assets, otherwise known as cavalry scouts, discovered such a force, it could never be known where they were at any precise time, because of the relative speeds of the horse-borne messenger and the moving enemy troops were too similar. During the two World Wars, and particularly the Second, this was not such an acute problem because modern communications and observation techniques, especially aerial reconnaissance and photography, reduced the time lapse considerably. There could be very little immediate action taken against forces more than a few dozen miles away, and knowing about their movements to an accuracy of a few hours was quite adequate.

What has happened in the last few years is that the potential reach and accuracy of weapons, particularly those promised by **Emerging Technology,** is so great that the time gap, in practical terms, has opened up again. Traditional methods of finding, plotting, and communicating the position of enemy units at distances of several hundreds of miles from headquarters are insufficient if one has, and needs, the capacity to destroy them as far away from the **Forward Edge of the Battle Area** as possible. The emphasis is now, therefore, on developing 'real time' observation and communication, which effectively means that a commander can follow the movement of enemy units as it happens. Real time, a concept borrowed from computing science, implies the sort of immediate knowledge about a target, deciding on an action, and launching it that in the past was only available in the face-to-face confrontation experienced by a platoon leader, rather than by a general commanding an entire army. (See also **Target Acquisition.**)

Recovery Capacity

Recovery capacity refers to the ability of a society, and particularly of its economy, to recover from a nuclear strike. Recovery capacity has become the prime object in nuclear targeting, replacing concepts like **Assured Destruction** and the idea of targeting for blatant population destruction. The most recent version of the American war plan, the **Single Integrated Operations Plan,** selects economic targets which would cause the Soviet Union the maximum problem in trying to regain its current productive capacity, while seeking to avoid unnecessary or **Collateral Damage.** The aim is to ensure that in a post-war world the USA would return to its current economic level faster than the USSR, thus ensuring future world dominance.

Although targets cannot be selected to avoid collateral damage or population destruction as easily as this plan suggests, because of the widespread effects of even an accurate and low-**Yield** nuclear weapon, the idea of targeting recovery potential is not meaningless. Quite often it is possible to identify a relatively small number of targets which have a **Bottleneck** effect on an economy, such that their destruction reduces overall economic performance disproportionately. Thus only a few nuclear warheads might be able to destroy the entire non-ferrous metals industry of a modern economy; a dozen warheads could interrupt the flow of oil or natural gas across an entire country. Identifying such targets involves complex econometric modelling, but in principle allows for a much subtler, and conceivably more humane version of **War-Fighting** than aiming simply to destroy a given percentage of the economic capacity.

Re-entry Vehicle (RV)

A re-entry vehicle, in the context of nuclear weaponry, is a **Warhead** carried by a ballistic missile. During the first stage, the **Boost Phase,** of the missile's flight, power is provided by rocket motors, which drop away when exhausted. The **Front End** or **Bus** then enters an exoatmospheric sub-orbital path during what is usually called the **Mid-Course Phase.** The warheads of the missile are ejected from the bus during this phase, and re-enter the atmosphere, hence 're-entry' in the name. Because of the high kinetic energy and the atmospheric ablation which RVs encounter, they have to be specially armoured against heat damage, while the bus, if it does not remain in orbit, burns up on re-entry.

Revolutionary War

Although the term revolutionary war is a common one, especially in Soviet and Marxist writings, it is not clear that it actually describes a specific phenomenon. What is a revolutionary war to one party may be an orthodox war, or even merely a 'police action' to another. The term originates with the wars of the French Revolution—hence the phrase 'the revolutionary and Napoleonic wars'. The point here is that by the time the wars which beset Europe from the 1790s until 1815 had become Napoleon's wars, they had been transformed into orthodox wars of acquisition. Earlier they were wars of ideological defence, fought by the mass population of France as armed citizens, rather than by organized armies subject to political control and with careful strategic goals.

There are two hallmarks of a revolutionary war. Firstly, such a war involves the whole population, directly or otherwise, rather than only the professional soldiers. Secondly, it is essentially defensive, in the sense of trying to protect a society that is undergoing or has undergone a revolution, although it may also be protecting itself against an external enemy, such as the old internal ruling class or, as is more common this century, a foreign colonial power.

Technically a revolutionary war is likely to take the form of **Guerrilla Warfare.** However, this is not an invariable pattern; the Vietnam War included some pitched battles between organized armies of both sides, for example. What *does* make revolutionary war different from ordinary war is that the war aims of counter-revolutionary forces are hard to define. It is not a matter of capturing cities or resources, nor of taking ground in the usual way. The aim has to be to defeat the revolution, but if the revolution is very popular and deeply seated, this is tantamount to destroying the population, manifestly as impossible as it is immoral.

Roll-Back (see Containment)

Rules of War

It is a common mistake to believe that war is an entirely lawless activity. In fact, in many ways it is one of the more regulated of human actions because of the reciprocal need of enemies to know which targets they can do what to. Only in a situation where one side is so overwhelmingly strong that it need fear its opponent hardly at all will the need to behave in such a way as to avoid reprisals for particular actions not be felt. Nevertheless, there are few written rules of war nowadays, the **Geneva Conventions** being notable but rare examples, because war tends to be between ideologically opposed nations whose decision-makers take rather different approaches to questions of moral obligation.

The rules of war really date from early medieval societies where war was not only fought for pragmatic rather than ideological goals, but also between professional soldiers, either mercenaries or members of Orders of Chivalry, whose ideas were directed more towards proper soldierly conduct than towards winning. Furthermore, they were all, by virtue of their common Catholic beliefs, imbued with a sense of morality and honour that led to understandings about 'what was not done' in warfare just as in peacetime. The traditional rules of war were not necessarily gentle—although they did deal with the right to have surrender accepted, or the need to allow envoys free passage—on the contrary, they could be very harsh. One commonly-accepted rule, for example, laid the responsibility for the massacre of a besieged town on its Governor, were he to refuse to surrender it to clearly superior forces, and not on the victorious besiegers. The extent to which any rules of war would apply in the conditions, even the pre-nuclear conditions, of **Central Front** battle, is very dubious.

S

SACEUR

SACEUR (the full title, Supreme Allied Commander Europe, is almost never used) is the senior military commander of the entire NATO land and air contingent on the European continent. His authority extends over six subordinate commands: Northern Europe, Central Europe, Southern Europe, the **Allied Command Europe** Mobile Force, the United Kingdom Air Forces, and the special NATO Airborne Early Warning Force. The main responsibility, however, is to prepare for air and land strategy on the **Central Front** where, in case of war, SACEUR would have direct final command authority over the air forces and army corps of all the NATO allies. SACEUR is always a US army general, partly because of the need to seek authority for the use of nuclear weapons from the US President, but also in token of the predominant US contribution to NATO. Once permission for the use of nuclear weapons was obtained, SACEUR would make the final decision on when and how to use them. The second-in-command can come from any of the other member states, but is often a West German or British army general because these two countries provide the bulk of the air and land power not provided by the USA. NATO has two other command positions at the SACEUR level: SACLANT and CINCHAN. SACLANT, the Supreme Allied Commander Atlantic, is an American admiral based at Norfolk, Virginia, and CINCHAN, the Commander-in-Chief Channel, is a British admiral operating from Northwood, London.

SACEUR's most important peacetime role is largely political. It is to maintain pressure on the NATO member governments for armament budgets, and to lay out clearly what the alliance's weapons and tactical priorities are. During peacetime the units of the individual country contingents operate quite independently of each other under their national commanders, and SACEUR is consequently not the equivalent of a supranational Chief-of-Staff. Neither does the holder of the post have a major bureaucratic function in the military side of the permanent NATO council structure, a role which is filled by the Chairman and President of the **NATO Military Committee.** Come wartime, however, the pressing need for unity in so diverse an alliance would arguably give SACEUR more direct command authority than any soldier has held in history.

SACLANT (see SACEUR)

Salami Tactics

'Salami' is used in defence discourse to represent the idea of taking very thin slices away from something, as in the customary manner of selling continental sausage. 'Salami tactics' usually refers to a **Scenario** in which the Warsaw Pact, rather than launching an all out attack on NATO with major territorial gains as its object, makes a very limited incursion. For example, it might expel Western troops from Berlin or, perhaps, push NATO forces 30 kilometres back from the border between the two Germanies in selected areas which most threaten its ally, East Germany. The problem that salami tactics would present NATO is that no individual act would seem to warrant the full, especially the nuclear, response of which NATO is capable. Would world war be risked in order to keep a foothold in Berlin? An analogy is often made to Hitler's policy in the few years before the Second World War. Neither the remilitarization of the Rhineland, which was German, nor the forced union (Anschluss) with Austria, which has close historical links with Germany, notably through the Holy Roman Empire, nor the taking of a German-speaking part of Czechoslovakia in themselves seemed to warrant a war for which the allies were, in any case, ill-equipped.

A second usage of the salami metaphor which is found in defence terminology relates to the problem of military budgeting in the UK. As the cost per unit of military equipment escalates rapidly, the **Ministry of Defence** periodically comes under pressure to reduce its overall expenditure. Typically this has been done not by identifying a whole area of military activity, for example, aircraft carriers or **Out of Theatre** capacity, and abandoning it, but by across-the-board small cuts. Most critics contend that this is a disastrous policy because it fails to reshape defence policy on rational grounds, and simply ensures that all the previous commitments are carried out inefficiently.

SALT: the General Process

The Strategic Arms Limitation Talks (SALT) between the USSR and the USA, which took place in the 1970s, are by far the most important and, with the exception of the Partial Test Ban Treaty of 1963 (see **Comprehensive Test Ban Treaty),** the only really successful **Arms Control** negotiations of the post-war period. Although the talks took place in two stages, **SALT I** and **SALT II,** they were seen as part of a continuing process. Some even expected them to lead to a SALT III treaty. (Had it ever been formally ratified, SALT II would have expired on 31 December 1985.)

The SALT process began in 1967 when the Johnson administration suggested talks with the USSR. The two superpower governments later agreed that talks should commence in September 1968. The Soviet invasion of Czechoslovakia in August 1968, however, led to a postponement, and it was not until the new Nixon administration had reviewed the whole **Strategic** arms situation that a new starting date of November 1969 was agreed. It is harder to date the end of the process. The SALT II agreement was signed in June 1979, but as it was never formally ratified it has also never formally expired. To the extent that both the USA and USSR were, roughly, holding to its terms, and accusing each other of breaches, it could be considered still to be a live process in mid-1987. As such it will have been a major concern of five US Presidents, through six presidential terms, and of four Soviet leaders.

The process may never end in any clear-cut way. Although he changed the name to Strategic Arms Reduction Talks **(START)** on coming into office, President Reagan's abortive early attempts to relaunch negotiations had to use the SALT II guidelines as a foundation. There is no formal name for the negotiations that have gone on patiently in Geneva ever since 12 March 1985, but they cover the same ground that a SALT III would have dealt with. Although periods like that between December 1983 and March 1985, when one country (in that instance the USSR) unilaterally pulls out of one series of talks and threatens never to come back may occur again, the probability is that arms control negotiations will be a permanent feature of international relations for the foreseeable future.

This does not necessarily imply that anything will happen to reduce the numbers of nuclear weapons. The main conventional arms reduction talks, the **Mutual and Balanced Force Reductions** negotiations, have been in nearly constant session since October 1973 with no agreements reached at all. Even so, many politicians believe that arms control negotiations are a good thing in themselves. In part this is sincere—the saying 'Jaw, jaw, is better than war, war' is not meaningless—but there is also the need to satisfy public expectations that governments will attempt arms negotiation. Thus President Reagan made no serious effort to arrange a summit meeting with any senior Soviet leaders until he came under fierce criticism during the 1984 re-election campaign for being the first President for many years who had never met the Soviet leader. Within a few months of Gorbachev's becoming leader in the following year, Reagan had met him in Geneva.

SALT I

The first set of Strategic Arms Limitation Talks began in November 1969 and ended when President Nixon and General Secretary Brezhnev signed two agreements in Moscow on 26 May 1972. These two documents were the **Anti-Ballistic Missile Treaty** and an Interim Agreement on Strategic Offensive Arms. The first, being a full treaty, had to be formally ratified by the US Senate, which happened by a vote of 88 – 2 in August. The interim agreement, which is what is usually meant when SALT I is referred to, was approved by votes of both the Senate (88 – 2) and the House of Representatives (307 – 4) in September.

The interim agreement had been produced relatively rapidly and easily, and had an easy political passage in the USA mainly because the difficult problems were intentionally shelved for later consideration in **SALT II.** The massive problems of equating such radically different force structures in coming to any real definition of **Parity** were largely ignored. The crucial disagreement between the USA and the USSR was over the Soviet desire to have British and French nuclear arsenals, and US **Forward Based** forces, included in the US totals. Furthermore, the USA wished to keep their bomber force out of the calculations, it being nearly three times as big, and vastly superior to, the Soviet long-range bombing capacity.

Rather than tackle any of these problems, a simple compromise was struck. The compromise was simply to freeze **ICBM** and **SLBM** forces at the level then existing or under construction. To be more specific, what this meant was that there was to be a freeze on launchers—either the silos for land-based missiles or the launching tubes for submarine missiles. This did not, of course, cover the actual size of the missile armouries, and, more importantly, it did nothing to constrain the number of warheads because there was no control over the process of fitting multiple independently targeted re-entry vehicles **(MIRVs)** to the agreed fixed number of missiles.

There were two other important aspects. The USSR agreed to a ban on any increase in their ultra-heavy missile force, the SS-9s and SS-18s, for which there was no US equivalent. Finally, it was agreed that the restrictions on launchers applied to the *total number,* so that changes could be brought about by increasing, for example, SLBM launchers if a suitable number of land silos were destroyed. The frozen levels were to be 2,347 for the USSR and 1,710 for the USA. The disparity was justified by treating it as compensation for the exclusion of US bombers and forward-based nuclear weapons, and of the French and British nuclear forces from the calculations.

In fact, the USA had no plans to enlarge its missile systems, the British and the French had no intention of allowing their nuclear weapons to be included in negotiations, the forward-based US systems were NATO allocated and incorporating them would have required NATO approval, and the USSR was not about to reduce its missile totals anyway.

SALT I was a success largely for the same reason that the associated ABM Treaty was—it did not require anyone to do anything which they did not want to do already.

Because it was an 'interim agreement' until a stronger and proper treaty could be negotiated, SALT I had an expiry date of October 1977. (The ABM Treaty, on the other hand, as a full treaty, is open-ended.) However, both sides later made a series of assurances that they would abide by the SALT I terms until SALT II was completed. As it happens, although both sides may have breached SALT II, it is generally accepted that they still abide by the more generous SALT I terms.

In one important way, the procedures determined for **Verification** of compliance with the terms of the agreement, SALT I set a precedent that was to cause major problems for later negotiations. Because of the intense Soviet objection to on-site inspection, it was agreed that **National Technical Means (NTMs),** that is satellites and other external electronic monitoring, were adequate to ensure that neither side would 'cheat'. Given the simplicity of the agreement, described by some American experts as 'limiting holes in the ground' the NTMs *were* adequate, but such a restriction has subsequently become a straightjacket in many other **Arms Control** arenas. (See also **SALT: the General Process.**)

SALT II

Negotiations on SALT II started in Geneva on 21 November 1972, less than seven weeks after the formal implementation of the **SALT I** interim agreement. The draft of the resulting treaty was signed by Presidents Brezhnev and Carter in June 1979 and, in the following month, was passed to the US Senate for ratification, as required by the US constitution. On 2 January 1980, realizing that he could never get Senate approval for SALT II, particularly in the light of the Soviet invasion of Afghanistan in December 1979, Carter withdrew it from discussion.

The treaty has never been ratified, although in March 1980 Carter announced that the USA would comply with its provisions as long as the USSR did so. Until 1986 there was no occasion on which either country openly admitted that it had breached the draft treaty, and many obvious cases of one or other country taking some action to comply with it. However, there has also been a series of allegations by both superpowers that the other has, in fact, broken one term or another. Legally the treaty has never been in effect, but it could be said that it has become a sort of 'international common law' obligation that, in general, has been complied with. How long that situation would prevail, and whether the uncertainty would be removed by a new agreement coming from the **Arms Control** negotiations that recommenced in 1985, or whether an uncontrolled **Arms Race** would ensue, was still not clear by mid-1987.

SALT II was, inevitably, a much more complex document than the interim agreement of SALT I, largely because the difficult problems from the first round of talks had been shelved for later discussion. Some of them continued to be ignored. Thus the question of American **Forward Based** weapons, that is, intermediate-range nuclear weapons which could hit the USSR only because they were stationed in Europe, had been subsumed by the now quite separate **Intermediate Nuclear Force** discussions. The British and French still resolutely refused to have their nuclear weapons counted with the US armoury and, as the USSR was beginning to wage its 'peace offensive' against Western Europe, it thought better of pressing this issue. Nevertheless, SALT II did tackle some serious problems. It accepted the fact of **MIRV** technology, and gave up the simplicity of counting each missile as one, regardless of how many warheads it carried. The draft treaty agreed not only on totals for missiles overall, but also on sub-category totals that went some way towards dealing with the very different needs of one superpower which had most of its warheads under the sea in the form of **SLBMs** and another which had most in **ICBM** silos. It brought the long-range bombing forces into the calculations, and even considered the new technology of **Air-Launched Cruise Missiles (ALCM).** What it did *not* do was to reduce the number of warheads either side had, or 'roll back' any technology, although it did constrict major new developments.

The full details of SALT II are too lengthy to enumerate here, but the highlights can be listed as follows:

A ceiling of 2,400 on ICBMs, SLBMs and heavy bombers;
a sub-ceiling of 1,320 on any one of the three categories;
a sub-ceiling of 1,200 on launchers for MIRVed ICBMs and SLBMs;
a sub-ceiling of 820 on MIRVed ICBMs.

The apparently conflicting sets of limits are to allow for various force mixes, which would be legitimate as long as they did not breach any one of the ceilings and sub-ceilings.

There were many other clauses—restricting the number of warheads on MIRVed weapons, restricting the number of new missile types that could be tested and deployed, controlling the size and nature of the bomber force on each side, and so on in great detail. The trouble with SALT II is that the US Senate refused to ratify it, and therefore, while it has served to guide and constrain the arms race, it has never had the authority which it ought to have had. For example, if ratified, the treaty would have required the superpowers actually to *reduce* their ICBM/SLBM and heavy bomber total to 2,250 by the end of 1981.

The key question of the 1980s as far as SALT II is concerned has been the extent to which the Reagan administration would honour the Carter pledge to abide by the terms of the unratified treaty. Despite his election campaign attack on SALT II as 'fatally flawed', President Reagan somewhat grudgingly observed the constrictions of the agreement until well into his second term of office. At the end of 1986 he openly allowed the **Strategic Air Command** to convert one more B-52 bomber for the purposes of carrying ALCMs than the treaty allowed. One interpretation of Reagan's actions is that he has been playing with the SALT limits to try to entice the USSR into a further arms control agreement; whatever one might think about the morality of this strategy, it indicates how powerful the agreement has been. What scope there is for progress towards a SALT III is unclear. However, such a treaty would certainly have to include quantifiable reductions in missile and warhead numbers, for the first time, because there seems little future, except in the field of **Ballistic Missile Defence,** of restraining the application of new technology to strategic weaponry. (See also **SALT: the General Process.**)

Sanctuarization

Sanctuarization is the description of a possible nuclear strategy which is only overtly admitted to by one member of the **Nuclear Club,** France, but may well underlie the thinking of all of them. De Gaulle developed France's independent nuclear force, the **Force de Frappe** or, as it is now sometimes called, the 'Force de Dissuasion' because he believed that no nuclear-armed nation would ever use its forces except in its own interests, and therefore the US guarantee of **Extended Deterrence** to Western Europe was invalid. For de Gaulle, and subsequent French thinkers, the point of owning a nuclear force was to make one's own country a sanctuary, an area against which no other power would dare use nuclear weapons. It was mainly to express this purpose of creating a sanctuary for France in the midst of a future war that de Gaulle stressed the idea that French missiles point 'à **Tous Azimuts**' (in all directions), rather than because anyone ever thought there could be a target other than the Soviet Union. The doctrine of sanctuarization, therefore, holds that nuclear weapons are not usable against a nuclear-armed power, because even a small nuclear force can present a retaliatory threat too awful to risk. (The French doctrine on the power of small nuclear forces, **Proportional Deterrence,** is complex, but not necessary to the general idea of sanctuarization.)

The USA cannot admit to sanctuarization as the basis of its nuclear policy because NATO's **Flexible Response** strategy depends on the American nuclear guarantee. However, even in the USA there are respected analysts and leading politicians who would argue that the arrival of nuclear **Parity,** combined with the dangers of uncontrolled **Escalation** from any nuclear use, in reality mean that the only function nuclear weapons fulfil is to deter the Soviet Union from attacking the US homeland. The British, in keeping with their general approach of never making any clear-cut statement on nuclear use policy (see **Declaratory Policy**), have never officially embraced sanctuarization. Nevertheless, much semi-official comment, and perhaps most expert opinion, takes it for granted that the *only* situation in which the UK would launch its own missiles is in retaliation for a nuclear attack on Britain (see **Moscow Option**). **Soviet Doctrine** on nuclear weapons is expressed in terms that do not fit Western debate, and it is difficult to speculate how far it is affected by thinking on sanctuarization. It is generally accepted that defence of the Soviet homeland is the single major object, which might well imply an unwillingness to risk retaliation by a nuclear attack on a nuclear-armed enemy.

The consequences of the nuclear club accepting sanctuarization in private, whatever they say in public, may be double-edged. On the one hand it would certainly reduce the risks attached to some versions of nuclear **War-Fighting** theory, in as much as even **Limited Nuclear Options** would not be executed against another member. On the other hand the implication that extended deterrence is vacuous might create something almost amounting to a nuclear 'free-fire zone' through much of central and northern Europe.

A major theoretical problem to the sanctuarization concept remains, however. It is one thing to say that possession of even a small **Independent Deterrent** nuclear force secures one against a *nuclear* attack. But does that make a country a national sanctuary in terms of conventional invasion? To claim that it does is to claim that there are grounds upon which a nuclear attack would be launched against a nuclear enemy short of that enemy having used such weapons itself. This, indeed, is official French policy; that the moment a Soviet soldier steps over the border they will use **Battlefield Nuclear Weapons** against the Warsaw Pact forces, to signal the danger of escalation, and that if it does not stop the advance, a full-blown **Counter-Value** nuclear strike would be launched against the USSR. The logic of this is fatally flawed; the status of nuclear sanctuary is immediately thrown away once the opponent is denied sanctuary.

Scenario

A scenario in strategic thinking is an imaginative account of a series of events, or of some crisis or problem, which might occur in the future. Because there is so little past experience of conflict approaching what a nuclear armed conflict would be like, the use of historical analogy is rare in modern strategic theory. Instead analysts concentrate on relatively simple imaginary situations to work out the implications of factors such as **Doctrine,** weaponry, **Declaratory Policy** and so forth. A scenario, for example, might depict a fully-fledged conventional onslaught by the Warsaw Pact on West Germany, as featured in novels like Tom Clancy's *Red Star Rising,* or a highly-technical precise phase of superpower interaction over a crisis in the Middle East. Essentially scenario planning emulates the traditional war-game techniques used by all military forces over the last century. War-gaming, if carried out with great care and attention to detail, as well as imagination, can be a vital tool. It is often said that the US navy fought no actions during the Second World War in the Pacific that its admirals had not practised in war-games during the 1930s. It is less clear whether the conditions of **Central Strategic Warfare** can be modelled in this way, and nuclear strategy scenarios almost inevitably involve a greater degree of cool rationality in their development than anyone thinks would be plausible in real life. Nevertheless, scenario-writing remains the nearest strategic thinkers have to a laboratory. (See also **Game Theory.**)

Schlesinger Doctrine

James Schlesinger became the US Secretary of Defense in May 1973, early in President Nixon's second term of office. He had previously worked for **RAND,** and was the first civilian professional strategist to hold the office. It was Schlesinger who presided over the development of the doctrine of **Limited Nuclear Options** and set in train the design of a nuclear arsenal that would be able to be used much more flexibly than the force structure **McNamara** had created for the **Assured Destruction** policy. Ultimately Schlesinger's doctrine was to be enshrined in President Carter's **PD-59.** The details of the doctrine revolved around the need to have a range of options with which to respond to a nuclear attack on the USA and, above all, featured a shift of emphasis away from **Counter-Value** strikes in favour of **Counter-Force** targeting. In many ways his ideas were not original even to the administration he served, because as early as 1970 President Nixon has publicly raised doubts about the **Massive Retaliation** doctrine, which he believed dangerously restricted the alternatives available to a President in the event of a Soviet strike.

Scowcroft Commission

The Scowcroft Commission was set up by President Reagan in 1981 to report on the future of the US strategic nuclear forces. Although it had a broad brief to report on the entire range of such forces, the original impulse behind the Commission was to allay Congressional fears about the **MX** missile and, in particular, the accompanying **Basing Mode** problems. The Commission strongly recommended the continuance of the nuclear **Triad** system by the USA, largely on the grounds that keeping all three legs, **SLBMs,** the **Strategic Air Command** and **ICBMs,** insured against a scientific breakthrough by the enemy in any one area. On the issue with which it was primarily set up to deal, the future of the MX missile, it was less positive than the President might have wanted. Although the building of MX, and its deployment in the old **Minuteman** silos, was supported, such a development was seen more as an intermediate step. The Commission clearly had some sympathy with Congressional interests in **Arms Control,** and recommended a move away from the heavy, multiple-warhead fixed missile. Instead General Brent Scowcroft, a previous National

Security Adviser, and his colleagues pressed for the development of the **Midgetman** missile. This is conceived of as a small, mobile missile with a single warhead. Both of the main distinguishing features of Midgetman are seen as improving international stability. Mobility would increase the **Survivability** of the missile in the face of an enemy's **First Strike,** while the single warhead, by moving away from **MIRV** capacity, would discourage a new **Arms Race.**

Sea Lines of Communication (SLOCs)

Sea lines of communication refers to the need for any maritime power fighting a war abroad to be able to protect sea routes for the transportation of supplies and reinforcements. In particular it is used to emphasize the problem NATO has in depending on **Sea-Lift** to move most of the 10 divisions of US reinforcements into Europe on which its **Central Front** policy relies. There would also be a need to ensure the supply of vital materials, particularly oil, to the NATO forces in Europe. The primary function of the NATO naval capacity, in the eyes of most strategists who are not enthusiasts for the **Carrier Battle Group** concept, is to protect these sea lanes, especially against Soviet submarine attack. As it is believed that **Forward Based** NATO troops might only be able to hold off a major Warsaw Pact invasion for about 10 days, the challenge of protecting the SLOCs becomes of vital importance.

It was the reliance on SLOCs that contrasted maritime powers, such as the USA or the UK, with continental powers like Russia or the German Empire, whose lines of communication were interior land links, during, for example, the First World War. The capacity to protect the SLOCs is what makes the real difference between being a maritime power and merely having a naval **Force Projection** capacity. The Soviet Union now has the latter, since it has developed a **Blue Water Navy,** but could not clear and protect extensive sea lines of communication over a protracted period to keep such a force supplied were the US navy to interfere. (See also **Maritime Strategy.**)

Sea-Launched Cruise Missile (SLCM)

Sea-launched cruise missiles are possibly the most numerous of all **Cruise Missiles,** and may be fired from both submarines and surface ships. The Soviet navy has been arming its ships with cruise missiles of varying degrees of technical sophistication since the late 1950s, and the US navy has made them very general equipment since the early 1980s. Most of the SLCM (pronounced 'slickum') inventories carry only conventional warheads, intended for ship-to-ship combat, but an undeterminable portion are nuclear-armed, and capable of striking targets over 2,000 kilometres inland (see **Dual-Capable System**). The main US SLCM, the Tomahawk, (from which was developed the US **Ground-Launched Cruise Missile**) comes in several versions, and there are about 1,000 in service.

Sea-Lift

Sea-lift, by analogy with **Air-Lift,** refers to the capacity to move troops, equipment and supplies rapidly to a conflict zone by ship. While the inability of, for example, the US air-lift potential to carry out many of its more dramatic plans, especially the emplacement of the Rapid Deployment Force (see **CENCOM**) is more often discussed, the traditional sea-lift capacity faces equally serious problems. A major reason for the sea-lift shortfall is that navies are prone to concentrate their attention on the more 'glamorous' business of acquiring combat ships, and to focus on what they see as their primary single-service tasks, rather than supporting army and air force needs which they regard as marginal (see **Inter-Service Rivalry**).

Sea-lift inadequacies are common, in fact, to all Western nations with a need or desire for **Force Projection.** The old-fashioned troop carriers of the two World Wars have not been replaced in naval inventories, and supply and amphibious warfare craft are in short supply in all navies. Of even more pressing concern is the fact that countries such as the USA and the UK have experienced a very serious shrinkage in the size of their merchant navies. Thus, in any sustained war, the ability of either government to rely on conscripting civilian ships and crews, as has been the pattern in the past, will be much reduced. The extent to which this is crucial is often not appreciated, but the number of merchant ships 'taken up from trade' by the Royal Navy to manage the comparatively minor task of moving one **Infantry** division to the Falkland Islands in 1982 indicates how vital the problem might become.

Second Decision Centre Thesis

This thesis, which appears under a variety of names and in more and less explicit versions, has been one of the major justifications provided by British governments for the UK's **Independent Deterrent** since the early 1960s. The argument is that the USSR might not believe in the firmness of the USA's **Nuclear Umbrella** covering Western Europe, and might conclude that it could safely use its nuclear weapons to intimidate or destroy European targets. But if Britain, and also possibly France, was known to have nuclear weapons and to be prepared to use them in the case of a Soviet attack on NATO, the uncertainty of what response it might face would deter the USSR. The argument is not based so much on a guarantee that UK weapons would be used in any particular circumstance as on the impossibility of the USSR ever being able to calculate *when* they might be used. The whole argument is an example of **Deterrence** through uncertainty—exactly the reason why the UK is so reluctant to announce a **Declaratory Policy.** Most probably the USA would act by itself or in co-ordination with the UK, but a second independent decision centre means that the USSR can never be sure, no matter what it decides about American policy, that it faces no nuclear threat. Unlike France, where the nuclear deterrent is justified in part precisely on the argument that no European country *can* trust the USA to risk **Central Strategic Warfare,** British governments have always been careful to insist that they do not doubt America, but that the USSR might do, and so a second centre of nuclear decision-making is vital to deterrence.

Second Strike

Second strike is no less ambiguous a concept than **First Strike,** but like the latter has an apparently simple meaning. A second strike in strategic language is a response to an enemy's nuclear attack, where that attack was the first use of nuclear weapons at that level of **Escalation.** The most usual place to encounter discussion of a second strike is in the more specific context of a 'secure second strike capacity'. In this case the meaning is clear. A secure second strike capacity means that one country has enough nuclear weapons which are invulnerable to attack (see **Vulnerability**) to 'ride out' an enemy first strike and still have a guaranteed ability to retaliate. (This generally means that secure second strike weapons are **SLBMs.**) Thus the key to the idea of second strike is that it is a response to an enemy's action.

The principal ambiguity comes from the following **Scenario.** Country A makes the first use of **ICBMs** in the war, and launches them against Country B. Country B responds with its secure second strike. Then Country A uses its remaining weapons (from *its* secure second strike force) to destroy what remains of Country B. In this case the second launch by Country A is both its second strike, and the use of its secure second strike capacity, even though it is actually the third launching of ballistic missiles in the war. Similar problems occur if movements over escalation boundaries are taken into account. If Country A uses **Theatre Nuclear Forces,** to which Country B responds with ballistic missiles aimed at Country A's heartland, and A responds with ballistic missiles, at which point are we talking about a 'second strike'? The only safe way to handle this ambiguity is to restrict usage of the term 'second strike' to a particular sort of attack—one made in response to the enemy's action, using weapons deliberately designed to survive a first strike, when all the strikes are, as pointed out above, at the same escalation level.

Selective Service

Selective service is the name given to the American system of **Conscription,** operated by the Selected Service Administration (SSA). The system is designed to cope with the fact that the USA could never use, let alone handle, the entire male age cohort, and therefore some way must be found of making a selection that seems to be socially just. This has never been achieved and the 'draft' has always occasioned great bitterness, especially during the Vietnam War era. It is a combination of lottery and selection. Every young man must register for the draft, and is given a number. If, when the services needs someone his number should be called, he must serve unless he can get exemption. However, the reasons for exemption, particularly those relating to being in full-time higher education, are socially inegalitarian. One consequence, for example, was that the black population of the USA was very much over-represented in the army during the Vietnam War, while young white middle-class college students were, in fact, quite unlikely to have to serve. After the Vietnam War the draft was suspended, and the USA has operated with **All Volunteer Forces,** but all men still have to register with the SSA in case there arises a future need to conscript.

Selective Strikes

Selective strikes are attacks with very precise and carefully delimited targets, chosen so as to maximize military value, or to send a very clear signal, while minimizing the risks of **Escalation.** As such selective strikes are at the heart of the more general concept of **Limited Nuclear Options.** The concept is not limited to nuclear warfare, however. In any offensive it is possible to distinguish between general attacks against any targets of military value that are within range, and carefully-defined targets which have been singled out. The essential point is that a selective strike is as much a matter of signalling to the enemy as of achieving purely military goals. The trouble is that the friction of war usually makes strikes much less selective than intended. Some military thinkers like to talk about **Surgical Strikes,** with the implication that such a use of violence is analogous to removing a tumour for the benefit of the whole body. Yet even with the **Target Acquisition** capacity of modern weapons this is seldom achieved, and **Collateral Damage** arises; for example a hospital might be bombed in a raid meant only to destroy a terrorist headquarters.

SHAPE

SHAPE (Supreme Headquarters Allied Powers Europe) is the headquarters of the entire European NATO military command (see **Allied Command Europe**) and is under the direct authority of **SACEUR.** Situated near Mons, in Belgium, it is essentially an administrative headquarters rather than a command location. Were NATO ever to go to war, SACEUR would change from being a largely political animal to a direct operational commander and, together with SHAPE, would move to some secret and highly-fortified **C³I** bunker. SHAPE used to be at Versailles, outside Paris, but had to move along with the rest of NATO's infrastructure when France removed itself from the unified military organization in 1966.

Short-Range Intermediate Nuclear Forces (see Intermediate Nuclear Forces)

Shot Across the Bows

This phrase is borrowed from traditional naval warfare, when a ship on **Blockade** duty would fire a cannon shot just in front of a blockade runner to make it clear that if it did not halt it would be sunk. In a modern context it refers to one possible strategy for an early use of nuclear weapons by NATO. The idea is that a very small and **Selective Strike,** possibly using only one warhead, would be made at that point in the **Central Front** battle where NATO's **Tactics** demanded the use of nuclear weapons. There would be a pause following this 'shot across the bows' to give the Warsaw Pact the chance to halt its advance, and perhaps a chance for **Intra-War Bargaining** to take place. If the Warsaw Pact chose, as it were, to ignore the warning, a serious use of **Theatre Nuclear Forces** would commence. (See also **Fire-Break.**)

Signals Intelligence (SIGINT)

SIGINT is the form of intelligence collection carried out by bodies such as the British Government Communications Headquarters (GCHQ) or the US **National Security Agency.** The activity consists of monitoring all the radio wavelengths used by an opponent (or at times an ally), transcribing, and attempting to decode its messages. The earliest use of SIGINT in a significant way was the British Admiralty's radio monitoring service which gave advance notice that the German Grand Fleet was putting to sea just before the Battle of Jutland in 1916. It was the failure of the US navy to take seriously signals intelligence reports of Japanese naval movements in December 1941 that led to their Pacific fleet being surprised at Pearl Harbor. It is not always necessary to be able to decode the enemy's radio communications to gain useful information. Changes in the volume, pattern and location of radio messages by themselves can give vital clues. (See also **Electronic Intelligence.**)

Silo

The silo is the predominant **Basing Mode** for **ICBMs** among the superpowers, although increasing attention is being given to mobile missile launchers. This is because the orthodox silo is becoming increasingly vulnerable. A silo is simply a hole in the ground walled with a thick concrete and steel shell, inside which a missile sits, sometimes on a sprung floor, with a heavy concrete cover that can be electrically opened. Silos were originally capable of considerable protection, and were built to withstand blast overpressures of more than 2,000 pounds per square inch (psi). (An ordinary brick building will collapse at between five and 10 psi.) However, combinations of high **Yields** and great accuracy (that is, very low **Circular Error Probable**) mean that most of the US silo-based missiles now suffer from problems of **Vulnerability** to the main Soviet ICBM force of **SS-18** missiles.

Single Integrated Operations Plan (SIOP)

The SIOP (pronounced sigh-op) is the US strategic nuclear war plan, supposed to cover all eventualities and contain options ranging all the way from a demonstration strike, or **Shot Across the Bows,** to an intentional 'city-busting' **Assured Destruction** bombardment of the USSR. The original SIOP was drawn up at the end of Eisenhower's presidency, when the utter confusion and incoherence of the existing nuclear war plans became apparent. Before its introduction each commander of a force that had any nuclear weapons attached to it drew up a separate target list and strike plan. As this involved not only the **Strategic Air Command (SAC),** but also **Tactical Air Force** commanders in various places around the world, the commanders of all navy fleets which had nuclear weapons on their aircraft carriers, the new ballistic missile submarine force (see **SSBN**) and even land troop commanders with medium-range rockets, the combined effect was absurd. Investigations found that dozens of vital targets had been put on the attack list by sometimes as many as three different commanders, and that there was no centralized priority list or agreed set of alternatives to suit various contingencies.

SAC, as the original main strategic bombing force, believed that they should have sole control over US nuclear war policy. This claim was hotly contested by the US navy, whose **Polaris** submarines would, before long, come to carry over 70% of all US nuclear warheads. With some difficulty, Eisenhower, aided by his own great experience of military bureaucracy, managed to force the creation of a single centralized targeting agency. This always has an SAC general as its chief, with a navy officer as deputy. It is responsible to the **Joint Chiefs of Staff** committee for maintaining a target list and preparing options from within it that can be presented for the President to choose from in any given situation. The SIOP is continually updated, and undergoes major revisions from time to time, as presidential directives change the emphasis of overall nuclear war policy. Thus the first SIOP, completed in 1960, was rapidly rewritten to fit the **McNamara** conceptions of nuclear strategy, and at least three subsequent revisions have taken place as first Nixon, and then Carter, developed the **Limited Nuclear Options** approach which concentrated on **Counter-Force** targeting.

The SIOP (the latest version is known as SIOP-5) is not one war plan, but many. For example, it is believed to contain more than 40,000 targets, yet the total US warhead inventory is only about 10,000; thus there can be no meaning to the phrase occasionally used of 'executing *the* SIOP'.

The targets are broadly defined into four categories:

1. Soviet nuclear forces and their supporting infrastructure, with about 20,000 targets;
2. **Other Military Targets;**
3. the Soviet leadership and key **C³I** bunkers; and
4. economic and industrial targets, of which there are at least 15,000 possible selections.

In the words of the official policy statement, the USA no longer targets population centres *per se,* although it would, of course, be quite impossible to avoid killing millions of civilian industrial workers and their families if many targets from the fourth category were selected (see **Collateral Damage**).

Because Polaris is officially NATO-dedicated, and thus comes under **SACEUR**'s control, the British Polaris squadron is also integrated into the SIOP, and Britain has officers posted to the joint targeting staff. It is widely believed, however, that Britain also has its own targeting plan for independent use, principally the **Moscow Option,** and it will certainly need to develop one if **Trident II** is finally deployed, with the dramatic increase in warhead numbers which it will bring.

Single Shot Kill Probability (SSKP)

Single shot kill probability is a measure of the general effectiveness of a **Weapons System.** It is essentially a combination of the accuracy of the weapon, whether it be a rifle, anti-tank rocket or **Anti-Ballistic Missile,** and the probability that a target, if hit, will be destroyed. It is the enormous increase in SSKP of modern weapons that has transformed the patterns of warfare today. Techniques of **Target Acquisition** and guidance of projectiles have brought the accuracy of a single round so high that it is sometimes said that, if one can see the target, one will be able to destroy it. The US army officially calculates that the SSKP for an anti-tank missile fired from one of its most modern tanks at an enemy tank 3,000 metres away is now 0.9 (that is, a 90% probability, as SSKP figures are rated on a scale from 0 – 1). (See also **Precision-Guided Munitions** and **Smart Bombs.**)

Skybolt

Skybolt was an American strategic weapon programme worked on in the late 1950s and early 1960s which, had it come to fruition, would have led to the British strategic nuclear force developing in a very different way. When it became apparent to the British government, in 1960, that their own **Intermediate-Range Ballistic Missile** programme, **Blue Streak,** would have to be cancelled, an agreement was secured with the Eisenhower administration to buy Skybolt. This weapon would have been mounted on the RAF's **V-Bombers,** thus extending their **Credibility** as a nuclear strike force for many years.

Skybolt was an unprecedented weapon and, indeed, nothing similar has since been built by either the USA or the USSR. Although carried on aircraft it was neither a **Stand-Off Bomb,** like the other British-built weapon, **Blue Steel,** nor an **Air-Launched Cruise Missile.** Instead, it was a true **Ballistic** missile which could be launched from an airborne bomber as far away from the target as desired. It would then behave exactly like a ground- or submarine-launched ballistic missile **(SLBM),** shooting up into an elliptical orbital trajectory in its **Boost Phase** and descending with great velocity, making it no easier to counter than any **ICBM** or SLBM. As long as the V-Bombers could avoid destruction on the ground, by a combination of dispersal and permanent airborne patrols, the UK could rely on the RAF to give it a secure **Second Strike** capacity.

However, Skybolt was cancelled in 1962 by the Kennedy administration, leaving Britain's strategic nuclear force with no long-term future. There have always been suspicions that the cancellation of Skybolt was avoidable, and that the decision was taken for political rather than technical reasons, either with the intention of sabotaging the British **Independent Deterrent** or, more probably, simply out of indifference to its fate. (Although the USA has always insisted that Skybolt was a technical failure, RAF officers attached to the development programme have often denied this.) The consequence of Skybolt's demise was that the UK had to accept instead the offer of the **Polaris** missiles which had been made earlier by President Eisenhower. The Polaris fleet has never been more than a **Minimum Deterrence** force, whereas Skybolt, if bought in sufficient quantities, could have maintained the strike capacity of **Bomber Command** at the level it had reached just as **Gravity Bombs** were ceasing to be a reliable instrument of nuclear **Deterrence.** Calculations vary, but some sources suggest that the V-Bomber force could, in 1962, have delivered at least 90 **Megaton**-level attacks on the USSR, and that Skybolt would have guaranteed the continuation of this capacity. With Polaris it has never been possible to guarantee the delivery of more than 16 such attacks, and perhaps fewer strategic strikes.

We are never likely to know the truth of the Skybolt affair, but it should be noted that the majority of the US strategic community at the time, led by **McNamara,** were strongly opposed to the British and French having their own nuclear forces. It was probably only the personal relationship between President Kennedy and the British Prime Minister of the time, Harold Macmillan, that even secured Polaris for the UK.

SLBM

A submarine-launched ballistic missile is the naval version of an intercontinental ballistic missile **(ICBM).** Carried on nuclear-powered submarines known as **SSBNs,** SLBMs have ranges varying from 2,000 kilometres for the first version of the US **Polaris** missile to between 9,000 and 11,000 km for the most advanced Soviet (Typhoon class) and American

(**Trident** D5) weapons. For much of their brief history submarine-carried missiles were much less accurate than land-based ICBMs, and could only be used for **Counter-Value** strikes against urban-industrial targets. The great value of this leg of the **Triad** was that submarines on patrol were virtually undetectable, so that the SLBM part of a nation's nuclear deterrent could ensure its **Second Strike** capacity. As these missiles were both **MIRVed** and given longer ranges the importance of the SSBN fleets, especially for the USA, increased greatly. With Trident D5's 11,000 km range there is an enormous volume of sea room for the submarines to hide in while staying in range of targets throughout the USSR. Historically the SLBM force has been more important to the USA, which keeps around 70% of its warheads in submarines, than to the USSR which, with much worse port facilities and ice-covered coast lines, has preferred to put its major strength into ICBMs.

Recent developments in satellite navigation systems have made SLBMs potentially as accurate as ICBMs, opening up the possibility of a secure force which could also be used for sophisticated nuclear **War-Fighting** or the carrying out of **Limited Nuclear Options.** The total power of a single modern SSBN is enormous. The American version of the Trident D5 (also referred to as Trident II) carries 24 missiles, each with up to 14 independently-targeted warheads; this amounts to an **Equivalent Megatonnage** of well over 100, and possibly over 150, MTE. If it is remembered that the **Mutual Assured Destruction** level plans of US Secretary of Defense Robert **McNamara** required only around 400 MTE, and that the USA will finally have a fleet of a dozen or more Trident submarines, the power of a secure SLBM armoury is obvious.

Britain and France also have SLBM forces, the former relying entirely on such weapons for its strategic deterrent, and China either already has, or is attempting to develop such systems. The attraction to smaller powers is, above all, the security that an underwater-based system provides against a superpower's **First Strike** removing the deterrent force.

Small Intercontinental Ballistic Missile (see Midgetman)

Smart Bombs

Smart bombs, and any other 'smart' weapon, are bombs that have some capacity to steer themselves directly onto targets. Thus their accuracy is not entirely dependent on the initial accuracy of the bomb aiming, which can be affected by such variables as wind speed. There are essentially two basic technologies for smart bombs. The first generation of such weapons depended on the target being illuminated by laser beams from an aircraft or ground observer. The reflections from the laser were picked up by sensors on the bomb which then controlled fins to angle the path the bomb took.

The next generation, which is currently coming into operation, is self-contained. With these weapons a sensing device detects emanations from the target directly. The most common form of this new generation relies on an infra-red detector which allows the bombs to home in on sources of heat. For this reason most second generation smart bombs are primarily anti-tank weapons, the heat of the tank engine making them perfect targets. (See also **Precision-Guided Munitions.**)

Social Defence

Social defence is a phrase used to cover not so much one policy as a collection of attitudes to defence. The idea comes from the left of the British political spectrum, and particularly from those opposed to nuclear weapons. It refers to any or all of the forms of defence that insist on doing without nuclear weapons, and as much as possible without any weapons that can be used for large-scale offensive warfare. An associated phrase—'a genuinely-defensive defence'—helps to illustrate the argument behind the concept. One way in which social defence is summed up is to argue that the best way of defending a country is to make it extremely difficult to occupy. Thus social defence can range all the way from passive civil disobedience, through total non-cooperation with occupying authorities, to large-scale **Guerrilla Warfare.** It is argued that no occupying power can afford enough troops to subdue effectively a population which thoroughly resists it and, at the same time, that a country which completely lacks the ability to attack another nation is much less likely to become an object of attack itself. The obvious problem with social defence is that the historical record, especially during the Second World War, shows that collaboration with an occupying enemy is at least as common as partisan warfare against it.

Sonar

Sonar (which is actually an acronym for Sound, Navigation and Radar) is a general label for the sound-acquiring detection systems used in underwater naval combat. There are two general types of sonar, which are similar to those found with radar. One is passive: a submarine, or submarine-hunting ship or helicopter, just listens for sounds transmitted through the water from the target submarine or ship. The alternative, active sonar, involves sending out noise 'blips' which create echoes. As the echoes return to the transmitting ship or submarine, a picture of the underwater environment can be built up which looks much like the picture on a radar screen. (See also **Hunter-Killer Submarine.**)

SOSUS

SOSUS is the acronym for Sound Surveillance System, a major American asset in the **Anti-Submarine Warfare (ASW)** that would be crucial in a war between NATO and the Warsaw Pact. It is a chain of underwater listening devices planted on the seabed at various strategic locations around the world, most notably across the approaches to the Atlantic Ocean, such as the **Greenland – Iceland – UK Gap** through which Soviet submarines would have to pass to attack America's **Sea Lines of Communication.** With these listening posts strung across such seaways, continuously monitored in land-based stations, the US navy is able to follow, even during peacetime, a large part of all Soviet submarine movements. Thus, in the event of a 'hot' war, US naval intelligence will have a fairly accurate idea of the size and deployment of Soviet submarines outside their territorial waters, and will be able to assess the main areas of threat. Combined with airborne monitoring by forces like the British maritime Nimrod squadrons and the ASW carrier forces it is believed that few, if any, undetected Soviet submarines would be able to approach the vital reinforcement convoys.

Southern Flank

Although the North Atlantic Treaty extends its geographical coverage as far south-west as Portugal and as far south-east as Turkey, the **Central Front** is cut off from the southern part of this area, mainly by the Alps. Not only is the southern region cut off from the northern, but it does not even constitute a single geographical military theatre itself. There are effectively three sub-theatres, between which NATO would find it very hard to co-ordinate and manoeuvre troops.

Italy forms one southern sub-region, and the bulk of the Italian army is concentrated in its north-east sub-Alpine area to protect against any invasion past Trieste, Yugoslavia, although as that country is not a Warsaw Pact member, and is outside the influence of Moscow, the threat may not be very obvious. Greece constitutes a second sub-theatre potentially subject to invasion; again a border is shared with Yugoslavia, and also with Albania—another non-Pact member. However, Greece does share a border with one Soviet ally, Bulgaria. Undoubtedly the most vulnerable part of the southern flank, and the one which would most benefit the Warsaw Pact to invade, is Turkey, which actually has a common border with the USSR, as well as with Bulgaria. The Turkish army, though large and well trained, is obsolescent in its equipment. The way in which Turkey provides an obstacle to Soviet control of the eastern Mediterranean, particularly in terms of naval strategy (Turkey controls the key Bosporus and Dardanelles straits and, with Bulgaria, Romania and the USSR, surrounds the Black Sea) increases the importance of its vulnerability.

As it happens, NATO's internal cohesion is at its weakest in these latter two sub-theatres because of the Greek – Turkish conflict of Cyprus and the Aegean Sea. A side-effect of this is that the USA has been prevented from giving Turkey as much military aid as the **Pentagon** would like by the powerful Greek lobby in the US Congress. A second factor leading to a lack of political cohesion and increasing the weakness of this potentially vital strategic area has been the ambivalent attitude of Greek governments since 1974 to NATO in general, and to the USA in particular. There is a strongly-held view among many strategic analysts that a third world war is more likely to come about because of US – USSR conflict in the Mediterranean than on the central front. It is, therefore, of great importance to NATO that the southern flank, which would pose problems enough just by its geography, be strengthened politically.

Soviet Doctrine

Soviet military **Doctrine** is only known in part to Western analysts, and much must be deduced from Soviet and Russian traditions. A few points are fairly clear. It is axiomatic to Soviet military thinking that their own **Strategy** must be based on observing the strengths and weaknesses of the opponent: it is in this way a reactive strategy. The way the Soviets analyse NATO leads them to the conclusion that surprise is all important in order to take advantage of NATO's greatest weakness, which is its mobilization time. This accords with long-term Soviet, and indeed Russian, preference for an offensive strategy. Secondly, speed is of the essence if Soviet objectives are to be achieved before any nuclear exchange can take place, and if the early stages of the war are to be conducted in such a way as to make NATO's use of nuclear weapons as difficult as possible. Thus it is likely that the Soviet Union would attach great importance to the use of **Operational Manoeuvre Groups,** to try to prevent the forming of a rigid 'front line'. If they can move very fast at a stage when an unmobilized NATO has not yet firmly established its defensive position on the **Central Front,** and succeed in inserting their armies inside the area covered by the mobilizing NATO units, it will be very difficult indeed for NATO to use **Battlefield Nuclear Weapons.** Finally the Soviet Union fears that NATO, given any time at all, will gain **Air Superiority,** and thus the doctrine stresses the need to open the offensive with an immediate attack on NATO air-bases and command and control centres (see **C^{3}I**), possibly using **Chemical Warfare** techniques, against which they need at the moment fear little retaliation.

A further reason for the emphasis on speed and surprise in Soviet doctrine is that a long drawn out war, even if it did not go nuclear, would pit the greater Western economic-military potential against the Warsaw Pact. The Warsaw Pact's main advantage is a greater short-term power of forces in being. Among other implications, this is leading some Western analysts to wonder if the supposed reliance by the Warsaw Pact on the **Echeloned Attack** approach, to deal with which most NATO doctrine and weapons acquisition for the central front is geared, is, after all, that important. It should be stressed that this focus on the offensive does not imply that the

Soviet Union has formed the *intention* of attacking Western Europe for gain; the point is that in a crisis situation where the Soviets have reason to fear that war will happen, they have every incentive to make **Pre-emptive War** in this way, believing that to be their best hope for victory.

The final point to make is that all Soviet experience, as well as doctrine, stresses the need for highly co-ordinated employment of all arms and services (see **Combined Arms**). The co-ordination would be at what the Soviets call the **Operational Level,** and what NATO would call the theatre level, so that, for example, all air-power across the whole central front will be directed from the **TVD** headquarters, as will reinforcements and the general decisions about axes of advance. This again is partly based on taking advantage of a perceived NATO weakness—the multiplicity of national armies and the lack of coherent planning and doctrine (see **Interoperability.**)

Soviet Strategic Thought

The central question which is discussed in professional military and strategic studies journals about Soviet strategic thinking is 'Does the USSR believe it is possible to win a nuclear war?' What makes this so vital is that Soviet military thinkers have never explicitly indicated that they share the concepts and logic of Western nuclear strategic analysis. For example, the US reliance on the doctrine of **Mutual Assured Destruction** has never been accepted by the USSR. Soviet strategists have always thought it better to guarantee their security by their own efforts to build all forms of weapons and to take all defensive measures, including **Civil Defence,** rather than rely on mutual vulnerability. Furthermore, it has always been part of Soviet thinking, at least at the **Declaratory Policy** level, that if a major war between the USA and USSR occurred, it would be fought to the utmost with every type of weapon available. Thus concepts such as limited nuclear war (see **Limited War**) are implicitly rejected: neither would there appear to be much, if any, faith in **Escalation Control** in Soviet circles.

Some schools of analysis believe that the USSR simply does not share the Western perspective that nuclear weapons are so categorically different from conventional weapons. Certainly what *can* be deduced about Soviet targeting doctrine would suggest that nuclear weapons are seen as having genuine military use, whereas it is fashionable, even among the military in the Western defence establishments, to decry this. At the same time there is little evidence that the Soviet Union envisages **Strategic** weapons as the major **Counter-Value** agents which much of Western thought identifies them as. The USSR studied the Anglo-American campaign of **Strategic Bombardment** in the Second World War, and was fundamentally unimpressed with it. In contrast, they see nuclear weapons much more as **Counter-Force** weapons and to be used on military-related targets even when they are used in a counter-value mode. That is, the targets would be what have in the past been thought of as 'war-related industries', not industrial capacity in general. None of this is to say that there is the slightest evidence that the USSR plans to initiate nuclear war, or that they think it could be fought in a way which would leave them relatively untouched. Rather it is a statement that they will be prepared to use whatever weapons are at hand, will pursue the nearest attainable thing to victory if a war does occur, and will not seek to limit war or to search for opportunities for **Intra-War Bargaining** such as are preached in American **War-Fighting** doctrines.

A second major question about Soviet strategic thought refers to the size of their arms build up. Put in Western terms, the problem is this: if the USSR is really as defensively, rather than offensively, oriented as they claim, and as NATO claims it is, why does it strive for such huge superiority in all areas of armaments? The generally agreed answer is that **Soviet Doctrine** has always stressed the need for massed forces of all arms, and that provision is made for a wide margin of error. Two aspects of Russian military history would tend to support this. The first is the experience of surprise invasion which twice, under Napoleon and Hitler, very nearly led to their defeat. The second is their long historical experience of having to rely on badly-trained and inflexible mass armies against superior, but smaller, invading forces. This has led, for example, to a very strong belief in the value of massive artillery bombardments—the Soviet army artillery in the closing battles of Second World War was the biggest ever deployed by any country. Considering these factors, the huge preponderance of artillery compared with NATO's, and the Soviet demand for far more **Divisions,** still largely staffed by ill-trained conscripts, appears more as a historical strategic pattern than as evidence of any military intent. The trouble with all these analyses, of course, is that available doctrinal publications are arguably written as much for Western consumption as for real internal use. In the West we simply cannot know in detail what doctrines inform actual Soviet strategic decisions.

Spasm War

Spasm war is a phrase used to describe one of the earlier **Scenarios** for a nuclear war. The spasm war theory was that the outbreak of a nuclear war would involve almost instantaneous launching of every missile in the armouries of the USA and the USSR. It stresses the uncontrollability of nuclear weapons interchange, and contradicts all theories of carefully calculated **Escalation.** While many analysts still believe that nuclear war would have this uncontrolled spasm character, more fashionable nuclear strategic thinking stresses control and rational targeting, as with **Limited Nuclear Options** or, at the opposite extreme to spasm war, the doctrine of nuclear **War-Fighting.**

Sputnik

Sputnik was the Russian name for the first ever artificial satellite, launched on 4 October 1957. It was very small, carrying nothing but an automatic radio beacon, but it demonstrated convincingly that Soviet missile technology was at least on a par with, and probably superior to, that of the USA. Sputnik's real importance was the psychological shock which it gave the Americans—not only was their self-esteem as the world's technological leaders challenged, but the implied threat to the USA itself from an advanced Soviet missile capacity was deeply frightening to them. In the wake of Sputnik, President Eisenhower pushed successfully for a major expansion of scientific education in the USA, as well as accelerating the American missile programme.

SS-18

The SS-18 is the largest of all the USSR's **ICBMs** and, given that the Soviet Union has always built much heavier missiles than anyone else, the largest in the world. The **SALT II** agreement placed a limit of 308 SS-18s, and deployment to this level has taken place in silos in the western USSR. There are several varieties of the SS-18, but the main one deployed, called the Mode 4 by NATO, is gigantic, with a **Throw-Weight** of eight tons and a range of at least 10,500 km. This allows it to carry 10 **MIRVed** warheads, each of 500 kilotons; the Mode 5 version may carry the same number of even more powerful 750-kt warheads. It is even accurate, compared with most Soviet missiles, with a **Circular Error Probable** estimated to be as low as 250 metres. A full strike with this set of missiles would be enormously destructive. It would allow the USSR to target three warheads on each of the US **Minuteman** silos. This was the basis for the American fear of the **Window of Vulnerability**; the **Pentagon** estimated that a Soviet **First Strike** might destroy 90% of the US ICBM force before it could be launched.

SS-20

The SS-20 is one of the most familiar weapon designations in the current strategic vocabulary because of the political furore surrounding its deployment by the USSR, which led to the **Euromissile** crisis. The SS-20 caused the West such problems because it was the first really successful **Theatre Nuclear Force** weapon the USSR had ever managed to produce. The Soviet Union had nuclear missiles deployed for European theatre purposes from the early 1960s, notably the SS-4 and, to a lesser extent, the SS-5. However, these missiles were very primitive and suffered from weaknesses in the two vital assets of any force meant for theatre purposes. The main disadvantage was that the missiles were all liquid-fuelled, meaning that they could not be kept at a state of high readiness for long periods. Thus the delay between deciding to use them and launch could easily be eight hours. Theatre use may well demand much more rapid reaction than this. A second disadvantage was their tremendous inaccuracy; the **Circular Error Probable (CEP)** of the SS-4 is estimated at 2,300 metres. Given that they were fixed-base missiles with a range of only 2,000 kilometres they may well have presented a greater threat to Warsaw Pact forces than to NATO, had they been used. A CEP so great meant that they could only be used against troop concentrations (they would be ineffective against a **Hard Target**), yet the eight-hour delay would make it likely that any suitable target would have moved before their launch.

Against this background it is easy to see why NATO reacted so anxiously to the replacement of the SS-4 by the SS-20. This new ballistic missile is solid-fuelled, so it can be kept at one hour's readiness. It is mobile, so it can be moved and dispersed, making it much less vulnerable. It has a range of at least 5,000 km, so it can be based as far east as the Urals and still hit targets in Western Europe. It has a **MIRV** technology **Front End,** with three warheads of about 150 kilotons, rather than one huge **Megaton**-level warhead as in the SS-4. Smaller warheads are far more suitable for theatre use; the SS-4, if used, might have carried huge amounts of **Fall-Out** over Warsaw Pact troops. Finally, the SS-20 is vastly more accurate, with a CEP as low as 400 metres—nearly six times more accurate than the SS-4. All this marked a quite dramatic improvement in the **Lethality** of Soviet theatre weapons, particularly when it is remembered that doubling the accuracy of a weapon is equivalent in effect to an eight-fold increase in megatonnage.

The USSR is thought to have deployed over 400 of these missiles since they came into operation in 1977. However, not more than about 270 of these really concern NATO directly, the remainder being sited in the Soviet Far East, targeted against the People's Republic of China. Nevertheless, this meant that NATO now had to face over 800 independently-targeted, and relatively accurate, fast-reaction warheads when before it might have felt that it could almost ignore the Soviet theatre nuclear threat. Given that NATO's strategy, far more than the Warsaw Pact's, depends on the early use of nuclear weapons, the arrival of what is at least a deterrent force was bound to cause enormous anxiety.

The removal of SS-20s from the European theatre, as well as the balancing US **Ground-Launched Cruise Missiles** and **Pershing IIs,** has been at the centre of **Intermediate Nuclear Forces (INF)** negotiations during the 1980s. There was every indication that agreement to dismantle these weapons would be reached by the end of 1987; the Soviet offer, made in July of that year, to scrap the Far Eastern-sited missiles raised hopes of a global INF ban. (See also **Double Zero Option, Long-Range Theatre Nuclear Forces** and **Zero Option.**)

SS-22 and SS-23

In the heat of argument about the Soviet **SS-20** missiles in Eastern Europe very little attention was paid to the other new missiles in the USSR's inventory, until, in the early months of 1987, an agreement on the reduction or even abolition of **Intermediate Nuclear Forces (INF)** seemed imminent. The USA, however, has been complaining of them since 1980, before their actual deployment. The SS-22 and SS-23 missiles are of much shorter range than the SS-20, and are genuine **Theatre Nuclear Forces (TNFs)** intended for **Battlefield** targets. It is unclear how many were planned to be, deployed, but the figure will be in the range of several hundreds.

In fact these missiles are replacements for older generations of weapons, just as the SS-20 itself is a replacement for the old and very inaccurate SS-4. The SS-22 replaces the old SS-12 missile, which was deployed at the army corps level and is thought to have a range of some 900 kilometres, while the SS-23 replaces the SS-1, known to NATO as the 'Scud', probably based at the divisional level and with a range of some 500 km. Little is known about the performance of the replacement weapons. They are likely be more accurate than their predecessors, but this may mean little given the notorious inaccuracy of all Soviet INF missiles before the SS-20. Unless they are much more accurate than is thought likely, they are liable to be of use only against fixed targets, such as air-bases and supply dumps, because the **Target Acquisition** problem would make them of no great value as counters to military formations.

The real problem for NATO presented by these new short-range missiles is that they fall across a gap in NATO's own armoury. Disregarding aircraft, NATO has no weapon with an equivalent range. NATO's short-range nuclear force comes mainly in the form of artillery weapons, with ranges only up to a few scores of kilometres, and of very old missiles of ranges up to only about 150 km. While the much longer-range NATO INF forces, represented by the **Ground-Launched Cruise Missile** and the **Pershing II** ballistic missile, are in place this matters relatively little. However, should the **Zero Option,** originally a Western European proposal, taken up by President Reagan in 1981 and, against NATO expectations, adopted by Soviet Secretary-General Gorbachev in 1987, become a reality, these forces would be removed. In such a situation NATO's lack of short-range nuclear weapons becomes of great concern to Western political leaders. Further suggestions by the USSR that they might be prepared also to remove their SS-22s and SS-23s from Eastern Europe, in return for NATO removing the existing, and ageing, Pershing IAs and not deploying any new weapons in the 500 – 1,000 km bracket, merely serve to focus attention on NATO's comparative weaknesses at the **Battlefield Nuclear Weapon** and **Conventional** levels. This development became known as the **Double Zero Option.** Whether NATO needs to match the Warsaw Pact in every category of weapons in order to exert some form of **Escalation Control** is unclear, but the absence of TNFs to balance the Soviet SS-22s and SS-23s is, at the very least, likely to put it at a disadvantage in **Arms Control** negotiations. Whether a global ban on shorter-range theatre nuclear weapons would be necessary to secure an INF agreement was another crucial element in the equation. (See also **Long-Range Theatre Nuclear Forces.**)

SSBN

SSBN is the technical acronym, standing for Submarine, Ballistic, Nuclear, used in the US and British navies for submarines, such as the British **Polaris** or American **Trident** and **Poseidon,** which carry part of the nuclear deterrent force. They are often referred to as the 'capitol' ships of a modern navy, representing the epitome of naval power much as battleships did before the Second World War, or the aircraft carriers during that war. The missiles they carry are known as submarine-launched ballistic missiles **(SLBMs).** The USSR has a variety of similar submarines, the largest class of which, the Typhoon, with a displacement of over 20,000 tons, is considerably larger than anything in a Western navy. The French also have a squadron of such submarines, which is in the process of being upgraded and may number as many as seven by the end of the century. It is thought that the fifth acknowledged nuclear power, the People's Republic of China, does not as yet have an operating SSBN fleet.

These submarines have the major advantage of being, with the current state of the art in **Anti-Submarine Warfare (ASW),** undetectable. They leave their home ports on long, secret patrols of three months or longer, during which they never broadcast so that the opponent has no idea of their position. Being nuclear-powered they need never surface, and the range of their missiles is such that they have a huge area of sea in which to hide. The modern US Trident II submarine, due to come into service in the late 1980s, will be able to carry up to 24 **MIRVed** missiles, each with a range of up to 10,000 kilometres. The USA, the UK and France all aim to keep about 50% of their ballistic missile submarines at sea at any one time, thus ensuring a secure **Second Strike** capacity. The US Trident fleet, when completed, could therefore, on an average day, have six submarines, each with 24 MIRVed missiles with up to 14 warheads, hidden underwater. That suggests a total of over 2,000 warheads, probably of about 300 kilotons each. The resulting **Equivalent Megatonnage (MTE)** would be around 900 MTE: Hiroshima and Nagasaki combined came to no more than one-tenth of a megaton using these calculations.

These submarines are likely to remain the most secure leg of the nuclear **Triad,** unless some as yet unforeseeable breakthrough occurs in ASW capacity. The most recent models, such as Trident, which the UK may also deploy during the 1990s, are far more accurate than their predecessors, enabling them to carry out virtually all the tasks usually reserved for the more vulnerable ground-based **ICBMs** and bombers.

SSN

SSN is the standard naval acronym for a nuclear-powered submarine which is *not* a ballistic missile submarine **(SSBN).** They are variously described as Attack Submarines, Fleet Submarines and **Hunter-Killer Submarines.** Their general purposes are to protect the surface fleet against enemy submarine attack, to seek out the enemy's SSBNs, and to prey on shipping. The USSR, the USA, France and the UK all deploy SSNs although, with the exception of the USA, traditional diesel-powered submarines are also used. Because of the political power of the nuclear submarine lobby in the US navy they no longer have any submarines that are not nuclear-powered, despite the proven superiority of diesel and electric submarines, which are much quieter, for many tasks.

Standing Naval Force, Atlantic

The full official acronym for the Standing Naval Force, Atlantic is worth noting as an example of the absurdity of some naval acronyms—it is STANAVFORLANT! It is a multinational NATO force of **Destroyers** and **Frigates** held at permanent readiness, and indeed normally at sea, able to answer any sudden emergency and, above all, ready to be deployed to any crisis spot to demonstrate NATO's resolve (see **Naval Diplomacy**). There is a second NATO naval standing force, STANAVFORCHAN, which is again a multinational force, this time of mine-sweepers, ready to clear the English Channel should the Warsaw Pact attempt to mine it, perhaps as a pressure tactic prior to the onset of a full-scale war.

Stand-Off Bomb

A stand-off bomb is, in fact, a relatively simple air-to-surface missile (ASM) which allows a bomber force to attack a target from a distance of some hundreds of miles, thus avoiding close contact with anti-aircraft defences protecting vital targets. The British **Blue Steel,** which used to be carried by the **V-Bomber** force, was a stand-off bomb, as are the Air-Sol Missile Portable (ASMP) weapons now being developed for the French air force, and what the US Air Force calls short-range attack missiles. A typical weapon like this travels at about three times the speed of sound with a range of 300 kilometres. Any bomber attack on the USSR will require such weapons because the Soviet anti-aircraft force is massive, well-equipped and subject to constant modernization.

Star Wars

Star Wars is the journalistic nickname for the US **Strategic Defense Initiative (SDI),** coined by the American media immediately after

President Reagan's initial announcement of the plan in 1983. It takes its meaning from the earlier and most ambitious plans which stressed the idea of space-borne **Laser Weapons,** and focused on a scheme to throw a defensive shield over the whole of the USA. The SDI Office (SDIO) naturally dislikes what it sees as a trivialization of a major strategic plan but, nevertheless, the nickname probably remains better known than the official title.

START

Strategic Arms Reduction Talks (START) was the title chosen by President Reagan for the round of negotiations which were to follow **SALT II.** Thus they were a continuation of the series of **Arms Control** talks which had begun under President Nixon in the early 1970s (see **SALT: the General Process** and **SALT I**). However,.the word *reduction* was included in the title to suggest a major development from its predecessors, which had been merely strategic arms *limitation* talks.

The START negotiations opened in Geneva in June 1982, but made little or no progress until they collapsed, in December 1983, in the wake of the Soviet walk out from the **Intermediate Nuclear Forces** talks. Following President Reagan's re-election in November 1984, and the rise to power of Mikhail Gorbachev in the USSR in March 1985, the two superpowers resumed talks in Geneva on strategic weapons limitation, alongside a series of Gorbachev – Reagan summits which had, by mid-1987, yielded meetings in Geneva and Reykjavik. The principal aim of these meetings is to recommence the START process, although the title itself has passed out of common usage, always having been an American rather than an international label.

Stealth Bomber

The American Stealth bomber is the next planned generation of long-range heavy bombers, after the B1-B which was coming into service in the mid-1980s. The Stealth bomber was originally commissioned by the Carter administration as a way of retaining the viability of the piloted bomber leg of the nuclear **Triad,** which was rapidly becoming too vulnerable to Soviet anti-aircraft defences. Stealth technology involves a variety of methods to make aircraft as near as possible invisible to radar and other, principally infra-red, surveillance. Little is known of the ultra-secret programme, but at its core appear to be design techniques which avoid straight or sharp edges, protruding control surfaces and other features which produce clear radar images. In addition effort is taken to reduce heat emission, and there are some suggestions that special paints have been developed which absorb rather than reflect radar beams.

A really successful stealth technology would, at least for a time, radically alter strategic calculations. A bomber force that could penetrate Soviet airspace without being detected would offer a much more flexible and controllable deterrent than missiles which can never be very accurate (see **Bias** and **Circular Error Probable**), cannot be recalled, and are not really testable in warlike conditions. On the other hand many analysts believe that the response to stealth aircraft would be a scientific **Arms Race** to develop more powerful detection methods, and the initial advantage would not last for long. The US Air Force almost certainly already has a fighter with stealth characteristics, believed to be called the F-19, in service at squadron strength, but the secrecy surrounding the stealth programme is so intense that no official admission of this has ever been made. Such aircraft, they are in fact fighter-bombers, could give NATO a vital lead in the early phase of an air war over the **Central Front,** by penetrating Warsaw Pact defences and destroying communication and radar bases to make sorties by ordinary fighters and tactical bombers much easier (see **Air Superiority**).

Stockholm International Peace Research Institute (SIPRI)

SIPRI is the leading independent centre for **Peace Research** in the West. Its work of collating and publishing detailed information on the stocks of nuclear weapons, and on weapons development and production is of particular importance. Furthermore, it probably contains more **Arms Control** experts than any other single centre, certainly outside the US government. Its annual surveys of arms control agreements, negotiations and implementation is regarded as highly authoritative. No aspect of peace research is ignored, however, and SIPRI's monitoring of arms sales is especially useful. Another important characteristic of SIPRI's work is that although the motivation of its researchers is clearly towards pacifism, its publications and analyses are widely seen as completely objective.

Strategic

Obviously the primary meaning of 'strategic' is that it is the adjective from the noun 'strategy'. However, it has come to have a quite separate meaning in modern defence and military language, as is demonstrated in usages like, for example, the Strategic Arms Limitation Talks **(SALT),** or the **Strategic Air Command.** Manifestly the SAC is neither more nor less capable of being used for strategic purposes than any other branch of the US services, and the Strategic Arms talks are not talks about anyone's strategy. The nearest one can get to defining this secondary meaning of 'strategic' is to say that it connotes a combination of 'long-range' and 'powerful'. Thus a strategic missile is one that can reach trans-oceanic targets, and has a high-**Yield** nuclear warhead. The SAC is a 'strategic' air force because it would be used to destroy vital targets very far away from its bases. This leads to another shade of meaning which is closer to the original. If one talks of a weapon as 'strategic' one is saying that it can be directly used to achieve the war aims that lie behind **Strategy,** rather than simply carrying out one of the thousands of moves (classed as **Tactics**) that collectively add up to a strategy. A tank, for example, cannot be regarded as a strategic weapon, because the aim of US strategy, to prevent war and if necessary to defeat the USSR, cannot directly be achieved by any number of tanks by themselves. The destruction of Soviet cities by **ICBMs,** on the other hand, could end a war directly. While this particular example demonstrates the absurdity of regarding nuclear warfare as serving any strategic purpose at all, it does serve to illustrate that there is some sort of distinction between weapons that form part of a complex of tactical interactions, and those that can directly carry out war aims.

Strategic Air Command (SAC)

The Strategic Air Command is that part of the US Air Force which operates part of America's nuclear deterrent, both with its bomber squadrons and the land-based **ICBM** force. When the USAF was set up as a separate service in 1947 the SAC took over the role and traditions of the US army's air forces that had played so major a role in the bomber offensives against Germany and Japan in the Second World War. For a very long time the SAC was, and may still be, the most politically weighty of the air force sections, and as its birth coincided with the commencement of large-scale **Atomic Bomb** manufacture it was essentially the cream of all the US services. As the Eisenhower doctrine of **Massive Retaliation** became the key to US defence policy, the SAC, the only service which could deliver nuclear weapons over the USSR, was paramount. Until 1961, indeed, the whole nuclear war plan of the USA was simply the SAC's own plan, with virtually no control from the central **Department of Defense.** As missiles were developed with the potential to replace the piloted bomber, the SAC ensured that they came under its control, although there is in fact no logical reason why an air force should control ICBMs. In the USSR the missile force is an entirely independent service of its own. The SAC's dominance was finally challenged by the US navy when the submarine-launched ballistic missiles **(SLBMs)** arrived on board the **Polaris** submarines. From that point the development of the war plan was centralized in the **Single Integrated Operations Plan (SIOP),** although the senior officer of the group that draws up the SIOP is always an SAC general.

Despite the advent of the ICBM, and the development of very powerful anti-aircraft defences in the USSR, the SAC has managed to retain, and is now entirely re-equipping, its piloted bomber force, which is one reason for the USA having a nuclear stockpile far larger than it could possibly need (see **Overkill**). The SAC currently has over 300 long-range bombers in some 16 squadrons, as well as five squadrons with some 60 medium-range bombers. The bombers, particularly the B-52, which was designed in the mid-1950s, are being equipped to carry **Air-Launched Cruise Missiles**—a move which has taken the USA beyond the warhead limit set by the **SALT II** treaty. They are to be replaced by the new and politically controversial B-1 bomber, and ultimately by the highly-secret **Stealth Bomber.** All three of these programmes are enormously expensive, and are regarded by most American defence analysts outside the SAC as quite unnecessary.

Strategic Arms Limitation Talks (see SALT: the Process in General, SALT I and SALT II)

Strategic Arms Reduction Talks (see START)

Strategic Bombardment

The first and only real strategic bombardment in history was the bombing offensive carried out by the RAF and the US army's air forces against Germany in the Second World War. The aim of a strategic bombardment is to destroy the enemy's industry, especially its war-related industrial plant, and thereby to shorten or even directly end the war. A strategic bombardment may also be aimed directly at killing the enemy population to destroy **Civilian Morale** and therefore the will to fight. The destruction of anything which will help this end, for example by hampering food supplies, is a legitimate strategic aim. Some early theories of nuclear **War-Fighting** stressed the strategic bombardment aspect, although the mass and indiscriminate destruction of nuclear weapons for some time seemed to make the killing of the population and the destruction of industry the actual war aims, rather than the means to them.

There is ample evidence that the Allied strategic bombardment in the Second World War was actually counter-productive. German morale did *not* collapse, crucial industry was *not* brought to a halt, and indeed productivity actually increased in the worst-bombed areas. The failure to stop industry was probably due to the sheer inaccuracy of bombing rather than an inherent flaw in the theory of strategic bombardment, and nuclear weapons are unlikely to suffer from this problem. The nearest experience we have to the desired effect of a successful strategic bombardment was, in fact, a **Blockade.** This was the almost total success of the American submarine offensive against Japan, which virtually stopped all oil imports by early 1945, so that the aircraft, ships and tanks of the Japanese forces were immobilized. Nothing remotely as effective as this was achieved by bombing in either Europe or Japan. (See also **Douhet.**)

Strategic Defense Initiative (SDI)

The Strategic Defense Initiative was announced in a speech by President Reagan in 1983, which was immediately dubbed the **Star Wars** speech by the media. It signalled a U-turn in US policy, which had previously rejected the strategy of building a **Ballistic Missile Defence (BMD)** when the USA and the USSR signed the **Anti-Ballistic Missile Treaty** as part of the **SALT I** talks in 1972. BMD had been given up by the USA because it was felt to be technically and economically impossible. The belief then was that the exchange ratio, which means the relative cost of building the capacity to destroy an incoming missile compared with the cost of adding to the attacking missile force, would always be unfavourable to the former.

Reagan's initiative is based on the assumption that modern technologies can invert this exchange ratio, and this is at the core of the disputes about SDI's viability. This entry cannot go into the technical details of the many alternative proposed systems, on which there is a voluminous literature. One or two points do need to be set out, however. To start with there are two different broadly-definable versions of SDI. One, that which President Reagan himself seems to have had in mind, would give blanket cover to the USA, or certainly would cover most of the major cities. Technically this would be an **Area Defence.** The alternative, and the only one that seems at all technically feasible, would be to build a system that protected a relatively small number of vital targets, which would principally be the US **ICBM** sites, some major command and control centres (see **C³I**) and possibly vital symbolic targets like Washington, DC. This **Point Defence** approach appears to be the lines along which the Strategic Defense Initiative Office (SDIO) is working.

The range of defensive systems advocated for SDI is large, and many depend on technology that has not yet been devised, let alone tested. Whether the system is designed for point or area defence, it is clear that it will be a 'layered' or 'tiered' system (see **Layered Defence**). This means that successive belts of weapons will attempt to hit incoming missiles at successive stages in their flight path. The most 'glamorous' aspect of SDI technology is undoubtedly the idea of using space-based weapons, probably **Laser Weapons** of some sort, in orbiting satellite battle stations, which would seek to hit Soviet missiles in either the **Boost Phase** or the **Mid-Course Phase.** The more likely system will simply be a modernization of the ground-based interceptor rocket system on which the earlier American and Soviet BMD (see **Galosh**) projects were based. Versions of these, including the use of **Kinetic Energy Weapons** in the form of 'smart rocks', that is, self-guiding solid projectiles, and possibly ground-based lasers, are within current technological capacity, although at a cost which may be prohibitive.

The single most important criticism about the technical feasibility of SDI relates not to the weapons themselves, but to the computing support they will need. The 'Battle Management' programme, as it is called, will have to be both enormous, and of extreme sophistication. Some estimates call for a programme that would contain 10 million lines of computer code. The problems of **Target Acquisition** and tracking, of allotting each target to a specified weapon, of dealing with system failures and endless forms of electro magnetic interference, will make the computing aspect paramount. It has been pointed out by many leading computer scientists that it is absolutely impossible to write a programme of this magnitude and be sure that it will work first time, yet by the very nature of the activity there can be no rehearsals.

The final aspect of SDI that needs consideration is its implication for international stability. There are those who take the obvious position that a defensive weapon cannot possibly be objected to, but most analysts argue that, to the extent that superpower stability has rested on **Mutual Assured Destruction,** a successful **Anti-Ballistic Missile** screen, by taking away the mutuality itself enforces an **Arms Race.** Certainly the recent American tendency to use SDI research as a bargaining weapon to get **Arms Control** concessions suggests that it is not genuinely neutral as a project.

Strategic Superiority

Strategic superiority is one of the more elusive concepts in modern strategic thought. Unsurprisingly it refers to a situation where one superpower has a significantly greater number of powerful **Strategic** missiles than the other. However, it is not just any imbalance which constitutes strategic superiority, because there are natural limits to the number of missiles that can be provided with suitable targets. Thus the much greater number of **SLBM** warheads possessed by the USA compared with the USSR is not thought to confer strategic superiority. Such missiles (at least until the introduction of the more-accurate **Trident** II) are only usable against citied, rather than missile fields, and thus the USSR need only consider the number of US cities, not the number of US missiles, to preserve **Parity.**

In order to have a meaningful superiority of strategic missiles there has to be some point in building more than one's opponents, and it is for this reason that **Arms Races** are triggered by strategic superiority considerations, rather than by sheer warhead numbers. The usual example of strategic superiority is the argument that the USSR has a considerably greater number of very accurate **ICBMs** with explosive power capable of destroying a **Hard Target** than the USA has. Calculations have been made that if one side has a ratio of such warheads to the other's missile silos of greater than about 2.5:1, it could destroy more than 85% of the opponent's missiles during a **First Strike.** The USSR has a land-based missile force estimated to be able to deliver this number of warheads over each American silo, and still leave a considerable part of its ICBM force unused. This is said to give it strategic superiority, because although the USA would still retain its submarine-based missiles, giving it a secure **Second Strike** destruction capacity against **Counter-Value** targets, it would be effectively disarmed in ICBMs.

The obvious question is 'Why would this matter?' To which the standard response is that the US President would not dare use the SLBMs against civilian targets for fear of an equivalent retaliation, and would not be forced to do so because at that stage in a nuclear confrontation no major US city would have been destroyed. The USA would, in real terms, be defenceless in a nuclear sense, and obliged to accept Soviet conditions for terminating hostilities. The USSR, however, would still have a large part of its highly-accurate missile force that could continue to be used for limited **Counter-Force** strikes with impunity.

If such an analysis makes any sense, it implies an instability in the nuclear arms balance which could cause renewed arms races with no obvious terminating points. Whenever a new missile was deployed it would be in the interest of the opposite side to increase its warheads by a factor of 2.5, or whatever the ratio became. Other measures, such as **Ballistic Missile Defence** or even improved **Basing Modes** for missiles, would have a similar **Force Multiplier** effect. An even more frightening consequence of taking strategic superiority seriously is that it might lead to the adoption of a **Launch on Warning** policy by the outnumbered force, or more generally reduce stability by increasing the attraction of making a **Pre-emptive Strike** if nuclear war ever seemed to be highly probable according to the **Use Them or Lose Them** doctrine.

In fact there are two powerful arguments which suggest to some analysts that strategic superiority is largely meaningless. Any first strike would require a guarantee of near perfection in execution before it could be risked. Such a strike that went even slightly wrong in timing, accuracy or effectiveness would immediately engender retaliation in kind by the surviving part of the defender's force. If, for example, just 50 American **MX** missiles were to be left undestroyed because the warheads targeted on them were inaccurate or ineffective, they could deliver 500 warheads over Soviet command and control centres (see **C³I**), probably removing the ability to make any use of their remaining ICBMs. As it is impossible ever to be sure that a high-technology system will operate effectively first time, and as nothing approaching a first strike can ever be rehearsed, it would be extraordinarily difficult to be certain that any superiority on paper could be realized in practice. The most graphic illustration of this is the quotation attributed to Henry Kissinger, when President Nixon's National Security Adviser, 'Will someone please tell me what one *does* with strategic superiority!'

Secondly, and perhaps more importantly, the idea of a **Surgical Strike** so perfectly executed that minimum **Collateral Damage** occurs is improbable. Most analyses suggest that even a strike aimed only at the US missile fields in the sparsely-populated Midwest would kill many millions of American civilians. Faced with that, the Soviet Union could not possibly be sure that the US President would not feel obliged to retaliate in kind, that is, with an SLBM strike against Soviet cities. Despite these counter-arguments, important sectors of the US defence establishment give priority to the avoidance of conceding strategic superiority to the USSR, and thus justify the MX missile programme.

Strategic Surrender (see Pre-emptive Surrender)

Strategy

There is no harder term to define, especially in a book which carries the word in its title, than strategy. There are three reasons for this. Firstly, 'strategy' is one of those words which most people can use more or less correctly, or can grasp the outline of what someone else means in using it, but never need to define. Secondly, it has a general usage—for example the term 'business strategy' and, indeed, the hundreds of courses in Business Schools on 'strategic planning'—which have nothing whatsoever to do with the contents of this book. Further examples are 'economic strategy', 'educational strategy', and perhaps even 'dictionary-writing strategy'. In these non-military contexts it is doubtful if the word means anything more than 'plan' or 'outline of aims and means'. Thirdly, *in* the military context 'strategy' and its adjective **'Strategic'** are also used very loosely, as with 'strategic weapons' or 'strategic warfare', where the meaning is often simply 'big', 'large-scale' or 'long-range'.

Strategy is the science of directing the overall use of a nation's power to achieve major long-term ends. In its specifically military context it refers to large-scale planning of broadly-defined force structures with which to pursue a nation's war aims. Thus, for example, it is possible to talk about a country adopting a **Maritime Strategy,** a strategy of **Horizontal Escalation** or one of **Flexible Response.** Also, **Mutual Assured Destruction (MAD)** is a strategy as is, in the NATO context, **Forward Defence.**

A strategy is an overall plan which links resources, **Doctrine** and goals, but which deals hardly at all with details. A simple example might be as follows. NATO's strategy for defeating the Warsaw Pact in a third world war is to hold the invading armies as long as possible, and as far east in Germany as possible, to give US reinforcements time to cross the Atlantic. Meanwhile American naval and marine forces will 'take the war to the enemy' by attacking the USSR elsewhere than on the **Central Front,** while preventing the Soviet use of **Theatre Nuclear Forces** by the threat of **Escalation.** Once reinforcements were in place an attempt to force the

Warsaw Pact back over the East German borders would be made, but they would not be pursued far into East Germany, because then the war aim—to return matters to the status quo ante—would have been achieved. (This is a very cursory and slightly controversial description of NATO strategy designed for purely exploratory purposes.)

The easiest way to grasp why the above is a description of a *strategy,* rather than something else, such as a doctrine of operational **Tactics,** is that the obvious question to ask at the end of the description is 'How?' The 'How' is the domain of detailed plans, of weapons design and acquisition, of training doctrine, of **Logistics,** of war-gaming and staff courses and so on. Although strategy is grander than these more mundane matters (and indeed the phrase 'grand strategy' is often encountered), it is in fact dependent on them. A single example will illustrate this. Forward defence, doing everything humanly possible to prevent the Warsaw Pact armies from penetrating anywhere on the front, is without doubt a major part of NATO's strategy, because a vital war aim is to avoid turning the whole of West Germany into a **Battlefield.** Forward defence implies a particular tactical doctrine, fixed defence, which is both unfashionable and, arguably, impossible today. The fixed defence tactics themselves require the development of particular weapons, for example cheap **Infantry**-borne anti-tank missiles, and also involve training and **Morale**-building plans. It just may not be possible to come up with a way of persuading infantry soldiers to stand up in the middle of a Soviet tank attack and fire a modernized bazooka. It might also not be possible to build such weapons which are accurate enough even if the soldiers can be motivated to fire them, and the logistics quickly to supply enough reloads over the entire front may be inadequate. Strategy is therefore inevitably tied to much smaller-scale problems, or at least it should be.

At the other end of the spectrum, it sometimes seems that strategies, and even the war aims from which they should derive, are entirely the product of technical feasibility. For example, mutual assured destruction is clearly a strategy. As initially designed under US Secretary of Defense Robert **McNamara,** MAD capability for the USA was defined as killing one-third of the population of the USSR and destroying two-thirds of its industrial capacity. These figures, however, turn out to be set not by some independent war-aim assessment, but by the fact that, given the urban structure of the Soviet Union, a relatively small number of warheads could achieve that level of destruction, but a much larger number could do little more. An even simpler example of the way strategy is determined by feasibility or capacity is the way in which the UK has moved more and more towards a strategy of fighting a submarine, rather than surface, naval war against the USSR. Its entire war effort is now concentrated inside European boundaries as it has become economically impossible for the UK to afford either 'east of Suez' bases or a major surface fleet.

Strike Command

Strike Command is the functional division of the Royal Air Force responsible for all direct combat action. For most of its history the RAF divided combat roles between **Bomber Command,** Fighter Command, and a series of others, the most important of which was probably Coastal Command for **Anti-Submarine Warfare (ASW).** Since the early 1970s the technological and economic imperatives of modern warfare have reduced the need for such specialization. There are far fewer aircraft types in a modern air force, especially one as financially constrained as the RAF. This has led to the development of **Multiple Role Combat Aircraft,** and to a greater unity in the combat roles of air-crew. At the same time the ever growing preponderance of maintenance, **Logistic,** intelligence and administrative tasks in a modern fighting force has increasingly overshadowed any need to differentiate within the overall function of combat. Despite this trend Strike Command continues to carry out discernibly different combat tasks, which are roughly differentiated between as army co-operation, tactical bombing, **Air Superiority,** and ASW roles. It was probably the abandonment of the **Strategic Bombardment** role, at the heart of RAF philosophy under the sway of Bomber Command, when the UK's nuclear deterrent was passed to the Royal Navy in 1968 that hastened the end of the separate combat commands.

Supreme Allied Commander Europe (see SACEUR)

Surface-to-Air Missile (SAM)

Although anti-aircraft guns still exist and have great importance in special roles, defence against enemy aircraft is now very largely the business of specialized surface-to-air missiles. These can vary from very small short-range weapons fired from a portable disposable rocket launcher to highly complex radar-guided missile systems. Missiles exist for differing needs, and the entire family of SAMs in a modern arsenal can cope with aircraft at all altitudes from a few hundred metres to 25 or more kilometres. Experience in modern wars, especially in Vietnam and the Middle East, but also the Falkland Islands conflict, has shown that well-trained troops can put up a very effective defence against attacking aircraft, and modern designs of SAM, like the British Rapier missile system, can be highly mobile but still based on very sophisticated radar guidance. As with anti-tank missiles, it is unclear as yet whether these relatively cheap weapons will make a massive inroad on the threat posed by the enormously complex and expensive **Weapons Systems** which they are designed to counter.

Surgical Strike

Surgical strike is a euphemistic term sometimes used by the military to describe plans to destroy vital targets by means of carefully-controlled force, so as to minimize **Collateral Damage.** It can refer to the use of conventional force, for example a precisely-planned bombing raid on a terrorist headquarters intended to spare all innocent civilians nearby. It can also be used in nuclear strategy, where a surgical strike might be considered as a **Limited Nuclear Option** against a **C³I** bunker or a set of missile silos. However, surgical strikes are rarely as clean and discriminating in their destruction as planned, and the idea of such an attack using nuclear weapons is patently absurd by any normal standards. The lowest **Yield** strategic nuclear warhead in any country's arsenal is of about 150 kilotons—at least 10 times the power of the Nagasaki bomb in 1945. Unless the target was in a most extremely deserted area, hundreds, and probably thousands, of people would surely become casualties.

Survivability

In the age of nuclear warfare, survivability has come to be an aspect of weapons and other military assets as important as the traditional qualities such as accuracy or troop **Morale.** Except for static assets like missiles, which can be placed in super-hardened silos, survivability is seldom a long-term aim in weapons design because it simply cannot be achieved. Aircraft and tanks can be 'hardened', up to a point, so that they can survive a near miss by a nuclear weapon long enough to carry out one mission, although their crews are unlikely to live much longer than that because it is so difficult to protect them against radiation sickness. In the earlier **Scenarios** and planning for nuclear war, survivability was restricted to the ability of a nuclear retaliatory force to survive a **First Strike** by the enemy—hence the positioning of **ICBMs** in silos and the deep-sea deployment of **SLBMs** which were expected to be entirely used up in an immediate **Second Strike.** What has become clear since the development of **War-Fighting** scenarios is the vital importance of ensuring the survivability of **C³I** assets, and the personnel to staff them, as well as the **National Command Authorities** for whose benefit they exist. However, estimates for the survivability of such assets are not optimistic, even over the short period of a few weeks which the most prolonged of nuclear war scenarios require.

Symbolic Deterrence

Symbolic deterrence is a phrase used by some strategic theorists where most would use the idea of **Minimum Deterrence.** Both refer to the relatively small deterrent forces deployed by France and, in particular, the UK. Compared to the impact of a major American strike on the USSR, the firing of the French and British forces would have relatively minor effect. Nevertheless, the majority of analysts, who would refer to the French doctrine of **Proportional Deterrence** or the UK's plan for the **Moscow**

Option as minimum deterrence, take the view that even the level of damage inherent in, for example, a **Polaris** strike by the British is so horrendous that no rational enemy would wish to risk incurring it. Symbolic deterrence is a term coined by those who wish to stress the huge asymmetry in a British attack which might inflict seven to 10 million casualties on the USSR, and the almost inevitable retaliation which could easily destroy the bulk of the British population while leaving the Soviet Union perfectly able to fight a second nuclear war against the USA. The point of calling small nuclear forces 'symbolically' deterrent is to stress their uselessness were **Deterrence** ever to fail. They do indeed represent possibly unacceptable levels of damage, but could not rationally be used were an enemy to call the bluff of the deterrer.

T

TACAMO

TACAMO is one of the many rather contrived acronyms in use in American military circles. It stands for 'Take Charge and Move Out', which itself gives little clue that it refers to a specific type of aircraft exercise. The TACAMO squadrons in the US navy are responsible for communicating instructions to ballistic missile submarines **(SSBNs)** in the event of nuclear war. Because radio communication with submerged submarines is extremely difficult, and otherwise depends on huge ground-based aerials using the **Ultra-Low Frequency** range, the TACAMO aircraft are vital. They follow slow flight paths trailing a very long aerial of about 9 kilometres, capable of sending simple and short commands to submarines. At any time there is at least one TACAMO aircraft patrolling each of the Atlantic and Pacific Oceans.

Tactical (see Tactics)

Tactical Air Force

A tactical air force is an air force command dedicated to co-operation with ground troops or, more generally, to carry out tasks, including air defence and **Air Superiority,** in a particular **Theatre of War.** Thus the air assets under the command of **SACEUR** in the NATO system are all tactical air forces, although only the US Air Force makes the distinction overtly, because they are all scheduled to operate in the context of the immediate **Central Front** warfare needs. In contrast, the US **Strategic Air Command** would be used to carry out quite separate raids, most probably nuclear, as part of a wider war strategy that might have only incidental impact on any immediate battle going on in Europe.

Tactical Nuclear Weapons (see Battlefield Nuclear Weapons)

Tactics

The term 'tactics' is really only definable in the defence context in contrast to **Strategy.** Very simply, tactics are the detailed means for carrying out the directions set to a military force within the strategy by which it is bound. A general's strategy may require a division to capture a town. The divisional commander makes the tactical decisions, such as which battalion to send by which route to take which intermediate point. The appropriate combination of armour and **Infantry** to be used, and the point in the battle plan when artillery should be employed to suppress enemy troops, are further typical tactical questions.

Tactics, in many ways, are the essence of military science. They deal with the more ascertainable and measurable variables of one's own and the enemy's military hardware and troop capabilities. It is possible, at least in principle, to develop a generally 'correct' tactical **Doctrine** to deal with the sort of combat, **Fire-Power** and **Force Ratios** expected in any particular context, and to train troops accordingly. Strategy, being both more wide-ranging and less reducible to military technicalities, cannot be turned into doctrine in quite this way. Both levels of analysis, strategy and tactics, are, of course, interrelated. It is not meaningless to talk about a division commander's strategy for taking a town, it simply means that tactical considerations are being confined to a much lower level of decision-making. To the extent that there is an absolute, rather than a relative, distinction it involves the technical and almost quantifiable, and therefore teachable, skills of dealing with practical conflict questions as suggested above.

Target Acquisition

Target acquisition is one of the two principal tasks of any long-range **Weapons System,** and is most usually used in reference to anti-aircraft or **Anti-Ballistic Missile** defences. The way these are set up usually involves two radar systems. One makes a general search for incoming threatening objects, whether they be missiles or aircraft. When such an object is identified it is said to be 'acquired', and its co-ordinates are passed to another radar set which tracks the threat continuously, in turn passing data to the defensive weapon itself in order for it to fire. More generally, target acquisition is the process of spotting and identifying enemy units or hardware which pose a tactical or strategic threat, and communicating this identification either to a decision-maker, to a **Weapons System,** or to another surveillance system. Target acquisition at long range is at the heart of the plans for **Emerging Technology** weaponry, and presents the most significant problems in the high-technology **Arms Race,** which is focused on **C³I** capacity.

Terminal Phase

The terminal phase of a **Ballistic** missile is, not surprisingly, the last part of its flight. This occurs when the **Re-entry Vehicle** or **Warhead,** re-enters the atmosphere and drops onto its target. By this stage the warhead is not under motorized propulsion and follows a predictable ballistic path. Although this phase is short, lasting only a few minutes, the warheads are at their most vulnerable; they are approaching any defence systems deployed around the targets, and they are less easily masked by **Decoys.** Furthermore, by the terminal phase it is much clearer to the defenders just what the targets are, and they can concentrate defences to protect their most vital assets. For this reason much of the effort in the **Strategic Defense Initiative** is being put into terminal-phase interception.

Theatre Nuclear Forces (TNF)

Theatre nuclear forces is a very common description for what are now more frequently described as **Intermediate Nuclear Forces (INF),** or, in media terms, **Euromissiles.** The term covers a range of **Ballistic** and **Cruise Missiles** that are deployed by NATO and the Warsaw Pact for relatively long-range nuclear strikes inside Europe. No TNF weapon can reach the USA, although NATO missiles could, in fact, carry out attacks on the western USSR as far as Moscow, and the USSR's missiles could destroy almost any target in Western Europe. The somewhat arbitrary upper limit of 5,500 kilometres has been used in **Arms Control** negotiations to define theatre nuclear weapons, and to distinguish them from tactical or **Battlefield Nuclear Weapons,** which are often defined as having a range under 500 km. The most important TNF systems are the **Ground-Launched Cruise Missiles** and **Pershing II** ballistic missiles under NATO command, and the Soviet **SS-20s.** (See also **Double Zero Option, Long-Range Theatre Nuclear Forces** and **Zero Option.**)

Theatre of Operations (see Theatre of War)

Theatre of War

A theatre of war, also known as a theatre of operations, is a major geographical area of conflict, or potential conflict, which is covered by one unified command authority. Within a theatre **Strategy** is determined, on the whole, independently of any other military considerations applying elsewhere in the world. Thus NATO covers one theatre, the European theatre of war, but in a third world war the USA and USSR might well carry on operations quite separately in, for example, a Pacific or even a Latin American theatre (see **Horizontal Escalation.**) Soviet military doctrine recognizes an intermediate level of command organization, known by the initials **TVD,** whereas NATO tends to make functional divisions within the main theatre, for naval, land or air operations, rather than geographical ones.

Thermonuclear

Thermonuclear refers to the explosion of **Nuclear Fusion** weapons, more commonly called H-bombs, or **Hydrogen Bombs.** As the title indicates, an initial atomic, or **Nuclear Fission** explosion is required to generate the enormous temperatures under which nuclear fusion can occur to release the energy of a thermonuclear explosion. 'Thermonuclear' is frequently found as an adjective attached to, for example, the noun 'war' to indicate the fully-fledged **Central Strategic Warfare** in which **Megaton**-level hydrogen weapons are used, in contrast to the use of a few relatively small weapons, possibly only **Atomic Bombs,** in a tactical confrontation.

Think-Tank

Think-Tanks probably originated in the American military establishment during the early **Cold War** period. The model is well established now in other spheres and other countries, but the first important think-tank was probably **RAND,** which is an acronym for the 'Research and Development Corporation'. RAND was created by the newly-independent US Air Force in 1947 (it had previously been merely a branch of the army) as a civilian research institute. It operated on a contract basis, and studied problems ranging from the highly technological to the social scientific. This wide-ranging area of expertise has, perhaps, been the hallmark of the most successful think-tanks. As the think-tanks became established they developed a less passive role, and much of the research they carry out now is conceived of 'in house' and 'sold to the client'. As a result the US services probably receive much better briefing on the whole range of military concerns than is to be obtained by the services of most other countries.

A good example is a less well-known but highly influential think-tank, the **Institute for Defense Analyses,** which conducts scientific and engineering research on the **Strategic Defense Initiative,** but also studies battlefield nuclear tactics and prepares reports on such diverse topics as British arms control policy and Soviet military thought. There are perhaps half-a-dozen major, and dozens of minor think-tanks in the USA involved with strategic policy, and two or three each in France and Germany. In the UK the 'defence analysis industry' is very small, and the **Ministry of Defence** does not give it adequate support, or recognize the need for advice from this quarter. The two or three research bodies that have the potential to become powerful think-tanks have no government support, and little acceptance, although one at least, the **International Institute for Strategic Studies (IISS),** is widely admired internationally.

Thor

Thor was a US **Intermediate-Range Ballistic Missile (IRBM)** in operation from the mid-1950s to the early 1960s. From 1957 until 1963 some 60 Thor missiles were located in the UK, operated by the RAF under a **Dual-Key** arrangement with the USA. They were needed at this stage because the USA did not have a proper **ICBM** capacity with which to hit the USSR from America, and thus tried to threaten the Soviet Union with shorter-range missiles stationed in the territories of its allies. For example, a similar missile, the Jupiter, was deployed in Turkey. These missiles very rapidly grew obsolete because of their extreme vulnerability to any Soviet IRBM strike. They were not based in **Silos** and, being liquid-fuelled, had a lengthy preparation period. Thus if the USSR were able to ready its own missiles undetected, easily possible at this period because there was no satellite surveillance, it could have destroyed the Thor and Jupiter sites in a **Pre-emptive Strike.**

Threat Assessment

Threat assessment is the military intelligence calculation of the danger presented by another country, or more specifically the threat posed by a particular action of that country. There is a more clearly defined distinction between these two assessments than is commonly realized. Assessing the threat of a particular action is usually quite unambiguous, because trying to analyse the motivations and intentions of the opponent hardly matters anyway. If, for example, the Soviet navy moves a squadron of ships to the west coast of Italy, it is largely a technical matter to assess the nature and degree of threat this poses to the US Sixth Fleet, which is based in Italian ports.

General threat assessment, however, ought to, but usually does not, involve a consideration of the reasons behind an opponent's armament programmes. Instead the assessment is usually made on a worst-case basis. For example, the threat posed by the USSR is assessed by counting up the hardware and personnel it has, and assuming that whatever this force structure *could* be used for is what it *will* be used for. Although it would probably be irresponsible for military staffs not to make this calculation, it is unfortunate that policy-makers tend to accept this restricted notion of how to assess a threat as the final word. The USSR has amassed a sufficiently large tank army in Europe to make invasion of West Germany at least a conceivably successful strategy. In military threat assessment this information is enough to establish that there is therefore a real threat of invasion. But is the capacity to do something evidence of such an intention? Suppose the Soviet Union is motivated by a great insecurity about its defensive capacity, and regularly over-arms by Western standards, so that its tank forces simply represent the same preparation as NATO's much smaller forces? This sort of essentially political calculation is a necessary part of any sound threat assessment on which strategies and procurement decisions can be based.

Naturally other elements besides personnel and hardware totals have to be taken into account, the primary one being the opponent's strategic doctrine. How a potential enemy thinks it can, or must, fight a war is a key element in assessing the threat which it poses, because it determines what *they* think they can do with their available capacity. For example, had the British and French militaries, in 1940, taken seriously the German doctrine of **Blitzkrieg,** they would have assessed the threat Germany posed to France more highly than indicated on the basis of hardware arithmetic, which correctly showed that the number and quality of French tanks was considerably superior to that of the *Wehrmacht.* Increasing attention is paid in the West to **Soviet Doctrine,** although it is less obvious that defence policy is based on a threat assessment derived from these studies.

Threshold Test Ban Treaty (see Comprehensive Test Ban Treaty)

Throw-Weight

Throw-weight is a measure of a missile's capacity to carry a load into the atmosphere on a **Ballistic** trajectory capable of reaching a target of, possibly, intercontinental range. The throw-weight which the missile can carry determines the size and nature of its **Bus** (or **Front End**), and therefore the size and number of the **Warheads,** penetration aids and **Decoys,** and the complexity of the guidance mechanism which the bus itself can contain. Throw-weight is usually a very small fraction of the total weight of the missile. For example, the new **Midgetman,** planned by the USA as a small mobile **ICBM,** will probably have a throw-weight of 450 kg against a total missile weight of perhaps 14,000 kg.

The throw-weight calculation is a function of the drive which the missile's engines can produce and the range over which the front end has to be projected. To some extent these are interchangeable: the same missile can 'trade' range for a heavier front end, but this is obviously not an alteration that can be done easily, or in any haste. Not surprisingly the bigger the missile is, the greater is its throw-weight. Land-based ICBMs, which are often three-stage rockets, can have significantly higher throw-weights than **SLBMs.** The planned US **MX,** for example, has a throw-weight of about 3,600 kg, compared with the **Trident** C4's 1,350 kg. In general, though, US missiles, even the ICBMs, have been smaller and less powerful than those of the USSR. The **Minuteman III,** which was still the main American ICBM into the late 1980s, has a throw-weight only one-third the size of the MX, while the biggest Soviet missile, the **SS-18** (which can carry 10 500-kiloton warheads), has a throw-weight of over 7,000 kg. This disparity (the USSR has some 500 of these missiles) is linked to the fear of some US analysts that a **Window of Vulnerability** might exist until the MX is fully deployed.

Time Urgent Target

A time urgent target is one that must be destroyed very rapidly in any nuclear exchange, and most probably in a **First Strike** or the earliest stages of a **Second Strike.** Typically they are targets such as missile silos. If a power is launching a nuclear first strike against an opponent, clearly the primary need is to destroy as many of that opponent's missiles and air and submarine bases as quickly as possible in order to minimize retaliation. If a second strike is being launched in retaliation much the same applies, but, in addition, enemy radar and anti-aircraft facilities may become time urgent targets so that a clear corridor through air defences can be established for piloted bombers or **Cruise Missiles.**

The speed with which time urgent targets must be destroyed presents a problem as it means that only ballistic missiles can be used. Cruise missiles would take so long to reach the targets that an enemy would have ample opportunity to retaliate before its own missiles could be destroyed. (This is a major reason why cruise missiles cannot, in general, be regarded as first strike weapons.) A further problem is that many time urgent targets are also **Hard Targets,** such as missile silos and command bunkers (see **C³I**), and even some air-bases with super-hardened aircraft shelters. To guarantee the destruction of such targets requires a combination of very high **Yield** and extreme accuracy. Consequently, until the advent of the third generation **SLBMs,** such as the American **Trident,** only land-based **ICBMs** have been regarded as suitable for time urgent targeting. As the USA keeps a preponderance of its warheads in SLBMs, while the USSR has most in ICBMs, this has led some American analysts to detect a **Window of Vulnerability** threat for the USA in recent years.

Total War

Total war is a way of describing confrontations such as the two World Wars. The expected nature of a **Central Front** war between NATO and the Warsaw Pact would place it in the same class. These are different from most wars in world history, according to this characterization, in two crucial ways. Wars are 'total' in the sense that the entire population of each combatant is affected both by being potential targets and because the entire economic and social efforts of the nations involved are required for the conduct of the war. In 'ordinary' war it would be possible to raise an army with perhaps only marginal extra taxation. The army could either be entirely made up of volunteers, or by conscripting a small and atypical part of the population. As weapons were not capable of reaching outside the

Battlefield, the bulk of the population would be entirely safe and might hardly be affected by the war.

In contrast to this the World Wars saw bombing of civilian populations, near starvation because of **Blockade** or submarine warfare, wide-ranging **Conscription** of nearly all men under middle age into mass armies, and the need for the whole adult population not in the services, including women, to be available for war-related industrial jobs. Thus warfare had become 'total', involving the effort and risk of every aspect of the social structure. The 'third world war', if such a disaster occurs, may be too short to be total in terms of military and industrial mobilization. However, the risk to the whole population from nuclear weapons would obviously qualify it for classification as total war. (See also **Limited War.**)

Tous Azimuts

Tous azimuts is a French military term used to indicate that its nuclear weapons are pointed 'in all directions'. This doctrine was first pronounced in 1967 by the influential French strategist General Ailleret, at the time that the **Force de Frappe** was first being deployed, and France was pulling out of NATO's integrated military organization. Obviously it has never been seriously contended that there is any power other than the USSR that needs to be deterred from making a nuclear attack on France, but the doctrine is an important part of French nuclear **Declaratory Policy.** In effect it strengthens the classic French doctrine that the possession of nuclear weapons can only credibly be intended for self-protection, and must not in any way be restricted by alliance obligations. For the French the purpose of possessing a nuclear capability is for the **Sanctuarization** of their own country. The independence which they have retained in their targeting strategy implies that no country can rely absolutely on the French refraining from using (or, indeed, not using) its nuclear force in concert with the rest of NATO. The other Western nuclear powers, the USA and the UK, at least technically dedicate their strategic weapons to NATO.

Triad

The triad, or nuclear triad, is the three-part structure of nuclear **Strategic** forces, consisting of the air force bomber squadrons, the land-based intercontinental ballistic missile **(ICBM)** force and the submarine-launched ballistic missile **(SLBM)** fleet. Three of the five members of the **Nuclear Club,** the USA, the USSR and France, deploy their forces in this triad, though with differing emphases. For example, the USSR has no truly intercontinental bombers, although it is sometimes argued that their Backfire bombers (TU-22s), might have such range with enough mid-air refuelling. In terms of numbers of warheads the land-based ICBM is the mainstay of the Soviet triad, while for the USA it is the SLBM fleet, which carries

70% of all US strategic warheads. France faces very different problems, being so much smaller a nuclear power, and it is arguable that the land-based leg, and perhaps even the air force leg, of its triad are kept in being for political and symbolic, rather than strictly military, reasons.

The justification for maintaining all three forms of nuclear capacity, when one might be seen as sufficient, is that the country in question has a protection against any suddenly-achieved **Strategic Superiority** which an opponent might develop. Thus, during the early 1980s, some Americans believed that the USSR had developed enough ICBM power and accuracy to destroy 90% of the US land-based missiles in their silos with a **First Strike** (see **Window of Vulnerability**). Were this true, the existence of the submarine force would have guaranteed the USA a secure **Second Strike** deterrent. At the same time, no country with the choice would risk putting all its warheads in submarines in case there was a breakthrough in **Anti-Submarine Warfare** techniques.

The least justifiable of the legs of the triad was, for a long time, the bomber force, because of its vulnerability both on the ground and while trying to penetrate the enemy's airspace. Recent development of the **Air-Launched Cruise Missile,** however, means that it can be fired from a considerable distance outside the range of air defence facilities. Potential developments in **Stealth** technology could make further additions to the viability of this element of the triad.

Trident

Trident missiles are the most recent development of submarine-launched ballistic missiles **(SLBMs),** and come in two forms. One is already deployed on US submarines, and is known either as Trident I or as Trident C4. This missile was a considerable advance upon the previous, second generation, SLBM, the **Poseidon,** and enormously more powerful than the old **Polaris** missiles still deployed by the British Royal Navy. However, it differs only in having greater range (about 7,500 kilometres, against Poseidon's 4,600 km) and a greater **Throw-Weight,** reportedly carrying 10 **MIRVed** warheads each of about 100 kilotons. Despite this greater power it is much the same size as Poseidon, and can be carried in the old Poseidon submarines. It was originally to be purchased by the British government, and could have been fitted into new submarines of the old Polaris design.

Early in the Reagan administration the decision was made to go ahead with Trident II, or D5, which is really more like a fourth generation of SLBM than a revised version of Trident I. It is much bigger, requiring the building by both the USA and the UK, which has decided to buy them instead of the C4, of a new and much larger model of submarine. The D5 has an even greater range, estimated at 9,700 km. This greatly increases the sea area in which the submarines can patrol, making them much less vulnerable to **Anti-Submarine Warfare.** Trident II also has the capacity to carry up to 14 MIRVs in the missile **Front End,** each possibly of as much as 400 kt. Thus, even though the UK denies that it will fully load the missiles, changing from Polaris with only three, fairly small, non-MIRVing warheads to Trident D5, which the British **Ministry of Defence** suggests will be deployed with at least eight MIRVed warheads, is a huge enhancement of strike power. The need to build a new generation of submarines has enabled the British to plan for engineering improvements that will probably guarantee two craft on patrol at any one time, compared with the Polaris squadron's restriction to one. This perhaps represents a 15-fold increase in striking power.

The real significance of Trident II is that it is the first submarine-launched missile that will have the accuracy to strike **Hard Targets,** because it is expected to have a **Circular Error Probable** not greater than 100 metres. Until this missile comes into service, in the late 1980s, only **ICBMs** will be able to achieve this degree of accuracy. Thus the USA is restricted to using only a small part of its warhead stock for carrying out **Limited Nuclear Options** or, more generally, both **First Strike** and nuclear **War-Fighting** tasks. The Trident D5 will make any country which possesses it much more able to consider the secure use of its nuclear armoury, without running the risk of retaliatory annihilation.

Trigger Thesis

The trigger thesis was, at one time, a more or less serious justification advanced to defend small strategic forces, such as those of France and the UK, against the charge that they were so insignificant compared with the Soviet armoury that they could not have **Credibility** as deterrents. The argument was that these small forces got their power not from what they could do themselves, but by their ability to 'trigger' a supporting strike by the USA. It was claimed that the USA might not actually wish to use its nuclear weapons in defence of an ally, and that the **Nuclear Umbrella** might not hold. But if the French or British, in their own defence, made a strike on the USSR the whole world would be plunged into an uncontrollable process of **Escalation.** The USSR might *believe* they were under attack from the USA. Alternatively the USA itself might think the risk that the USSR *would* think that the Americans were attacking was so great that they would have to launch a **Pre-emptive Strike** against the Soviet Union anyway (see **Catalytic War**).

The trigger thesis was probably never actually held by any government as real policy, but it made some sort of macabre sense in the early **Spasm War** days of nuclear strategy. Now that Western strategies are based so firmly on the hope of being able to exercise very precise control over escalation processes, there is even less room for the theory.

Tripwire Thesis

The tripwire thesis was a way of describing the role of conventional troops, particularly US forces, on NATO's **Central Front** during the 1950s and early 1960s. This was the period when NATO was committed to the strategy of **Massive Retaliation.** According to this, any serious incursion by the Soviet Union on the 'free world' would have been met not by prolonged defensive war, but by immediate nuclear retaliation on the USSR. Because there was a problem of **Credibility** with this theory (would the USA really engage in **Central Strategic Warfare** just because the Warsaw Pact had invaded a few miles into Western territory—a scenario referred to as **Salami Tactics**), some symbol of intent was required. It was argued that a relatively thin screen of troops would serve this purpose. They would put up some real resistance, so the Warsaw Pact would have to make their intentions and determination quite clear by defeating them. This would signal the appropriate moment to commence nuclear war, and it was hoped that the Warsaw Pact would realize this. At the same time the presence of the troops, and even more so their dependants, would leave the US President no alternative but to activate the **Nuclear Umbrella** to protect American citizens in Europe.

Truman Doctrine

The Truman Doctrine, announced in 1947, was the more or less formal American acceptance of its self-imposed responsibility to protect the 'free world' from communist or other totalitarian takeover. It was enunciated to legitimize the American intervention in the Greek civil war between Albanian-backed Greek communists and the Western-oriented democratic government of the period, after the UK had indicated its own inability to continue in that role. In essence the doctrine, which has never been repudiated by subsequent American administrations, committed the USA to fight, or provide military assistance, wherever a democratic government was under pressure either from an external enemy, as in the **Korean War,** or from internal insurgency, as was partially the case in Vietnam. Later Presidents, notably Kennedy, with his commitment to 'pay any price' and 'oppose any foe to assure the survival and success of liberty', have elaborated the doctrine. The main problem is that the initial concentration on the protection of 'freedom and democracy' has often been taken to embrace the protection of any non-communist status quo, as in El Salvador during the 1980s, or even the encouragement of anti-communist 'freedom fighters', as in Nicaragua. Thus the Truman Doctrine has been used, but not in its original spirit, by the supporters of a **Global Strategy** in the USA.

TVD

TVD is a term which refers to the main organizing units of Soviet military posture, and is usually translated into English as standing for 'Theatre of

Military Operations'. It highlights a difference between NATO and Warsaw Pact doctrine, because **Soviet Doctrine** demands a much more centralized control at a high level, the TVD headquarters, than does NATO's more diffused command structure. The whole of the **Central Front** will be one Soviet TVD. In general the USSR concurs with the NATO identification of potential zones of conflict in Europe: the **Northern Flank,** the central front and the **Southern Flank.** In addition Moscow probably recognizes one Asian TVD, and possibly a purely naval TVD for the North Sea. Unification is of paramount importance, however, so coastal activities off Europe would be seen as being part of the relevant mainland TVD, not as separate naval operations. (See also **Theatre of War.**)

Twin-Track

In 1977 the West German government initiated a Western European request to the USA and to NATO planners to upgrade NATO's **Theatre Nuclear Force (TNF)** capacity. This was a reaction to the USSR's deployment of a new generation of **Intermediate-Range Ballistic Missile,** the **SS-20.** Both the USSR and NATO had already, for some considerable time, deployed **Long-Range Theatre Nuclear Forces** in Europe. The Soviet Union relied on missiles, the SS-4 and SS-5, while NATO depended on medium-range bombers, mainly the American F-111, but also, until the early 1980s, the RAF's **V-Bomber** squadrons.

The early enthusiasm for the modernization of TNFs waned, however, when Western European governments discovered there was intense, if minority, opposition in their own countries to this rearmament step; this initiated what became known as the **Euromissile** debate. There would have been a technical need to modernize an ageing force in any case, and NATO's planners had indeed been working on the problem since at least 1972. In an attempt to carry out this modernization in a politically acceptable way NATO announced, in December 1979, what came to be known as the 'twin-track' (or 'dual-track') policy. Under this they would simultaneously begin to deploy new weapons, and actively pursue an **Arms Control** agreement reducing or banning all theatre nuclear weapons. Actual deployment of the NATO weapons, **Ground-Launched Cruise Missiles** and **Pershing II** ballistic missiles, would thus be geared to Soviet action.

If the Soviet Union agreed to remove its SS-20, SS-4 and SS-5 missiles, NATO would not go ahead with its plans (see **Zero Option**). Alternatively, the full number might not be deployed if the USSR agreed to limit its own deployments. Such an agreement was almost reached during the famous **Walk in the Woods** episode before the USSR walked out of the **Intermediate Nuclear Forces (INF)** negotations, in Geneva, in 1983. In the end the arms control 'track' faltered, and NATO began deploying its new weapons in Europe. However, it was only in 1986 that the Netherlands finally agreed to allow deployment of its share of the weapons, and full equipment of the batteries may never actually happen—particularly if the renewed interest in INF negotiations during 1987 was to prove fruitful. (See also **Double Zero Option.**)

U

Ultra-Low Frequency (ULF)

Potentially the most crucial single problem in nuclear strategy is to ensure safe and efficient communications between the **National Command Authorities (NCAs)** and the **Strategic** nuclear forces. The most difficult of all these problems is communicating with the missile-carrying submarines **(SSBNs)**, because their very value lies in being nearly impossible to detect. As soon as they come near the surface to pass radio signals, they risk detection. Radio waves do not normally pass through water, so the NCAs cannot quickly and safely communicate launch instructions, movement orders or other tactical and strategic information.

However, one form of radio emission can travel through water for great distances—signals carried at ultra-low frequencies. Unfortunately it is an unavoidable law of physics that such signals not only require enormously long land-based aerials, but transmit information extraordinarily slowly. All nations whose navies include SSBN squadrons have ULF transmitters, which take up hundreds of square miles of land. Even then only extremely simple messages can be transmitted, as it can take half-an-hour to transmit a three-letter code group to a submarine. These transmitter stations, whose localities can hardly be hidden, are highly vulnerable to disruption by nuclear attack, and thus the ULF systems go only a small way to solving the problem of **C^{3}I** with the SSBN forces. (See also **TACAMO.**)

United Kingdom Atomic Energy Authority (UKAEA)

The United Kingdom Atomic Energy Authority was set up in 1954 to cover all matters relating to the production and design of nuclear facilities, weapons, and material. Originally it had three divisions: for research, weapons and industrial energy production. However, the nuclear weapons division at Aldermaston has always operated effectively as an autonomous institution, it being decided from the outset that it should be quite separate from the civilian research centre at Harwell. Depending on the political situation in the USA, the UKAEA has usually had very good relations with its US counterpart, the **Atomic Energy Commission** and its successors). As the use of nuclear energy has increased, functions of the UKAEA have tended to slip away to other organizations, such as the Central Electricity Generating Board or quasi-private bodies like British Nuclear Fuels. Meanwhile the AEA itself has come to be more a co-ordinating agency than a direct controller, except in the field of civilian nuclear research.

Use Them or Lose Them

'Use them or lose them' is a term, particularly common in US strategic circles, to describe the predicament that NATO could very well find itself in with regard to its short-range tactical or **Battlefield Nuclear Weapons** in the event of a **Central Front** war. The problem is that a very large number of these low-**Yield** battlefield weapons, particularly the thousands of nuclear artillery shells but also some missiles, have very short ranges. The Lance missile, for example, has a range of only 130 kilometres, while the typical artillery weapon range is 30 km or less. Obviously these weapons have to be kept very near the front line if they are to be used at all. Yet a successful attack by the Warsaw Pact, particularly if achieved with surprise, could penetrate this far into NATO territory in a matter of hours, and certainly within a couple of days. Consequently there is a strong risk of these weapons being overrun and captured unless they are used very early in the conflict.

Although NATO is committed to first use of nuclear weapons if necessary (see **No First Use**), it naturally does not want to be forced into their use prematurely. This has led to the suggestion that these weapons, several thousands of warheads for which are stored in Europe, can serve no valid military purpose at all. To some extent NATO recognizes this and has withdrawn part of its stocks. The USSR has never built such weapons, preferring to rely on longer-range missiles such as the SS-4 and SS-5.

V

V-Bomber

The V-bombers were the mainstay of Britain's nuclear deterrent until that role was assumed by the Royal Navy with its **Polaris** squadron in 1968. There were, in fact, three different aircraft, the Victor, Valiant and Vulcan—hence 'V'-bombers. The initial design work for all three was carried out in the early 1950s, although they came into squadron service at different times. The most successful, and the one that became the main combat aircraft of the V-bomber force was the Vulcan, which remained in service for NATO theatre uses until 1982, when it was replaced by the Tornado. At its height, in 1963, the V-bomber force numbered between 140 and 180 aircraft each carrying a **Thermonuclear** one-megaton **Gravity Bomb,** or **Blue Steel** stand-off bombs. The Valiant was the least successful aircraft, having to be withdrawn from service in 1961 because of metal fatigue. The Victor was soon transferred to duties such as photo-reconnaissance, or refitted for mid-air refuelling. The Vulcan was widely regarded as the best medium-range jet bomber in the world, and might have continued to carry Britain's deterrent force for decades had the US **Skybolt** programme not been cancelled.

Verification

Verification is the process of checking whether or not a party to an **Arms Control** agreement is abiding by the terms. It has largely been disputes on how to verify such agreements that have made it difficult for the USA and USSR to find common interest in arms control. Usually the demand for effective verification measures has been made by the West and denied by the Warsaw Pact. Concern over verification has also been used by groups inside Western political systems who are unenthusiastic about arms control to pressure their governments into refusing some Soviet offers. So, for example, many conservative Senators in the USA used verification arguments against the ratification of **SALT II.** The Thatcher government in Britain defended its refusal to support the USSR's proposal of a **Comprehensive Test Ban Treaty** on the grounds that long-range verification was impossible. However, most commentators believe that the real reason for rejection was Britain's need to carry out tests in the USA to develop the UK's own warheads for the **Trident** programme. Verification can either be carried out externally to the suspected country, by scientific instruments, which are usually called **National Technical Means,** or by on-site inspection by observers from the accusing country or from neutral states.

Vulnerability

Vulnerability principally refers to the extent to which a missile system, **ICBM, SLBM** or **Intermediate-Range Ballistic Missile (IRBM),** is at risk to destruction in a nuclear **Pre-emptive Strike.** It has no specific technical definition, but the vulnerability of a missile battery can be seen as a function of its detectability, the ease with which it can be destroyed by a nuclear explosion of any given **Yield** and accuracy (see **Bias** and **Circular Error Probable**), and the speed with which it can be launched. There are really three approaches to reducing vulnerability. One is to hide the weapons system by deploying it underwater in submarines. The ballistic missile submarines **(SSBNs)** are generally regarded as the least vulnerable alternative, because **Anti-Submarine Warfare** techniques have not yet developed to the point where they can be detected reliably. The traditional alternative has been to reduce vulnerability by making missiles exceptionally **Hard Targets** by putting them into super-hardened **Silos** capable of withstanding immense blast overpressures. However, improvements in missile accuracy and **Equivalent Megatonnage** performance have made even these silos highly vulnerable. Consequently most development plans are now aimed at reducing vulnerability by making missile launchers mobile. By being able to take the missiles out of the identifiable military base and disperse them throughout the country, as with the **Ground-Launched Cruise Missiles** stationed in Britain, the enemy will not only find it difficult to locate them, but will be presented with a large number of widely-scattered targets. Both the American **Midgetman** plans, and the newest Soviet missile project, the SS-25, rely on mobility rather than silo deployment. The quest to reduce the vulnerability of US ICBMs, which had led to fears of a **Window of Vulnerability,** forms the backdrop to the **Basing Mode** arguments over the **MX.**

Walk in the Woods

The 'walk in the woods' is one of the more unusual phrases to have entered the history of **Arms Control** negotiations. During the protracted and ultimately fruitless Geneva negotiations on **Intermediate Nuclear Forces,** held between November 1981 and November 1983, there was only one moment when agreement might have been reached. Just days before the talks were due to recess the senior negotiators for both the USA and the USSR took the unusual step of dodging their assistants one evening after a formal dinner, and literally going for a walk by themselves in the woods. In this private conversation they managed to strike a balance which would have substantially reduced the number of warheads each side planned to deploy. Their initiative was met by outright denunciation from both governments and, shortly thereafter, the meetings were suspended by the USSR, not to resume again, at least in that form. Both sides went on to commence deployment of missiles according to their original plans.

War in Space (see Anti-Satellite Warfare and Strategic Defense Initiative)

War-Fighting

War-fighting has come to have a special meaning in nuclear strategy, and one of increasing importance over the last decade. The idea is a continuation from the **Limited Nuclear Options** which were developed as part of US nuclear policy from the early 1970s. Advocates of war-fighting claim that no nuclear **Deterrence** posture can be credible if it is not supported by the weapons and doctrines actually needed to fight and win, in some sense, a prolonged nuclear war. The contrast is with the older school of thought that nuclear war could have no winners, and that the only available course, were deterrence to have failed, was to indulge in a single massive response in the **Spasm War** tradition.

War-fighting theorists point out that a nuclear war would not necessarily involve an immediate massive exchange, with the war over in hours. Instead it might follow a slow and controlled path of **Escalation** and involve periods of cease-fire (or **Fire-Breaks**) for **Intra-War Bargaining.** In such a **Scenario** the ability to select targets, and to preserve **C³I** links would be vital. Furthermore, unless there is a 'theory of victory', as one of the war-fighting exponents, Colin Gray, has called it, there will be no way of knowing what to do if such a war does break out.

The doctrine of war-fighting, which has never quite gained official recognition, is deeply unpopular among the more orthodox proponents of **Assured Destruction** deterrence for two reasons. Firstly, it is claimed that rational preparation to fight a nuclear war makes it more probable that such a war will break out, because it will come to be seen as a possible policy option in a way that an all out spasm war of the orthodox deterrence theories could not. Secondly, the demands in terms of sophisticated weapons and C³I facilities are extremely expensive, whereas a simple secure **Second Strike** capacity need not be. Those who are tempted towards developing war-fighting doctrines and **Weapons Systems** understandably scorn the established wisdom that a nuclear war can have no winners. Their greatest problem is in explaining how the carefully controlled escalation processes can be expected to work in the unprecedentedly catastrophic conditions of *any* **Central Strategic Warfare.**

Warhead

A warhead is very simply the 'business end' of a missile or other **Delivery Vehicle,** whether it be a submarine torpedo or **Smart Bomb.** It is not very often used by professional strategists and scientists, perhaps because it is too blunt, if honest, and lacks the neutrality of terms such as **Re-entry Vehicle** or even 'mission package'. What is important is to distinguish between the complex machinery, possibly in several stages, that propels, guides, and delivers the explosive or nuclear device, and the device itself. In most forms of weapons where the warhead/delivery vehicle distinction applies the latter is more complex, much bigger, and even more expensive than the warhead itself. Furthermore, there is an increasing tendency to try to diminish the sheer destructive power of the warhead, in order to reduce **Collateral Damage,** and as a consequence the importance of accuracy and **Survivability** (against counter-weapons) has been emphasized, making more demands on the delivery vehicle than on warhead design. (See also **Bus** and **Front End.**)

Warsaw Pact

The Warsaw Pact or, to give it its full title, the Warsaw Treaty of Friendship, Co-operation and Mutual Assistance, was set up formally in May 1955, officially as a response to the rearmament of West Germany when that country became a member of **NATO.** The Warsaw Pact (WP) members are Bulgaria, Czechoslovakia, East Germany, Hungary, Poland, Romania and the USSR. The Pact has a complex governing structure, with a ministerial-level Political Consultative Committee, a Committee of Defence Ministers, a Military Council and a Joint High Command. Few of these bodies meet very regularly, and effective control comes from the USSR's Ministry of Defence via the Joint High Command, where the Commander-in-Chief and Chief of Staff are always very high-ranking Soviet officers.

The primary use of the WP forces and structure is to support the various Soviet military deployments, especially the Northern Group of Forces in Poland, the Central Group in Hungary and, above all else, the **Group of Soviet Forces, Germany.** The non-Soviet Warsaw Pact (NSWP) armies are attached in one way or another to these groups of forces, and exercise with them. However, unlike NATO, this peacetime organization, with its attempt to appear as an alliance of equal partners, is not intended to be the command structure for war. If war was to break out all NSWP forces would come under the direct command of their associated Soviet force groups, and would serve as reserves and lines-of-communication troops to aid what would be in name, as well as fact, the USSR's war. There is considerable doubt about the political reliability of NSWP forces and, with the exception of some East German divisions, little direct use is expected to be made of the East European forces in front-line combat. Each NSWP army has a Soviet military mission attached which supervises and, to some extent, controls training and liaison, with the whole of the Warsaw Pact commanded and organized from Moscow.

Weapon of Last Resort

This phrase is more usually heard in British discussions of nuclear weapons than anywhere else and is, indeed, the way that the UK's nuclear force is usually described by the **Ministry of Defence.** The point that is being made is that the British **Independent Deterrent** has no **War-Fighting** role, as it would now be called, and is not to be seen as part of strategic **Escalation.** Instead, it exists only for use 'in the last resort', which would presumably be to retaliate against a nuclear strike on the UK. However, as the UK does not give clear **Declaratory Policy** guidelines for the conditions under which Britain *would* launch its missiles (so as to increase the deterrent effect through the uncertainty) it is unclear what the last resort is to be taken to mean. As, furthermore, the British nuclear force is officially dedicated to NATO, and is supposed to come under the targeting plans developed by **SACEUR,** and as these NATO plans are based on the possible early use of nuclear weapons, the term may have no meaning in the real world at all. (See also **Moscow Option.**)

Weapons System

Weapons system is a general concept, made necessary by the complex technology of modern warfare, which intentionally has a very wide usage. The whole of a modern military aircraft, including its radars and navigational equipment as much as its missiles, bombs and guns, can be regarded as one weapons system, because on a technical level they are so highly interdependent. Similarly it would be pointless to treat only the **Warheads** or **Re-entry Vehicles** on a long-range missile as the 'weapon', given their uselessness unless they can be dropped onto the target. Therefore strategists, as much as engineers or arms controllers, increasingly talk about weapons systems in describing the whole of mechanical units.

Western European Union (WEU)

The Western European Union is based on the Brussels Treaty of 1948, which is also fundamental to the **NATO** agreements, and was set up in 1955. It was formed, following the collapse of plans to create a **European Defence Community,** with seven members—Belgium, France, West Germany, Italy, Luxembourg, the Netherlands and the UK. Meetings of both Defence and Foreign Affairs Ministers from these seven countries are provided for. It also has a parliamentary assembly consisting of delegations from the national parliaments. The WEU, which is neither an operational nor policy-making body had largely fallen into disuse until its reactivation in 1984, as a result of a French initiative, and some now see it as a way in which the position of the **European NATO** members can be strengthened. As yet it is unclear exactly what such a body could do, although as a focal point for expressing a non-technical European outlook on the collective defence of Western Europe, it may have some value. Meetings of ministers under the auspices of the WEU in 1986 proved useful as a forum to discuss Europe's response to the Reagan – Gorbachev proposals made at Reykjavik in October of that year. However, there are already many fora in which European ministers meet and, in any case, it is not clear that a geographically-restricted 'pressure group' within NATO is appropriate.

Window of Vulnerability

The window of vulnerability refers to the fear expressed by some American strategists that the USSR would be in a position during the early 1980s to launch a relatively safe **First Strike** against the USA. Their fear was based on the way that the USSR had built up a strong force of very heavy

ICBMs with large and accurate **MIRVed** warheads. Because the US ICBM force, which in any case only accounts for about 25% of its total warhead stock, was deployed in missile silos, prone to problems of **Vulnerability,** some estimates released by the **Department of Defense** suggested that a Soviet first strike might destroy 90% of the missiles before they could be launched. The only way to avoid this would have been to embrace the extremely risky **Launch on Warning** strategy, which the USA has always denied having adopted.

The fact that the USA would still have most of its warheads safely at sea in the **SSBN** fleet did not invalidate the window of vulnerability thesis. It was argued that these submarine-based missiles might be useless. If the Soviet Union designed and executed its strike carefully, it could destroy the American ground-based strike force without hitting any cities, and thus limit **Collateral Damage.** However, the remaining US missiles, being relatively inaccurate, could only be used for **Counter-Value** or 'city-busting' strikes. The President, it was suggested, would be unable to risk using them for retaliation, because the USSR would be effectively holding the American cities which they had spared in the first strike as **Nuclear Hostages.** The Soviet Union would still retain a considerable number of unused and accurate land-based ICBMs, relatively secure in their hardened silos, as well as the whole of its own submarine fleet. It would be able to carry out further strikes at will, with the USA effectively disarmed.

There were two major flaws in this theory. Firstly it would be impossible to rely on the success of such a first strike, because wholesale launching of nuclear missiles has never been tested, and great doubts surround the measures of accuracy, normally quoted in terms of **Circular Error Probable** and **Bias.** Unless the USSR's strike really *was* at least 90% successful they would put the Soviet heartland at great risk from the surviving US missiles. The command and control of nuclear war is so uncertain that no rational strategist could depend on such a plan, yet it was put forward precisely as a rational **War-Fighting** doctrine, and was used to justify the US rearmament programme which featured the **MX** and **Trident** developments. Secondly, the civilian deaths from such a strike, far from being minimal, could be as high as 20 million using worst-case assumptions. A death toll far lower even than that could not intelligently be distinguished from a deliberate counter-value strike, and it is most improbable that the American **National Command Authorities** would believe that they had suffered too little to justify the use of the submarine ballistic missile force.

Y

Yield

The power of a nuclear explosion is measured as yield, which covers the entire range of energy released *and* its impact. Yield is expressed in terms of megatonnage, where one **Megaton** (mt) is the equivalent of detonating one million tons of a conventional explosive such as TNT. Nuclear weapons with lower yields are expressed either as a decimal of one mt, or in kilotons (kt, with 1,000 tons of TNT equivalent to one kt). Thus the warheads on the British **Polaris** missile are usually said to have a yield of 200 kt or 0.2 mt. In practice, yields of less than one megaton are understated if this direct comparison is made. Instead it is usual to express effective yield as **Equivalent Megatonnage (MTE).** Using this method of calculation the MTE of a one mt detonation is, naturally enough, one, but kiloton-level yields are proportionately higher, and yields above one megaton proportionately lower, than the nominal value would indicate.

Z

Zero Option

The zero option is an **Arms Control** policy suggested by President Reagan in 1981 as a solution to the **Euromissile** conflict which had arisen when the USSR started deploying the **SS-20** theatre nuclear missile in 1977. When NATO responded by announcing that it would install the **Pershing II** and **Ground-Launched Cruise Missiles** in Europe, arms control propositions began to be made by both sides. Formal talks covering the whole **Intermediate Nuclear Forces (INF)** issue began between the USSR and the USA in Geneva in 1981.

Various levels of deployment were suggested by both sides. The basic Soviet position was that they had a need for deployment (to counter Chinese nuclear forces) that NATO did not, and should not be required to make concessions. The 'zero option' was the USA's starting point; NATO would stop short of INF deployment if the USSR agreed to remove all of its SS-20s. This was never even considered as a serious bargaining position by the USSR; indeed, they later rejected much more pragmatic suggestions of equal, but small, numbers of warheads. The zero option was not, in fact, entirely popular with **European NATO** members either. Although the actual decision to construct and deploy new missiles by NATO was a response to the SS-20, NATO's planners had already come to the conclusion that their theatre forces, at that time mainly aircraft-delivered bombs, were in need of modernization. The Geneva INF negotiations eventually collapsed, in November 1983, following a Soviet walk out, shortly after the **Walk in the Woods** initiative by the leading negotiators from both sides.

The zero option surfaced again as a potential solution to the need for European theatre arms reduction in late 1986, at the Reykjavik Summit between Reagan and Gorbachev. The US President astounded Western European governments, none of whom had been consulted, by repeating his offer of the zero option. While it might seem 'fair', the Warsaw Pact would actually gain most from the removal of this category of weapons. Until the SS-20 was deployed they were clearly at a disadvantage in **Theatre Nuclear Forces (TNF)**; to move to a position where neither side had such weapons would effectively remedy this previous disproportion. Furthermore NATO, dependent on possible first use of nuclear weapons (see **No First Use**) far more than the conventionally-superior Warsaw Pact, would be in an overall weaker position on the **Central Front.** The Warsaw Pact also held a clear advantage at the lower-range end of the TNF category, between 500 and 1,000 kilometres, in the form of its **SS-22 and SS-23** missiles. In an attempt to clear this hurdle General Secretary Gorbachev proposed a **Double Zero Option,** in which all missiles with ranges between 500 and 5,500 km would be removed from Europe and, conceivably, from the rest of the world as well.